Advanced High-Frequency Radio Communications

For a complete listing of the *Artech House Telecommunications Library*,
turn to the back of this book.

Advanced High-Frequency Radio Communications

Eric E. Johnson
Robert I. Desourdis, Jr.
Greg D. Earle
Stephen C. Cook
Jens C. Ostergaard

Artech House
Boston • London

Library of Congress Cataloging-in-Publication Data
Advanced high-frequency radio communications/ Eric E. Johnson... [et al.].
  p. cm.
Includes bibliographical references and index.
ISBN 0-89006-815-1 (alk. paper)
1. Wireless Communication systems.   2. Shortwave radio.
I. Johnson, Eric E.
TK5103.2.A38   1997
004.6'6—dc21   97-483
                                                                            CIP

British Library Cataloguing in Publication Data
Advanced high-frequency radio communications
1. Shortwave radio
I. Johnson, Eric E.
621.3'84151

ISBN 0-89006-815-1

Cover design by Deborah Dutton and Joseph Sherman Design

© 1997 ARTECH HOUSE, INC.
685 Canton Street
Norwood, MA 02062

All rights reserved. Printed and bound in the United States of America. No part of this book may be reproduced or utilized in any form or by any means, electronic or mechanical, including photocopying, recording, or by any information storage and retrieval system, without permission in writing from the publisher.
  All terms mentioned in this book that are known to be trademarks or service marks have been appropriately capitalized. Artech House cannot attest to the accuracy of this information. Use of a term in this book should not be regarded as affecting the validity of any trademark or service mark.

International Standard Book Number: 0-89006-815-1
Library of Congress Catalog Card Number: 97-483

10 9 8 7 6 5 4 3 2 1

We dedicate this work to our many loved ones, here and gone, and to those men and women whose creativity, insight, and devotion to their science and technology have made this book possible.

Contents

Preface   xiii

Acknowledgments   xvii

Chapter 1	High-Frequency Communication System Engineering		1
1.1	Wireless		1
	1.1.1	Legacy	1
	1.1.2	Lessons Learned? High-Frequency Standards Development	3
	1.1.3	Automated High-Frequency Communications	6
1.2	High-Frequency Radio in the Global Information Infrastructure		10
	1.2.1	Introduction	10
	1.2.2	Joining the Internet	12
	1.2.3	Conclusion	19
1.3	High-Frequency Systems Engineering		19
	1.3.1	Purpose	19
	1.3.2	The Systems Engineering Process	20
	1.3.3	The Customer-Supplier Relationship	22
1.4	High-Frequency System Design		25
	1.4.1	Requirements Definition	25
	1.4.2	Formation of Candidate Designs	26
	1.4.3	The Comparative Design Approach	28
	1.4.4	Design by Link Budgets	29
References			32
Chapter 2	High-Frequency Sky-Wave Channels		35
2.1	Overview		35
2.2	Sky-Wave Propagation in a Benign Ionosphere		35

		2.2.1	Ionospheric Plasma	35
		2.2.2	Ionospheric Layers	38
		2.2.3	Sky-Wave Propagation	41
		2.2.4	Ionospheric Measurements	60
		2.2.5	Link Performance Prediction	71
	2.3	Sky-Wave Propagation in a Disturbed Ionosphere		85
		2.3.1	Global Phenomena	85
		2.3.2	Latitude-Dependent Phenomena	91
		2.3.3	Summary	112
	2.4	Noise		114
		2.4.1	Background	114
		2.4.2	Atmospheric and Man-Made Noise	115
		2.4.3	Galactic Noise	119
	2.5	Radio Interference		121
	References			123
Chapter 3	High-Frequency Antennas			129
	3.1	Introduction		129
	3.2	Narrowband Antennas		132
		3.2.1	Vertical Polarization: The Whip Antenna	132
		3.2.2	Horizontal Polarization	133
		3.2.3	Antenna Tuners	133
	3.3	Broadband Antennas		135
		3.3.1	Vertical Polarization	136
		3.3.2	Horizontal Polarization	138
		3.3.3	Multimode Antennas	144
	3.4	"Small" High-Frequency Antennas		145
		3.4.1	Overview	145
		3.4.2	Receive-Only Antennas	146
		3.4.3	Small Transmit Antenna Design	149
	3.5	Environmental Effects		151
		3.5.1	Overview	151
		3.5.2	Ground Effects	152
		3.5.3	Antenna Mounting Effects	157
	3.6	Conclusions		160
	References			162
Chapter 4	Digital Modems for High-Frequency Radio			165
	4.1	Introduction		165
		4.1.1	Background	165
		4.1.2	Channel Distortion	167
		4.1.3	Additive Noise	168
		4.1.4	Cochannel Interference	168
	4.2	Traditional Design Approaches		169

		4.2.1	Background	169
		4.2.2	FSK Modems	170
		4.2.3	Parallel-Tone Modems	170
		4.2.4	Single-Tone Modems	172
	4.3	Relative Merits of Contemporary Designs		172
		4.3.1	CCIR Channel Performance	172
		4.3.2	Performance in Impulsive Noise	174
		4.3.3	Performance With Cochannel Interference	176
		4.3.4	On-the-Air Trials	179
		4.3.5	Other Attributes	180
	4.4	Enhanced High-Frequency Modem Design		182
		4.4.1	Design Issues	182
		4.4.2	Enhanced Parallel-Tone Modems	184
		4.4.3	Enhanced Single-Tone Modems	186
	4.5	Conclusions		187
	References			188
Chapter 5	Automatic Link Establishment Signal Structure			191
	5.1	Introduction		191
	5.2	Linking Performance		192
		5.2.1	Minimum Required Performance	192
		5.2.2	Measured and Simulated Results	194
	5.3	ALE Word Format		194
		5.3.1	Word Types	195
		5.3.2	Valid Sequences	198
	5.4	Coding		198
		5.4.1	Overview	199
		5.4.2	Golay FEC Coding	202
	5.5	Word Synchronization		203
		5.5.1	Transmitter Word Phase	203
		5.5.2	Synchronization at the Receiver	205
	5.6	Addresses		206
		5.6.1	Character Sets	206
		5.6.2	Individual Station Addresses	207
		5.6.3	Multiple Station Addresses	209
		5.6.4	Special Addressing Modes	209
	5.7	ALE Frame Structure		212
		5.7.1	The Calling Cycle	212
		5.7.2	Message Section	213
		5.7.3	Conclusion Section	214
		5.7.4	Sounds	214
	5.8	Conclusions		214
	References			215

Chapter 6		Automatic Link Establishment Protocols	217
	6.1	Overview of ALE Operations	217
		6.1.1 ALE Operational Rules	219
		6.1.2 Listen Before Transmit	219
	6.2	Channel Evaluation and Selection	219
		6.2.1 Channel Evaluation	220
		6.2.2 Channel Selection	225
	6.3	Link Establishment	226
		6.3.1 Overview	226
		6.3.2 Available State	227
		6.3.3 Linking State	227
		6.3.4 Linked State	227
		6.3.5 One-to-One Calling	227
		6.3.6 One-to-Many Calling	237
	6.4	ALE Data Structures	247
		6.4.1 Channel Table	247
		6.4.2 LQA Table	247
		6.4.3 Self-Address Table	247
		6.4.4 Other Address Table	248
		6.4.5 Scalar Variables	248
	6.5	ALE Orderwire Functions	248
		6.5.1 Link Quality Analysis	249
		6.5.2 LQA Report	251
		6.5.3 Noise Report	254
		6.5.4 Advanced Link Quality Analysis	255
		6.5.5 Channel and Frequency Commands	257
		6.5.6 Power Control	258
		6.5.7 Mode Control	259
		6.5.8 Scheduling Commands	263
		6.5.9 Automatic Message Display	265
		6.5.10 Data Text Message Mode	266
		6.5.11 Data Block Message Mode	269
		6.5.12 Cyclic Redundancy Check	271
		6.5.13 User-Unique Functions	273
		6.5.14 Time Exchange	274
	6.6	Future Developments	274
		References	275
Chapter 7		Linking Protection	277
	7.1	Introduction	277
	7.2	Linking Protection Technique	280
		7.2.1 Review of the Link Establishment Procedure	280
		7.2.2 Vulnerability of Unprotected ALE Stations	280

		7.2.3	Overview of the Linking Protection Procedure	282
		7.2.4	Transmit Processing	283
		7.2.5	Receive Processing	284
		7.2.6	Resolution of Ambiguity	286
		7.2.7	Linking Protection Processing Example	287
	7.3	Protection Interval Analysis		290
	7.4	Performance Evaluation of Linking Protection		291
		7.4.1	The Simulator	292
		7.4.2	Simulations for Protected Versus Nonprotected ALE Links	292
		7.4.3	Measurement-Simulation Comparison	296
		7.4.4	Summary of Linking Protection Performance Evaluation	297
	References			297
Chapter 8	The High-Frequency Data Link Protocol			299
	8.1	Introduction		299
		8.1.1	Background	299
		8.1.2	History	300
		8.1.3	HFDLP Overview	302
		8.1.4	HFDLP Operation	304
	8.2	Data Transfer Protocol		305
		8.2.1	Data Transfer Protocol Operations	305
		8.2.2	Data Transfer State Machine	310
	8.3	Message Management Protocol		310
		8.3.1	Message Management Protocol Operations	310
		8.3.2	Message Management State Machine	315
	8.4	Link Management Protocol		316
		8.4.1	Link Management Protocol Operations	316
		8.4.2	Link Management State Machine	318
	8.5	Timeouts		320
		8.5.1	Turnaround Times	320
		8.5.2	Response Timeout	320
		8.5.3	Link Timeout	321
		8.5.4	Action Following Timeout	322
	8.6	Broadcast Mode		322
	8.7	ARQ Circuit Mode		323
	8.8	Full-Duplex Operation		323
	8.9	Adapting Data Rate		323
	References			324
Chapter 9	High-Frequency Radio Networking			325
	9.1	Introduction		325

9.2	Automated Networking Technologies		328
	9.2.1	Finding Indirect Paths	328
	9.2.2	Monitoring Connectivity	331
	9.2.3	Using Indirect Paths	332
9.3	An Example Implementation		332
	9.3.1	High-Frequency Node Controllers	332
	9.3.2	Network-Layer Protocol Overview	337
	9.3.3	Automatic Message Exchange	339
	9.3.4	Store-and-Forward Operation	342
	9.3.5	Null S&F Functions	342
	9.3.6	Connectivity Exchange	343
	9.3.7	High-Frequency Relay Management Protocol	343
	9.3.8	High-Frequency Station Status Protocol	345
9.4	Automated High-Frequency Network Management		346
	9.4.1	Background	346
	9.4.2	High-Frequency Network Management Requirements	348
	9.4.3	Applicability of SNMP for High-Frequency Network Management	349
	9.4.4	HNMP: The High-Frequency Variant of SNMP	351
References			356
About the Authors			359
Index			365

Preface

High-frequency (HF) communications has been an integral part of worldwide information transmission since the dawn of radio and has kept pace with the Information Age. For example, to quote from Section 1.2.2, "The HF e-mail system provides secure, error-free automatic delivery of e-mail messages and binary files such as images and graphics." As shown in Chapter 1, HF radio communication systems have evolved along with the global information infrastructure to provide line-of-sight all the way to round-the-world connectivity capable of supporting a variety of compressed multimedia voice and data services. This book describes standardized state-of-the-art and advanced techniques used by automated HF radio equipment to provide primary and backup communication links and networks within this infrastructure.

I became involved in the development of military standards for automated HF radio more than three years ago. Although only a small "cog" in the standards-development "engine," I soon believed that a need existed to present the standards in a more palatable form. I hoped to offer an understanding of the standards in a less terse format, while still retaining the necessary detail and accuracy. Of course, this vision would have gone unrealized had it not been for Dr. Eric E. Johnson, who was at that time the central figure in the day-to-day drafting and technical review of the HF automation standards. I came to know and respect Dr. Johnson, and found that he was quick to share the vision of a book largely devoted to automatic link establishment, or ALE, and the evolving data link and networking standards.

For several years, Dr. Johnson had approval for an AFCEA (Armed Forces Communications Electronics Association) course, but had not had time to put together the needed course materials. With the publication of this book, a textbook has been developed that is ideally suited for this course. In addition to the general information on HF radio, the book contains the only exposition to date on the U.S. military and federal standards for automated HF systems. These standards have swept the world, and Dr. Johnson has received frequent requests from engineers implementing these standards who need help in understanding them. These engineers are increasingly called on to build communication systems that include automated HF radios. They need to understand how this technology works, and how to integrate it into their systems. Dr. Johnson wished he could just refer them to a book, but none existed until now.

I knew it was logical to precede any discussion of automated HF communications with a complete description of the HF channel. I wanted this description to discuss the significant signal, noise, and interference characteristics from a phenomenological perspective. In this way, the reader would gain an appreciation of the need for the many intricacies of the various operating, addressing, and information exchange protocols that constitute automated HF communication networks. For this reason, I solicited the participation of Dr. Gregory D. Earle. I have known Dr. Earle for some years and have a great deal of respect for his first principle's knowledge of the ionosphere and ionospheric processes as well as sky-wave radio propagation. Dr. Earle agreed to the project at once and became one of the key "founding fathers" of our work, sharing our vision of a complete presentation of both theoretical and practical HF radio. We believed from experience that only such a complete understanding permits system integration and test with a minimum of surprises.

While engaging Dr. Earle in the book project, I sought an additional contribution from another friend and colleague, Mr. Jens C. Ostergaard. Mr. Ostergaard worked for some years managing an HF communications network in Greenland and has since provided technical leadership for the former Greenland Meteor Burst Test Bed. This U.S. Air Force and Danish test bed program produced some of the most complete high-latitude meteor-burst measurements and analysis produced in the West. Since the lowest operating frequency of this test bed was 35 MHz, and because these links provided a diagnostic tool for the high-latitude environment, Mr. Ostergaard volunteered to contribute relevant material to Chapter 2 on HF sky-wave channels. Mr. Ostergaard worked with Dr. Earle to author other portions of Chapter 2 and, to our great benefit, sought the support of Dr. Leonard S. Wagner for material describing the disturbed channel. In addition, Mr. Ostergaard provided important material describing noise and interference in the HF band.

I asked Dr. Stephen C. Cook to add important material on HF modem technology. Dr. Cook offered his HF '95 Nordic Shortwave Conference paper entitled "Advances in High-Speed HF Radio Modem Design," which provided immediate and up-to-date material for Chapter 4 of this book. This contribution provides an understanding of what could be expected from advanced HF modem types in *realistic* channels. The paper also included some useful results regarding noise and interference that were moved to the corresponding section in Chapter 2. Given his overall HF systems experience, he also volunteered to write material on HF systems engineering presented in Chapter 1.

Given the wide variability in link power budgets encountered in practice, RF systems engineering are founded on proper selection of the transmission equipment, including antennas, as well as modem design to meet performance requirements. Because significant performance swings are often attributed to antenna effects, I sought the support of Mr. Malcolm J. Packer to contribute Chapter 3, "HF Antennas." Mr. Packer provided tutorial material describing HF antennas, presented popular HF antenna types, described antenna design trade-offs, and discussed the effects of the antenna environment on its electrical performance. Nearly 300 pages of additional material provided by the authors had to be deleted from the manuscript to meet publication requirements. This material included

- Chapter 1: Historical perspective on the impact of radio standards and interoperability on the disappearance of Amelia Earhart; a description of a shipboard HF e-mail implementation; and a sample specification for an HF e-mail system
- Chapter 2: Quantitative descriptions of HF ground-wave and meteor-scatter propagation and material describing the mathematical formulation of the standard MUF calculation and the first-principles IONORAY algorithm for prediction sky-wave propagation
- Chapter 3: Tutorial describing antenna design and performance parameters as well as a description of numerical modeling tools and techniques
- Chapters 5 through 9: Detailed descriptions of various bit fields described in the protocol standards documentation, an alternative codec implementation approach, and several related tables and figures

These materials can be made available to the interested reader from the editor, Robert I. Desourdis.

This book was written for seasoned HF radio engineers who wish to learn the evolving technology of automated HF stations as well as newcomers who need to come up to speed quickly regarding their knowledge of modern HF radio systems. These two types of readers are often tasked with designing, evaluating, or implementing HF radio systems. For this reason, the book has been written to address the system designers problem and proceed from the medium (HF channels) through the link and network layers needed to integrate HF radio into the global information infrastructure. This "layered" architecture theme is apparent in the HF automation standards and therefore provided a natural theme that we followed throughout the book.

Robert I. Desourdis, Jr.
February 1997

Acknowledgments

In a decade of remarkable collegiality among the top engineers from the U.S. HF industry, it has been my pleasure to work with quite an outstanding group: Gene Harrison (of course), Bill Beamish (of "Midnight Oil" fame), Gene Teggatz (the "Camel" who proudly wears the tent!), Dan McRae, Stan Adams, Chuck Lynn, Dan Roessler, Bill Jackson, Steve Elvy, Lars Stromberg, David Young, Chad Seymour, and many others. The cast of collaborators from governments here and abroad is no less illustrious: Joe Whitney, Simon Rosenblatt, Dave Peach, Bob Adair, Chris Redding, and the whole crew at NTIA ITS, Stephen Cook, Jason Scholz, and the merry band from HF '95.

Eric E. Johnson

I would like to acknowledge the assistance of my colleagues at UT Dallas and at Science Applications International Corporation (SAIC), who contributed to this work through many thought-provoking discussions, as well as the understanding and patience shown by my wife Nancy throughout the publication process. I am grateful to Bob Desourdis for inviting me to participate in this work, and for undertaking so many of the most arduous tasks himself. Finally, I am indebted to NASA, the U.S. Air Force, and the National Science Foundation for supporting my research in space sciences over the years, thereby providing me with the opportunity to gain a deeper understanding of radio propagation through the ionosphere.

Gregory D. Earle

The work presented in Chapter 4 embraces far more than the author's experience; it summarizes the accumulated knowledge of the Defence Science and Technology Organization, Australia's Radio Networks Discipline. I would like to acknowledge specifically the assistance of Martin Gill for provision of information on his 52-tone modem and his integrated ARQ systems and Tim Giles for assistance in describing his work on impulsive noise and cochannel interference tolerance. Thanks also go to Steve Morris for the 3200-bps single-tone modem results, Marcel Scholz for his new results on HF noise, Mark Preiss for editorial assistance, and to the rest of the team for all their contributions over the years. Finally, I would like to express my gratitude to the reviewers whose comments have significantly improved the material presented in Chapter 4.

Stephen C. Cook

I wish to acknowledge and express my appreciation to several persons and institutions who made my contribution to this work possible. I wish to thank the Geophysics Laboratory, Ionospheric Interactions Branch, U.S. Air Force, for their invitation to me to serve as a guest scientist with the division and for providing access to the GL (now the Phillips Laboratory) facilities near Boston, and most importantly, the Greenland Meteor Scatter Test Bed in Sondrestrom and Thule, Greenland. I also want to acknowledge and express personal appreciation to John Rasmussen, former chief GL/LID, and Paul Kossey, deputy director PL/LID, who supported my long-term involvement in the test bed program and provided kind tutoring. I also acknowledge and thank Alan Bailey, Don DeHart, Patricia Bench, and Eric Li for their devotion to the test bed effort, as well as Sergeants Anthony Coriaty, Carlton Curtis, and Douglas Carter for their "spirited" support under freezing conditions in Greenland. Finally, I acknowledge and express appreciation for many productive discussions with and suggestions from Paul S. Cannon of the Royal Aerospace Establishment in the United Kingdom and Jay Weitzen of the University of Lowell, and to R. Desourdis for the opportunity to participate in the creation of this book.

Jens C. Ostergaard

This work would not have been possible without the creative genius of Gene Harrison (the "Father of ALE") and Fred Leiner of the Mitre Corporation for the initial conception and development of automatic link establishment (ALE). In this regard, I hope this book yields a greater understanding of their contribution to the field of HF radio communications.

Dr. Eric Johnson (the "Stepfather of ALE") played a dominant role in the detailing and enhancement of the automated HF standards and is the principal book author, having contributed more than five chapters to this work. The practical success and widespread acceptance of ALE as a viable (and protected) HF signaling scheme is largely due to the intelligence, engineering excellence, and communication skills of Dr. Johnson. I am indebted to Dr. Johnson for his drive and determination in making the first book to detail automated HF standards and ALE a reality. It could not have been done without his generous commitment made more than two years ago and faithfully maintained since that time. His ceaseless efforts to mature Gene Harrison's work into an industry-accepted standard (MIL-STD-188-141A), with critical help from Joseph Whitney and Rich Besselman of Science Applications International Corporation (SAIC) provided under enlightened leadership by the U.S. Army, are greatly appreciated. My small contribution to their great effort has been entirely my privilege.

I wish to thank my other excellent coauthors on this work. Dr. Greg Earle provided established theoretical and empirical understanding of the ionosphere and sky-wave channels. Dr. Earle's intelligence, patience, understanding, and communication skills are undoubtedly a great benefit to his students at the University of Texas. He was always quick to respond to editorial changes, and accepted the seemingly endless rewrites of previously accepted passages and several "just-one-more-thing" requests.

Dr. Stephen Cook was asked to contribute to this work in great haste and at the last minute. I greatly appreciate his immediate response and devotion to this book. More importantly, I appreciate Dr. Cook's major contribution to the field of advanced HF modems and his generous contribution to this effort. His contribution was particularly appreciated because I could not constantly heckle him for input as I could the other authors. His presence "Down Under" left us only the Internet to exchange outlines, schedules, and material, and it performed quite well.

Finally, Mr. Jens Ostergaard provided the expertise gained from years of operational HF experience in Greenland, combined with the wisdom regarding high-latitude effects gained from conducting a long-term measurement program. I am indebted to Mr. Ostergaard for his contribution and his seasoned understanding of high-latitude effects on HF communications.

I want to thank Dr. Leonard Wagner for applying his wealth of scientific knowledge regarding the magnetized ionosphere and its effects on propagating radio waves. I also greatly appreciate Dr. Wagner's quick response to what must have seemed like endless requests for more material as well as the inevitable further requests for explanation and clarification from someone far less knowledgeable then himself in this field. Similarly, I want to thank Mr. Malcolm Packer for his proven antenna expertise combined with the speed with which he contributed the HF antenna chapter. My old colleague from SAIC proved again that his talent for his chosen field was only surpassed by his willingness to help a friend.

I want to express my gratitude to Richard Formato, my friend and colleague, who reviewed Chapter 2 (HF Sky-Wave Channels) and Chapter 3 (HF Antennas). Richard provided excellent review of these materials and suggested many substantive as well as editorial changes. I also want to thank Joseph Katan of the Naval Undersea Warfare Center in New London, Connecticut. Mr. Katan provided the scaled ionosonde data and ionogram interpretation reports needed to perform the work reported in Section 2.2.4.2.

I want to express my appreciation to those companies who contributed materials for a planned chapter on state-of-the-art systems and their suppliers. Although I had too little material from enough suppliers to write this chapter in the time available, I nevertheless want to thank those individuals and their organizations who provided material. In this regard, I want to thank Dr. John Goodman, who not only provided material describing his company but also offered to contribute in general. I thank him for his offer and hope that I am able to accept this offer, as well as materials submitted by other responsive organizations, in future editions of this work. I regret that both the length of the book and the limited time available prevented my better utilization of such a premiere authority as Dr. Goodman.

This book would not have been possible without the devotion and patience of Mr. Mark Walsh, who signed these experts to complete this work. Mark's commitment to a quality product surpassed his devotion to schedule, which was greatly appreciated by yours truly and, ultimately, by the authors and contributors. I must also thank Ms. Kimberly Collignon and Ms. Roberta DeAvila, as the kind but firm "extractors" of the manuscript from an all-too-distracted editor. I thank Mr. Michael Holt for his professional direction in preparation of the final manuscript. I also thank the anonymous reviewer, who was often asked to analyze multiple chapters in the shortest possible time in anonymity.

I must express my gratitude to Ms. Paula Marcangelo, my office administrator at SAIC, who managed all of the paperwork needed to launch this work. In addition, she persistently contacted many members of the HF Industries Association (HFIA) to solicit their input for the book. I thank Paula, who saved me much aggravation in the early stages of this effort and always offered her kind assistance irrespective of other pressures in the office.

I also thank my previous employer, Science Applications International Corporation, for the opportunity to learn the material that ultimately fostered the idea and realization of this book. I would like to thank Arthur Hurtado, Jeffrey Wright, Barry Shay, Susan Moore, and Linda Bunton, my new colleagues at the Telecommunications Division of the Advanced Technologies Division Group of CACI International, Inc., in Fairfax, Virginia, for their vital support during the final stages of book production.

Last, but definitely not least, I must thank my wife Betty and daughters Danielle Marie, Nicole Elizabeth, and Amanda Michelle, as well as my mother Margaret, for their understanding and patience regarding the need to sequester myself at home to compile this work. Betty also implemented authors' editions in several chapters and reviewed grammar in several sections. She hopes, as I have come to understand, that I "will not" venture into publishing a work of this size and scope again in the near future. We'll have to "see about that."

Robert I. Desourdis, Jr.

We are indebted to all of those engineers and scientists who, although not mentioned in these acknowledgments, nevertheless deserve credit for advancing our knowledge of the medium and achieving interoperable HF communications. This major accomplishment should establish a model for the industry—government standards process for all similar far-reaching technology development efforts in the future. We regret that many important professional and amateur advances in HF communications could not be included in this work. Although many of these advances are significant but have not received general acceptance, we thank these contributors for their excellent efforts and encourage their further devotion to the field. We are, of course, indebted to our families, friends, and coworkers who endured our commitment to the completion of this work.

All

High-Frequency Communication System Engineering

1.1 WIRELESS

1.1.1 Legacy

Harold Cottam left his berth to get the latest news. From his wireless set, he passed some messages to Cape Cod and inquired from a nearby ship if Cape Race had messages for his ship, the SS *Carpathia*. Cottam received quite urgent news, a CQD from the nearby ship, then within 60 miles of the *Carpathia*. The CQD was the regulation international call for help, soon to be replaced by SOS. The message continued [1]: "CQD from MGY. We've struck a berg. Sinking fast. Come to our assistance. Position, Latitude 41.46 North, Longitude 50.14 West, MGY." Cottam notified Captain Rostron, and the *Carpathia* made for North 52 West. It was just before 12:30 A.M., April 15, 1912, and the *New York Times* was printing the following dispatch [2]:

> Cape Race, Newfoundland.
> Sunday night, April 14.
> At 10:25 o'clock tonight the White Star Line steamship *Titanic* called CQD to the Marconi wireless station here and reported having struck an iceberg. The steamship said that immediate assistance was needed.

Wireless operators despaired as the *Titanic*'s signals weakened, because they knew it was a sign that the engine rooms were flooding. At 1:27 A.M., senior operator John George Phillips left the wireless on the upper deck when it too began to flood and ran to the stern. He had stayed with his equipment for almost an hour after Captain John Phillips had announced "Every man for himself."

Harold Bride, the junior wireless operator, had quipped shortly after the collision that Phillips should "Send the SOS; it's a new signal and it may be your last chance to send it." Later, he would put an overcoat and lifebelt on

Phillips, who was working the key. Bride would avoid the suction of the sinking ship and be pulled from the water into a collapsible boat, where he found Phillips dead. Ultimately, Bride would recover in the SS *Carpathia*'s sick bay and alternate the key with Cottam until the ship could make New York. When the gangplank was dropped, Guglielmo Marconi was one of the first to go aboard. He wanted to interview the wireless men personally. Emerging from the ship and noticeably in grief, Marconi said [3]:

> It is worthwhile to have lived to have made it possible for these people to have been saved. Just now all the world is thinking of this greatest of sea disasters, I feel that I must speak of it, but I do it reluctantly. I know you will understand me if I say that all those who have been working with me, entertain a true feeling of gratitude that wireless telegraphy has again helped to save human lives. I also want to express my thanks to the press for the hearty approval it has given my invention.
>
> I am proud, but I see many things that will have to be done if wireless is to be of the fullest utility. It will be necessary to compel all ships to carry two operators, so that one may be on duty at all times.
>
> Some of the ships failed to hear the *Titanic*'s call for help because they were receiving news bulletins from Cape Cod. With two operators, one could be working the news, the other—on any ship equipped properly—could be listening for distress signals, which would not interfere with the long distance news messages.

What would have happened if Cottam had gone to bed that night without hearing the CQD? Would the toll have been higher than 1517 lives and the $7.5 million luxury liner, the largest "unsinkable" ship then afloat?

In the days following the disaster, misinformation was disseminated by overeager journalists, and wireless operators alternately raised and lowered the hopes of readers and listeners. As Dunlap writes [3]:

> Land stations, numerous ships at sea and hundreds of amateurs tried to reach the *Titanic* or ships rushing to her aid. There was a babel of dots and dashes. That great confusion resulted from jamming the air with wireless messages in a frantic effort to get more news is found in a dispatch of that day, which read:
>
> "It was practically impossible to get any reliable information by wireless because of the great number of wireless stations breaking into the field, and because of the work of amateur operators. It appears that the disaster to the *Titanic* had no sooner been flashed over the

sea than about every wireless instrument along the coast within range began to transmit with no thought of others, and so the net result soon became a hopeless jumble, from which distorted and inaccurate messages were patched up in haphazard fashion and announced to the anxious world."

This tangle was responsible for the messages reporting the *Titanic* was en route to Halifax under her own steam at six o'clock at night, when, as a matter of fact, she had been sixteen hours under the surface of a sullen ocean.

Later, in discussing triangulation from radio beacons, Marconi states that "there must be regulation of wireless and assurance that all vessels carry equipment before that feature comes up." He went on to say [4]:

> Wireless, however, should not be regulated to death, as it easily could be. But, it simply must be governed in some manner, and the one body fit to do the regulating would be an international board. It's a bigger job than any one nation could handle. All must be considered and must join in the proceedings.

Much has changed since these events, and much has remained the same. The radio compass and beacon stations, now surpassed by the satellite global positioning system (GPS), can safely negotiate ports and avoid other ships and even icebergs. Continuous operator monitoring of vessel radio equipment and fast maritime network operating procedures have evolved since the *Titanic* slipped below the waves. Spectrum allocations have reduced, albeit not eliminated, interference between operators "breaking in" on each other. Automated systems allow a single user to "come up" on many frequencies, thus increasing channel congestion and the need for better coordination and consideration between radio neighbors.

1.1.2 Lessons Learned? High-Frequency Standards Development

Although more than 70 years have passed since Marconi foresaw the need for standards and interoperability, individual teams of scientists and engineers continue to make significant, yet uncoordinated, contributions to the advancement of wireless. High-frequency (HF) radio researchers in industry brought the power of automation to bear on the problems of challenging sky-wave propagation in the form of proprietary systems in the early 1980s. The merger of the microprocessor and radio led to a number of innovative solutions to traditional problems in radio communication systems, such as channel selection, unattended operation, and interoperability. These systems rapidly and automatically scanned a set of assigned frequencies, listening for modem tones that carried their assigned call signs. The resulting automatic link establishment, or ALE, systems were received with delight by the radio operators: Not only did the radio take on the task of finding propagating frequencies, the speaker stayed mercifully silent except when traffic addressed to the station was under way. Of course, radios for use in automated stations had to be controllable by an ALE or other radio controller if the benefits of automation were to be achieved. Sales of these automatic HF radio systems to government and other HF users soared.

Within a few years, however, a problem surfaced: the differing ALE schemes were incompatible. They did not use the same modems, nor the same encodings for call signs. As a result, a station being called by a radio from another manufacturer would silently continue to scan the channels, with its operators entirely unaware of the attempt to reach them. The government was, of course, not pleased with this state of affairs. The National Communication System, which is chartered to ensure that federal agencies can communicate with each other, contracted with the Mitre Corporation for consultative work, which led to the federal standard for ALE. After surveying the available ALE schemes, a Mitre engineer, Gene Harrison, proposed a second-generation ALE scheme that captured the best ideas of all of the existing schemes, plus some clever new ideas of his own [5–7]. His work was documented in a proposed FED-STD-1045 [8] in September 1986, which was subsequently adopted by the military community as well for their revision of the military HF radio standard, MIL-STD-188-141 [9].

In the next couple of years, the "Mitre scheme,"[1] as it came to be known, was thoroughly debated, analyzed, and simulated by parties friendly to the interoperable approach, and those who wished to maintain the status quo of noninteroperable proprietary systems. Analysis, simulation, and eventually on-air testing confirmed that the Mitre scheme exhibited superior linking performance over the first-generation systems, and was confirmed as the technique to be standardized in MIL-STD-188-141A and FED-STD-1045. The military standard had the advantage of a shorter coordination cycle, and was published first, in September 1988. Publication of FED-STD-1045 followed in January 1990. The ALE technology standardized in MIL-STD-188-141A and FED-STD-1045 is now in wide use throughout the world.

Building on the immediate success of the ALE standards, work began on follow—on standards to develop and standardize, in advance, the innovative capabilities foreseen by Harrison in his visionary "Stairway to Heaven"

---

[1]The Mitre scheme drew ideas from the first-generation automatic link establishment schemes of many U.S. manufacturers and benefited from the participation of a wide cross section of engineers from industry and academia as it was transformed into the current second-generation standards.

(see Fig. 1.1). The Stairway showed how increasingly more capable HF radio systems could be built up by successively adding more sophisticated features. This entire family of features (along with some additional features not foreseen in the Stairway to Heaven) has now been standardized in a pair of military standards, MIL-STD-188-141A and MIL-STD-187-721C [10]. The latter standard is a *planning standard,* which is not mandatory for military users of automated HF radio, but instead illustrates the planned evolution of the mandatory standard 188-141A. The suite of capabilities standardized in 187-721 was developed through a thoroughly stimulating interaction with some of the best engineers in the U.S. HF industry. It constitutes a significant break with the usual standardization process, in which standards are developed after several industry solutions have been developed and tried in the marketplace. In the latter procedure, users of equipment using nonstandard techniques are left with dead-end noninteroperable technology (such as Betamax in the video world).

The planned evolution of the HF military standards will incorporate lessons learned in the first implementations of the 187-721 techniques into subsequent revisions of MIL-STD-188-141. Thus, even if the ideas in MIL-STD-187-721 are modified during their first implementation, the best that industry can do will be incorporated in that first implementation and then in the standard. In this way, all users get the benefit of excellent design and interoperability—at least in theory.

Parallel federal standards (1046-1051) are under development, and FED-STD-1052 [11] was recently published to standardize the MIL-STD-188-110A

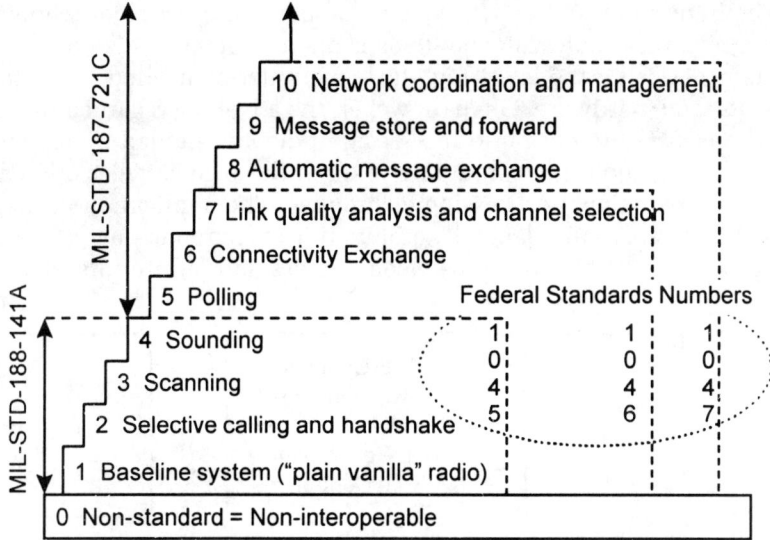

Figure 1.1 The Stairway to Heaven.

HF data modems and the HF data link protocol (also included in MIL-STD-187-721C). At the time of this writing, the military standards community is beginning work on a third generation of HF automation standards. It's truly an exciting time to be working in the HF business.

1.1.3 Automated HF Communications

1.1.3.1 Overview

An ALE-capable station consists of an ALE controller, a controllable single sideband (SSB) radio, and transmission equipment such as antennas and couplers. More advanced automated stations also include high-speed HF data modems, networking controllers, and so on, as shown in Figure 1.2. Figure 1.3 shows a functional breakdown of such an automated HF station, where the functions are shown in layers along the lines of the ISO Open Systems Interconnection Reference Model (OSI-RM). In this figure, only the functions at the network layer and lower are depicted.

In addition to automated controllers, great progress in the development of HF transmission equipment has been made in recent years. In fact, state-of-the-art professional equipment can be modeled very accurately as linear elements with frequency translation functions. Older equipment is less ideal and its imperfections need to be taken into account during both the system design and equipment selection processes.

The importance of key HF system components, particularly antennas, to the success of these automated systems is often underestimated because of the customer's desire for the latest high-tech computer-controlled or digital signal processor-based hardware. Often, however, the advantages gained from careful application of traditional radio link-design practices—aided by modern computer analysis and design techniques—exceed those advantages offered by use of state-of-the-art signal processing techniques. The application of these techniques, though not as intellectually stimulating to computer-savvy trained engineers, can make the difference between success and failure in system design.

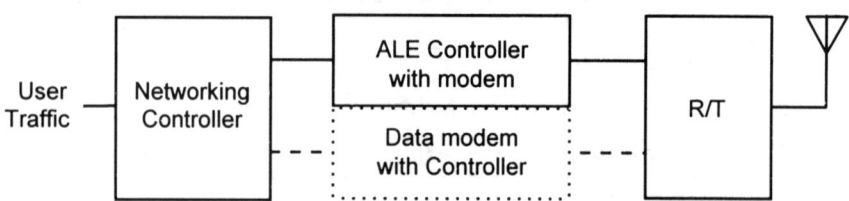

Figure 1.2 Automated HF station block diagram.

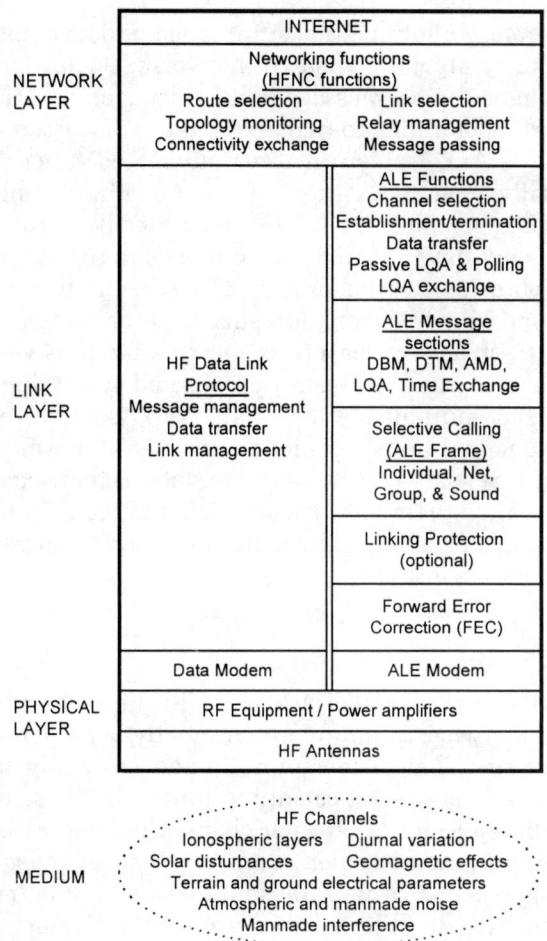

Figure 1.3 Hierarchical layers of an HF radio system.

For example, if the antenna system places a null in the direction of propagation, thereby attenuating the peak received signal strength 50 dB below the background noise, few if any signal processing algorithms will be able to recover the transmitted information. Likewise, choice of a nonpropagating frequency renders the rest of the system useless. Automatic evaluation and selection of operating frequency is one of the key features of ALE technology.

To many users, ALE is the most visible aspect of modern automated HF systems, and a detailed discussion of the best known ALE technology is the subject of two chapters of this book. However, the worldwide HF engineering community is currently applying the entire range of data communication network techniques to create automated HF networks. Such topics as automatic network topology monitoring, message routing using any available media, and automated network operation and management are discussed in Chapter 9.

The essence of ALE is seen in three automated functions that are not present in manually operated stations: (1) automatic channel selection, (2) automatic channel evaluation, and (3) selective calling. Automatic channel selection has distinct aspects at ALE transmitters and receivers: Receivers rapidly scan many channels listening for ALE calls, whereas transmitters select the channel for a transmission based on the results of automatic channel evaluation. Channel evaluation, an active measure as compared to passive channel prediction, employs measurement of HF energy propagated by the channels of interest, either by specialized equipment such as a "chirp" sounder or by measuring the effects of the channel transfer function on the waveform of interest (such as the ALE modem). Selective calling refers to stations that, except when linked to another station, mute their speakers and listen for calls from other stations. The selectivity in calls is achieved by including station addresses in transmissions that seek to establish links.

1.1.3.2 Scanning

All MIL-STD-188-141A and FED-STD-1045 ALE-equipped stations, when operational and not otherwise committed, continually scan a preselected set of channels, or *scan set*. They listen for calls and are ready to respond. The channels in the scan set are repeatedly scanned in the same order with a *minimum* dwell time (T_d min.) on each channel defined as the reciprocal of the scan rate (several channels per second). When a transmitter wishes to capture a scanning receiver, it will transmit a signal that is recognized by the scanning receiver. The duration of this scanning call must be sufficient to ensure that, if the receiver is indeed scanning for calls, it will land on the channel carrying the scanning call before the transmitter ceases emitting.

The minimum scan period (T_s min.) is simply the product of the number of channels (C) times the minimum dwell time on each channel (T_d min.); that is, T_s min. $= C \times T_d$ min. However, when a scanning receiver detects ALE signaling on a channel, it may extend its dwell time on that channel. This extended dwell time is needed for the station to determine whether or not the signal is a valid call, and if so, whether or not it is itself the destination of the call. This channel dwell time may be increased to the duration of two ALE words ($T_{drw} = 2T_{rw}$), where an individual word duration is abbreviated as T_{rw}. Therefore, when a transmitting station initiates a call to a second station, it must first compute the duration of its scanning call (T_{sc}) to maximize the probability that the destination station will be able to read the scanning call.

It must therefore assume a receiving scan period (T_s) based on the probable *maximum* pause (T_d) to read words on each channel, or T_{drw}. Of course, a net manager may adjust the value of T_d to optimize performance in a network composed of multiple ALE-equipped stations. These considerations suggest some of the complexities involved in the important scanning process.

1.1.3.3 Selective Calling

Referring to Figure 1.3, selective calling in an ALE system involves the exchange of ALE frames among stations. This selective calling capability supports all higher level ALE functions, including link establishment, data transfer, and so on. The general structure of an ALE frame (Fig. 1.4) consists of one or more destination addresses, an optional message section, and a frame conclusion, which contains the address of the station sending the frame.

The fundamental ALE operation of establishing a link between two stations proceeds as follows: (1) The calling station addresses and sends a call frame to the called station; (2) if the called station "hears" the call, it sends a response frame addressed to the calling station; and (3) if the calling station receives the response, it now knows that a bilateral link has been established with the called station. The called station does not yet know this, however, so the calling station sends an acknowledgment (or ACK) frame addressed to the called station. At the conclusion of this three-way handshake, a link has been established, and the stations may commence voice or data traffic, or simply note that they *can* communicate and drop the link.

In addition to the ability to establish individual links, the ALE standard also describes net and group calls, and sounds as follows:

- A *net call* is addressed to a single address that implicitly names all members of a *prearranged* collection of stations (a net). All stations belonging to the net that hear the net call send their response frames in prearranged time slots. The calling station then completes the handshake by sending an acknowledgment frame as usual.
- A *group call* works similarly, except that an arbitrary collection of stations is named in the call. Because *no prearranged net address has been set up*, each station must be individually named. Called stations respond in time slots, determining their slot positions by reversing the order in which

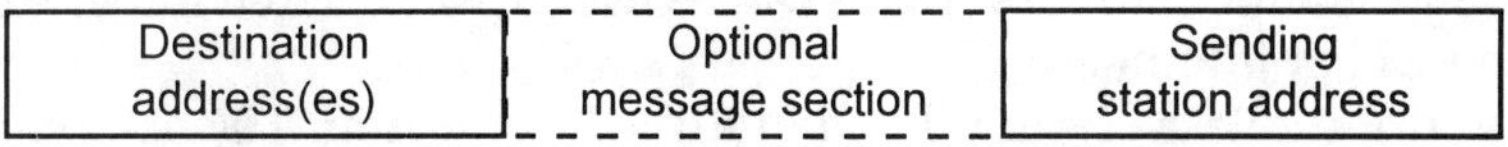

Destination address(es)	Optional message section	Sending station address

Figure 1.4 ALE frame structure.

stations were named in the call. The calling station sends an acknowledgment as usual.
- Finally, a *sound* is a unidirectional broadcast of ALE signaling by a station to assist other stations in measuring channel quality. The broadcast is not addressed to any station or collection of stations, but merely carries the identification of the station sending the sound.

This ALE process is vital in the establishment of HF links, irrespective of their propagation, noise, and interference environments. Once the link is established, however, information transfer over the link is controlled by the operator for voice traffic or the data link modem and protocol for digital communications as was shown in Figure 1.2.

1.2 HIGH-FREQUENCY RADIO IN THE GLOBAL INFORMATION INFRASTRUCTURE [12]

Daily advances in the international information infrastructure bring concepts from science fiction into the everyday fabric of society. Only a few years ago, the rapid, asynchronous communication provided by electronic mail (e-mail) became a compelling argument for buying a computer for the home. Today, home computer users contemplate personal "agents" that will nightly scour information sources to produce a customized electronic "newspaper" in time for breakfast.

Underlying this explosion in new capabilities is a dense web of networks including the global telephone network and the Internet (the latter relying largely on the services of the former). However, some communities of potential users of these information resources are not well served by the existing networks. Many of these users could make good use of HF radio technology for on-ramps or bridges to the information superhighway.

1.2.1 Introduction

In building the global information infrastructure, communication planners have emphasized terrestrial high-bandwidth service to urban users. Satellite links are assumed for customers who are either mobile or too distant from the high-capacity backbones for terrestrial service. However, automated HF radio provides a viable low-cost alternative for many of these users. This section presents an overview of how HF radio systems can be usefully integrated into the rapidly expanding global Internet.

Due to the relatively low bandwidth available from HF data channels, most scenarios for routing Internet traffic through HF subnetworks involve special circumstances:

- *Voice or data to remote locations:* HF is currently in use to provide relatively low-cost voice service to locations too remote for economical landline or line-of-sight radio service. With the addition of modern HF automation, such remote sites can be linked into HF subnetworks, with multiple gateways into the information infrastructure to improve the robustness of connectivity to these sites as shown in Figure 1.5.
- *Voice or data to mobile platforms:* For communications to mobile platforms beyond line of sight, HF provides an economic alternative to satellite communications. ALE has been shown to largely alleviate the link-level connectivity problems that formerly plagued HF. Automated HF node controllers (HFNCs) will integrate individual voice and data terminals, as well as the networks aboard larger platforms, into the high-bandwidth, low-cost stationary infrastructure. For example, shipboard LANs (see Fig. 1.6) may be linked within a task force using UHF, VHF, and HF radio (as appropriate for each link), with long-haul trunks carried by an optimized mix of satellite and HF radio.
- *Emergency connection to severed networks:* Natural or man-made disasters can sever segments of our backbone networks. A backup network of automated HF radio stations can quickly detect and bridge such faults to carry high-precedence traffic into and out of emergency areas (see Fig. 1.7). Bandwidth limitations will require precedence and preemption mechanisms to optimize use of the HF links.
- *Connection to rapid-deployment networks:* From disaster areas to combat theaters, the transportability, low cost, and long range of HF radio make

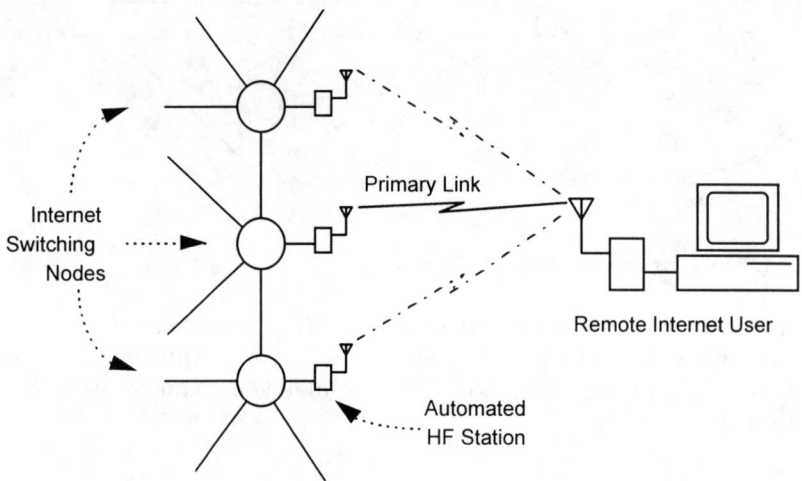

Figure 1.5 Internet entry from remote locations.

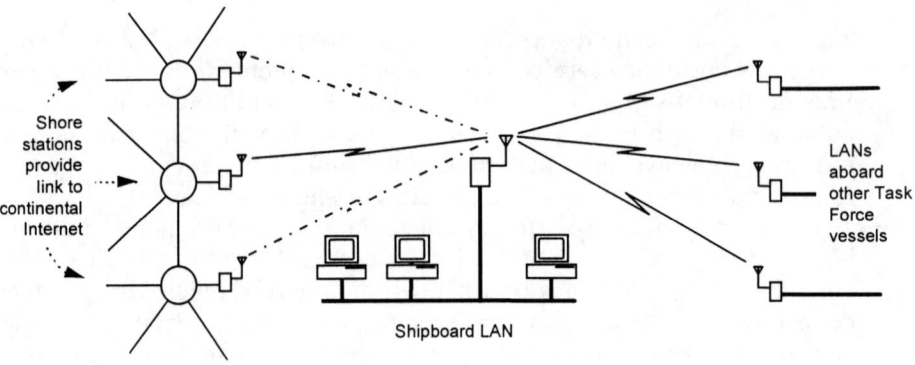

Figure 1.6 HF networking among mobile platforms.

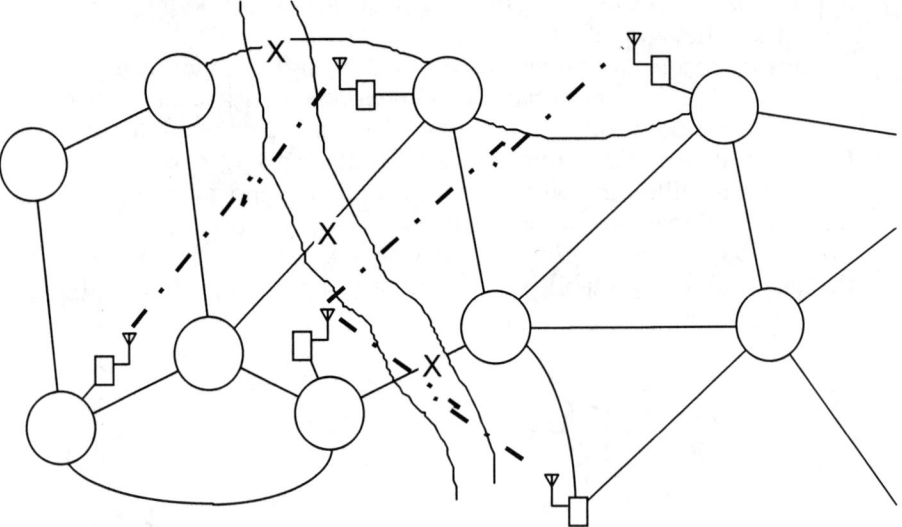

Figure 1.7 HF links connect partitioned networks.

it a primary quick-response medium. With an automated HF link from field networks into the Internet (see Fig. 1.8), deployed teams can use familiar communication tools such as e-mail to ease the transition to operations in the field.

1.2.2 Joining the Internet

The usual uses of HF radio as a backup medium or to extend the terrestrial communications plant to mobile or remote users require that HF networks

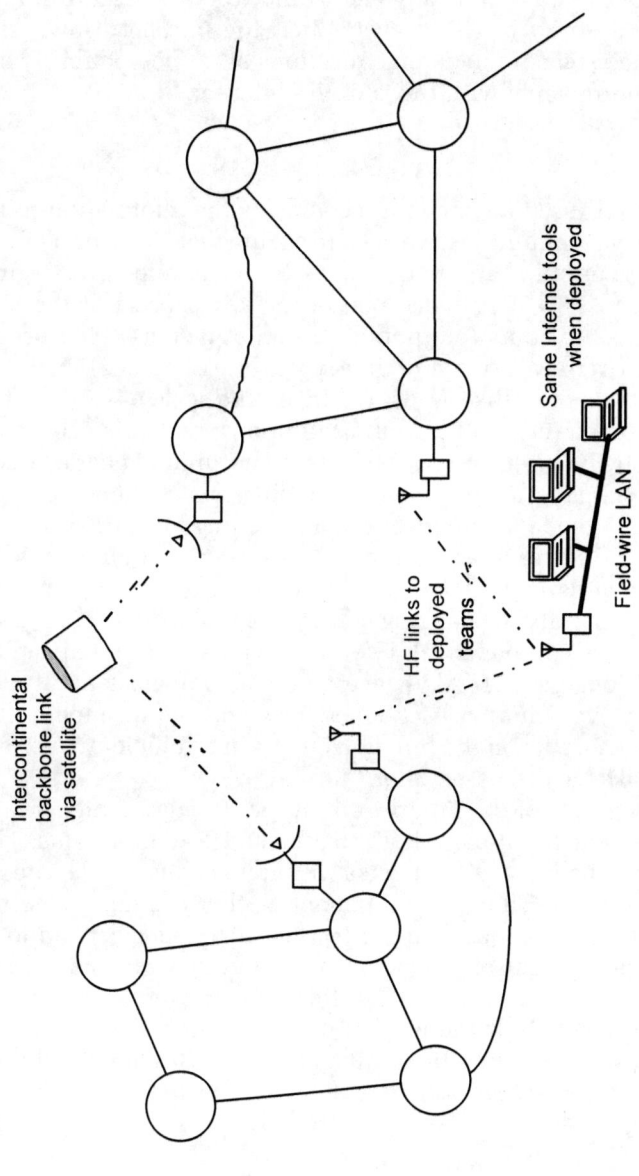

Figure 1.8 Extending the Internet to rapid-deployment teams.

interface with other media. As discussed in Section 9.3.1, a level 4 HFNC is specifically designed to provide the interfaces and protocol conversions required to link HF radio into "any media" networks. In Figure 1.8 we blithely assumed the feasibility of HF gateways into the Internet. However, we have not yet examined some fundamental questions of the compatibility of HF capabilities and requirements with those of the Internet.

1.2.2.1 Compatibility

If HF radio is to be used to extend transparently the information infrastructure to these new user communities, we need to ensure that HF technology is compatible with the assumptions implicit in the Internet architecture. Beginning with the characteristics of HF systems, one of the key aspects of the HF medium that distinguishes it from more popular Internet media is that propagation is highly variable over a wide range of time scales (see Chapter 2) such as multipath effects on the scale of milliseconds, fading on the scale of seconds to minutes, diurnal variation on the scale of hours, and ionospheric disturbances and sunspot activity on the scale of days to years. The unique characteristics of HF technology are largely the results of addressing this challenging environment, including unique modems, interleavers, and coding for shorter-term variations as discussed in Chapter 4, and adaptive frequency and antenna selection for the longer-term variations as described in Chapter 3. The ability of automated HFNCs to adapt rapidly to changing conditions is steadily improving the reliability of HF links (see Chapter 9). However, the data rates that can be reliably achieved over long-haul HF sky-wave channels are substantially lower than those expected over Internet backbones (or even dial-up modem links); this will impose some limits on the functions that can be efficiently performed over HF Internet links, as discussed later.

The architecture of the Internet emphasizes issues at higher layers in the OSI Reference Model [13] than those that make HF radio unique. The essence of the Internet is technology that links disparate subnetworks into a seamless network of networks. The key component of this technology is the Internet protocol (IP), which provides *datagram* service to higher layer end-to-end protocols. Because datagrams sent via a subnet are not guaranteed to emerge from that network in order, or without duplication (or even to emerge at all), the upper layer protocols bear the burden of providing a user's expected quality of service. Thus, the existing IP architecture is already prepared to cope with the vagaries of HF propagation.

1.2.2.2 Performance Limitations

Given that HF networks and the Internet are compatible on this most fundamental level of interoperability, we must examine issues of performance and congestion that arise in HF subnetworks due to the restricted bandwidth of HF links. Although HF modems with data rates of 9600 bps or more are currently in development, achievable throughput over HF links is currently closer to 1200 bps, an order of magnitude less than the rates achievable over wireline modems. For example, a pair of 28.8-Kbps (v.34bis) modems operating within a metropolitan area and using v.42bis data compression may be expected to achieve a user data throughput on the order of 16 Kbps for text transfer. HF data modems, operating over a mid-latitude sky-wave path (Boulder, Colorado, to San Diego, California), were able to achieve about 1-Kbps user data throughput [14].

The data rates required by Internet users vary over several orders of magnitude, depending on the application. Electronic mail typically requires only a few thousand bytes per day, while a World Wide Web server must handle data rates of a million bytes per second. Clearly, users of multimedia applications will be disappointed with the throughput of an HF connection to the Internet, while other users may be satisfied. The key to expanding the class of potential satisfied users of HF Internet links is to improve the throughput of user data.

Three possible avenues for increasing the usable data bandwidth of HF links are as follows: (1) increase the allocation of HF spectrum per link, (2) improve the data efficiency of the modem (in bits per hertz), and (3) increase the information content (reduce redundancy) in the modem bit stream. The first option is beyond the scope of this book; the second option is addressed in Chapter 4, which leaves for this chapter an evaluation of the compression of Internet traffic to improve the response time to users within the constraint of relatively low actual HF link data rates.

The lower data rates of HF links compared to wireline modems, local-area networks (LANs), and so on, allow more time per channel data bit for compression. Although this increased time will not result in commensurate increases in compression, it does allow for somewhat better compression than the high-speed technique commonly used in wireline modems. The figures given later illustrate the improvement in compression achievable when increasing computational effort is applied. All are variants of the Lempel-Ziv algorithm. Timings were collected on an IBM RS/6000 model 580 workstation running AIX 3.2.5.

The compression of text files (Fig. 1.9[a]) is much the same as that for PostScript files (Fig. 1.9[b]). The lowest compression ratio in every case is achieved by the Lempel-Ziv-Welch (LZW) algorithm, which is widely used (e.g., in GIF files and the UNIX compress command) due to its high compression speed. For HF applications, we can achieve 50% better compression by instead using a more aggressive implementation of the Lempel-Ziv algorithm such as that in the GZIP utility (which produced the third point in each case in Fig. 1.9). Graphic images (GIF) and QuickTime movies (MOV) are stored in a compressed

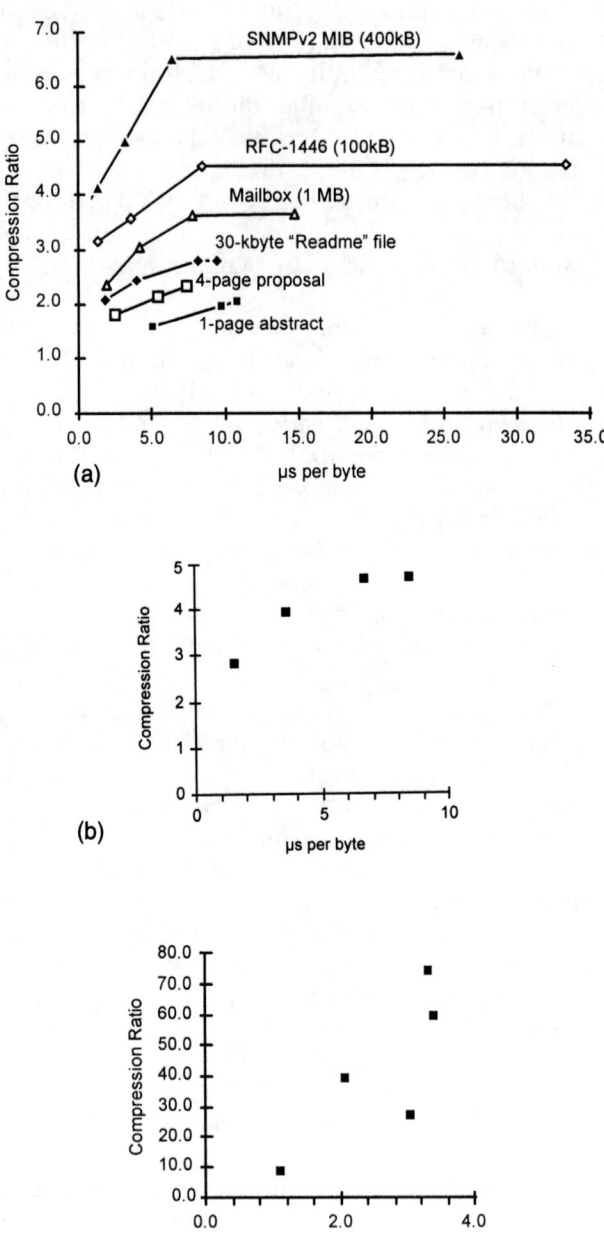

Figure 1.9 File compression examples: (a) Text, (b) PostScript, and (c) Audio (.WAV).

format by default, so no additional compression by networks is usually feasible. However, given the potential payoff for HF users, it might be worthwhile to remove the LZW compression and recompress using more powerful techniques for improved image throughput on HF links.

From these compression results, it is apparent that Internet access via HF is most likely to prove satisfactory for applications that are text based or PostScript based, because the additional compression achievable partially mitigates the lower data rates of HF versus wireline modems. Audio files can also be compressed remarkably well (see Fig. 1.9[c]). However, applications that require rapid transfer of large photo or video files are not likely to work well over 3-kHz HF links.

1.2.2.3 HF Interface to Internet

An inexpensive technique for connecting an HF subnetwork to the Internet is shown in Figure 1.10. Here, a desktop computer (labeled "Gateway") executes off-the-shelf IP software that routes packets among any connected data links.

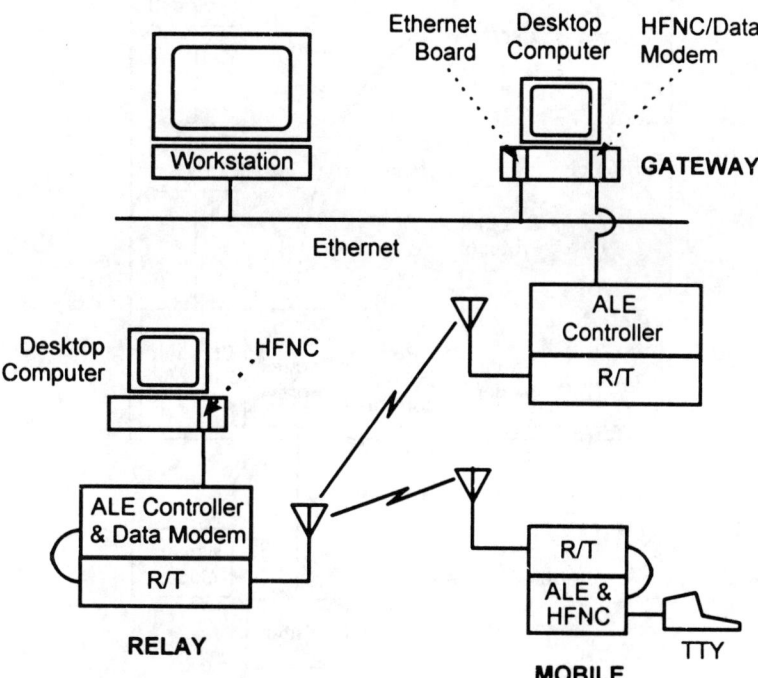

Figure 1.10 HF interface to the Internet.

18   Advanced High-Frequency Radio Communications

The figure shows both an Ethernet board and an HFNC present in this computer, so it serves as the gateway between all nodes reachable via the Ethernet (probably the entire Internet) and all nodes reachable from the local HF station.

The internal structure of such an HF Internet gateway is depicted in Figure 1.11. In this figure, functions and components that are readily available in desktop computers are shown with solid-lined boxes. Most of the remaining

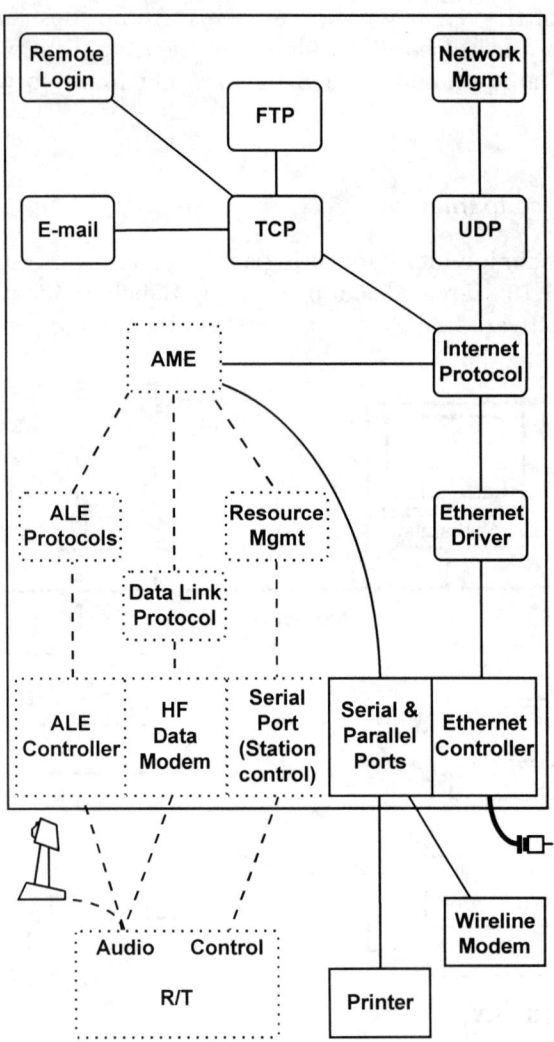

Figure 1.11 HF Internet gateway.

HF-specific components are also available off the shelf, so the changes required to produce an HF Internet gateway from existing hardware and software do not appear to be extensive.

1.2.3 Conclusion

HF radio appears well suited to provide connectivity to and within the Internet for applications that can tolerate the relatively low bandwidth currently available from HF modems. These applications certainly include text-oriented applications, for which powerful compression algorithms mitigate the lower throughput of HF compared to wireline modems. Audio files can also be compressed substantially, but photo and video files will be more difficult to accommodate without incurring very long file transfer times and producing significant congestion of HF networks.

Due to the compatibility between the standards developed for HF automation and the Internet standards, the integration of HF into the international information infrastructure should be relatively painless. Much of the software required for the end-to-end protocols and internetwork routing is commercially available, with documented interfaces.

New developments may be required only to implement the HF-specific protocols and algorithms for routing and station control. This software can be targeted to the inexpensive desktop and portable computers that currently run the higher layer protocols and applications, and will benefit from the mature development environments available for these machines.

PC cards are currently available that supply both an Ethernet interface and a fax/data modem in a credit card size package. The day may not be far distant when a PC card implementation of the automated HF technology described here will connect a palmtop computer to a LAN and an HF transceiver to form a complete pocket-sized HF Internet gateway.

1.3 HF SYSTEMS ENGINEERING

1.3.1 Purpose

The field of *systems engineering* has gained increasing attention since its recognition as a discipline in the late 1940s. This recognition has been stimulated by the increasing cost and technical complexity of development and acquisition programs. Some of this attention is no doubt due to large program failures that could possibly have been avoided, or at least mitigated, by proven systems engineering practices [15]. This section presents an introduction to HF systems engineering and alternative approaches for developing a system design and implementation.

The purpose of systems engineering is to prevent failures through a unified approach that begins with the complete definition of all system *requirements* used to derive a final system configuration shown to satisfy those requirements before detailed development commences. The definition of the terms *system* and *systems engineering* depends on the application. The U.S. Department of Defense definition [16], derived from MIL-STD-499A [17], arguably provides the broadest and most widely accepted interpretation:

> System engineering is the application of scientific and engineering efforts to (1) transform an operational need into a description of system performance parameters and a system configuration through the use of an iterative process of definition, synthesis, analysis, design, test, and evaluation; and (2) integrate related technical parameters and ensure compatibility of all physical, functional, and program interfaces in a manner that optimizes the total system definition and design; and (3) integrate reliability, maintainability, safety, survivability, human, and other such factors into the total engineering effort to meet cost, schedule, and technical performance objectives.

In its simplest terms, systems engineering includes both technical and management functions and is often referred to as a *front-end process.* That is, the majority of systems engineering tasks are completed in the initial phase of the program, termed *realization* by M'Pherson [15], when about 5% of the program's funding is expended [16]. The purpose of this section is to outline some of the key elements of the systems engineering *process* as it applies to the design of HF communication systems [18].

1.3.2 The Systems Engineering Process

The systems engineering process can be considered as a sequence of design stages followed by implementation and *through-life* support phases as shown in Figure 1.12. Similarly, the HF system design process can be considered as a number of iterations through an archetypal procedure illustrated in Figure 1.13 [19]. Note that the output from each stage of the process provides ever-increasing detail about the system to be created. In this regard, each design process is fueled by either a *needs statement* or a *requirements specification.* The output of each process is either a more detailed requirement specification for the next design phase or, ultimately, information for creation of the system to be implemented.

The first stage of the system design phase is the generation of key performance parameters and system configuration from the operational needs statement. During this initial stage, dubbed *requirements analysis,* these needs define

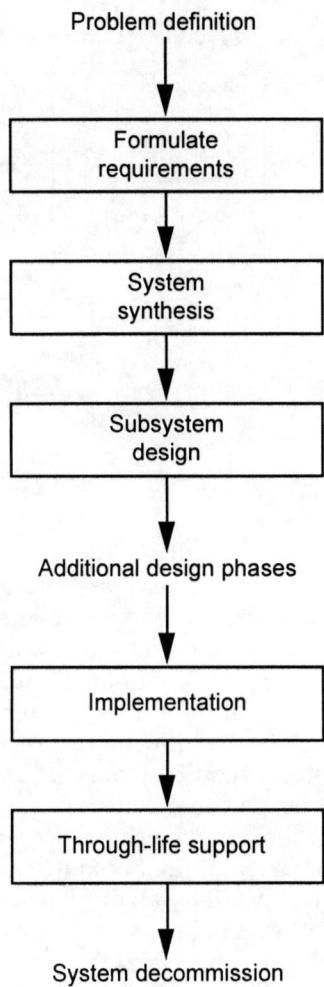

Figure 1.12 HF systems engineering process.

the top-level functions to be performed by the system. The outcome of this process is a complete set of system requirements documented in a system specification, in which both preferred and mandatory requirements can be defined. In the case of larger, more complex systems, this document would be augmented by system support specifications and a number of subsystem or equipment specifications. Subsequent system design processes partition the system requirements by function or discipline until a level is reached where specific hardware items or software routines are identified and specified.

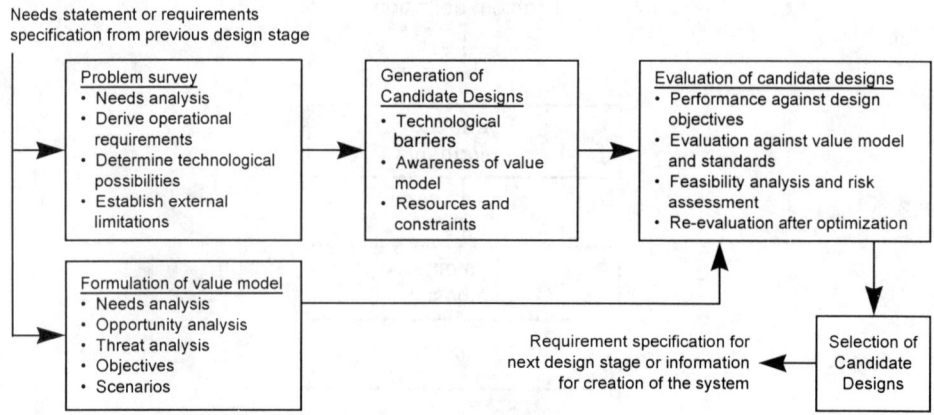

Figure 1.13 An archetypal design process.

The partitioning process, called *functional decomposition,* usually involves the use of analysis and simulation to translate and allocate all the system-level functional requirements to equipment-level specifications [20]. Examples of these specifications include error budgets, power consumption, reliability, physical size limitations, communication traffic loading, and information transfer latency. Note that nonfunctional requirements are necessarily included in this process and *constrain* the form or content of the ultimate design. One such nonfunctional requirement is the limitation on space available for the antennas imposed by physical constraints at the chosen antenna site. Thus, the common perception of an equipment specification as consisting of a single page of predominantly functional requirements is dangerously inadequate. The resulting system may fail to satisfy all user requirements. This failure is likely if inadequate attention is paid to such factors as availability, reliability, maintainability, supportability, survivability, ergonomic factors, safety, and internal and external compatibility [15]. Hence, it is not only desirable, but essential, for the specification to encompass all of these subjects.

Systems engineering integrates the design and management of the system acquisition process. In fact, the number and content of subsequent design stages are determined more by the nature of the contractual relationship between the customer and the supplier than by technical considerations. Thus, the next section investigates possible customer-supplier relationships and their impact on the system design process.

1.3.3 The Customer–Supplier Relationship

The nature of the relationship between purchasers and suppliers of communications systems has been evolving considerably in recent years along with government and industry procurement policies. A wide range of procurement philosophies is currently in use. Several samples of alternative procurement approaches are described next along with their relative advantages and disadvantages.

1.3.3.1 System Design by Customer

The traditional acquisition approach was that an organization requiring a communication system would undertake the systems design itself and then call for vendor bids to supply their specifications within certain contracting conditions. As communication systems became more complex, consultants were used to help with the systems design effort. These organizations were characterized by an extensive track record in systems design, knowledge of state-of-the-art equipment, and cognizance of the system design trade-offs. Most importantly, they understood the important trade-off between economic and schedule risks incurred when using customized equipment to *meet* requirements versus the lower technical risk of employing exclusively nondevelopmental items to *approach* requirements.

When the system design is complete, contracts are let for the implementation and commissioning phases. Although the delivered system meets the agreed specification, it may fail to satisfy the customer's real needs. The shortcoming of this customer-supplier relationship occurs because of an incomplete mapping of user needs into the system specification. In fact, the probability of this failure is increased as more levels of contractors are added between the eventual users and the equipment suppliers.

1.3.3.2 System Design by Prime Contractor

One approach to overcome the multiple contractor problem is to select a single prime contractor for overall system design, implementation, and commissioning. Thus, the original customer-developed specification addresses only the highest level of user requirements. As an example, the contractor may be required to install an HF system that can "provide, store, and forward message traffic services between two fixed points with a delay of less than one minute." It is then the contractor's responsibility to define the customer's concept of "message traffic" and then supply a system that can realize this concept. This approach has the advantages that (1) system design is performed by the most capable of the two parties and (2) if the system fails to satisfy the user's needs, it is the prime contractor's responsibility to rectify the situation.

A potential disadvantage of this approach, however, is that the customer is still required to assess whether the user's needs statement can really be met by the proposed system. This disadvantage is a major source of technical risk for an HF radio system. In particular, the variability of the ionosphere makes it impractical to verify performance parameters over the full range of propagation conditions expected during the 11-year solar cycle. To minimize this risk, computer modeling and simulation are used to forecast system performance and verify that user requirements are met for a full range of expected conditions. Once these efforts are completed, the primary challenge remaining is to determine if the forecasting results adequately represent the user's operating scenarios. Thus, the customer-supplier interface in this procurement approach is no longer an engineering specification, but a requirements specification supplemented with detailed test and evaluation criteria. Of course, the contractor counters the increased responsibility inherent in this design approach by raising the system cost.

1.3.3.3 Shared System Design by Customer and Contractor

It is common practice to use a combination of the preceding two methods to procure HF radio systems. In this case, a functional specification is prepared defining system operation together with a range of technical performance requirements that must be met. A typical system acquisition strategy adopted for large projects is founded on partitioning the system design into a number of phases, and then contracting each phase to two or more competing companies. This approach ensures competition, provides alternative solutions, and often gives the customer a greater sense of control. This approach is effective and appropriate when the technology being procured is stable and the system is not urgently needed.

However, this combined approach exacerbates a frequently encountered problem when procuring a large complex system: Acquisition time often extends to 10 or more years. When dealing with communications technologies, obsolescence can occur even before the specifications are complete. Thus, the system is found to be of limited value when finally installed. It is therefore increasingly common for this approach to fail in providing a system that meets user needs *on schedule.*

1.3.3.4 Partnership Approaches

A number of novel acquisition schemes based on *partnership* approaches are being developed to overcome system-design failures. The safeguards of the systems engineering approach, rigid specifications and associated contractual agreements, are replaced by a much looser partnership agreement between the eventual user and the selected supplier. The two organizations work closely together to perform the design functions and to formulate implementation and support functions. These partnerships often draw on the commercial information technology industry for inspiration. In this regard, they adopt strategies that reflect the rapidly changing nature of hardware and software seen in the industry.

In place of the classical linear systems engineering approach, systems design, acquisition, and integration become a continuous process. In this process, the system is continually upgraded to reflect the latest developments in the field. *Incremental* acquisition and *evolutionary* acquisition are names often associated with the establishment and maintenance of a capability using these partnership approaches. In certain systems, product lifetimes are so short and the hardware so reliable that through-life support reduces to simply replacing faulty units with new ones. Operational support, such as system tuning and training, take on major importance. The partnership approach requires greater levels of trust between the parties. The difficulties associated with building this trust can be rewarded by lower overall system costs, shorter lead time before the introduction of capability, and greater probability of a successful outcome.

1.3.3.5 Choice of Customer–Supplier Relationship

The preferred customer-supplier relationship for a particular job depends on the acceptable lead time to provision the desired capability and account for the rate of change of technology. A useful approach for the acquisition of HF radio systems is to (1) procure the high-cost, stable-technology items such as the radio transmitting and receiving equipment and the radio station infrastructure using a functional specification, then (2) use an evolutionary approach for such elements as the equipment to support source encoding, error correction, and modulation. In this way, much of the project expenditure is carefully controlled, and yet the system can still gain advantages from improvements in emerging technology and systems.

1.4 HF SYSTEM DESIGN

As already mentioned, a complete HF system design usually comprises a number of phases. The exact number and content of each is determined by the procurement philosophy and the size and anticipated cost of the project.

1.4.1 Requirements Definition

Regardless of the exact procurement strategy selected, it is ultimately necessary in the systems engineering process to interpret the customer needs and prepare a requirements specification. This specification describes the top-level system functional requirements, including key interfaces to other systems, and essential design constraints. This document has been given various names, including operational specification, functional specification, or, simply, the system specification. Regardless of the name, the purpose of this specification is only to describe the purpose and objectives of the system, not to specify system components or how the design is to be performed. Moreover, functional decomposition should be avoided at this initial specification level.

Customer organizations often rely on the traditional procurement approach to generate detailed and restrictive system specifications in the belief that this approach reduces project risk. This approach not only reduces the design options available to system contractors, but also can motivate the contractor to focus on the creation of overly detailed specifications rather than defining the user's real needs. Instead, a more modern approach focuses on development of the most abstract specification that fully encompasses the user's need statement. This approach is consistent with MIL-S-83490 Type A System Specifications:

> Type A specifications shall state all necessary requirements in terms of performance, including test provisions to assure that all requirements are achieved. Essential physical constraints shall be included. Type A specifications shall state the technical and mission requirements of the system as an entity. Specifications shall include requirements for specific functional areas, interfaces between functional areas, interfaces with other systems, and application of any known specific existing equipment.

Systems specifications are commonly prepared in prescribed formats such as that described in MIL-STD-490A. This methodology is useful both because its format is comprehensive and because its procedures minimize the risk of ignoring a critical issue. At this top level of specification, however, the requirements document is simply a statement of needs elicited directly from the user. The document does not contain derived requirements from functional decomposition or any other results from subsequent stages of the systems engineering process.

1.4.2 Formation of Candidate Designs

Once the requirements document is completed, HF system design proceeds with generation of multiple candidate designs that each meet the functional system specification. Trade studies are then conducted to evaluate various performance-related aspects of these designs until a preferred design is selected.

The outcome of this phase of HF systems engineering is the determination of key system parameters, including (1) the number and location of transmitting and receiving stations; (2) the minimum required effective radiated power (ERP) and number of transmitters per station; (3) infrastructure requirements, such as power supply, air conditioning, receiving site noise limits, and buffer zone size; (4) radio link description, including modulation, coding, protocols, radio access scheme, etc.; and (5) expected cost and implementation schedule. The selection of these features for the preferred design completes a critical stage of system acquisition.

Two approaches have evolved for the definition and evaluation of candidate designs and selection of a final design. The first approach bases definition of candidate architectures by comparison, or analogy, with existing systems. This approach draws on decades of systems engineering and operational experience, thus minimizing technical risk and the need for a fully capable engineering team. This *comparative* approach is particularly applicable in the replacement or upgrade of an aging transmission system. In this case, it is necessary to first assess how well the current system meets current requirements, which may have changed since the design of the original system. The results of this assessment are then used to judge the performance of one or more alternative designs in meeting these original requirements.

The second design approach employs the classic *first principles* of radio engineering based on achieving a specified HF link power budget. This link budget may be specified in several ways depending on the known or fixed design parameters and the unknown parameter to be determined. For example, the required transmitter output power P_t in dBW is given by

$$P_t = N_r + \text{SNR} + L_p - G_t - G_r + L_t + L_r, \tag{1.1}$$

where N_r is the noise power at the receiver input in dBW; SNR is the required signal-to-noise ratio in dB; L_p is the HF path loss between transmitting and receiving antennas defining the HF link; G_t and G_r are the transmitting and receiving antenna power gains in dBi, respectively; and L_t and L_r are the combined transmitting and receiving system losses in dB, respectively.

The link budget approach is typically more time consuming and expensive than the comparative approach. It often requires significant radio system-design knowledge, expertise with propagation and noise models, and greater familiarity with contemporary HF communications equipment. This method is appropriate for the design of large systems whose expense would easily justify the additional design cost. It is arguably the only option for those systems planned for upgrade using the latest digital techniques, because there is no operational experience from which to evaluate the performance of alternative components. It is therefore useful in practice to exercise both the comparative and link budget methods whenever possible to give confidence in the ability of the candidate designs to meet the user's needs.

1.4.3 The Comparative Design Approach

As a design example, consider upgrading a broadcast transmission facility for an analog voice service. This upgrade is typically undertaken for two reasons: (1) Listeners at the periphery of the broadcast area are complaining that the signal is often "noisy," and (2) the current transmission equipment has reached the end of its economic life and is becoming difficult to maintain.

In general, a "noisy" signal implies that the SNR at the demodulator input is inadequate to meet user speech intelligibility needs. In this case, it might be suspected that noise or interference power at the receiving sites has increased since installation of the original system. This increase could be the result of greater congestion of the limited HF spectrum caused by (1) changes in ionospheric conditions that enable interfering HF links with an increasing number of distant transmitters, (2) an increase in the number of HF users and user groups requiring assigned spectrum, and (3) a decrease in the number of frequencies employed per user for ALE. In addition, man-made noise levels near the affected receiving sites may have increased due to the introduction of industrial, residential, vehicular, or other sources of electromagnetic interference. Otherwise, the link could be operated during a sunspot minimum, so the received power via the sky-wave signal could have decreased. In any case, interference and noise measurements can be performed during careful site surveys at proposed station locations. This measurement campaign would include recording and analysis of received signal levels on the frequencies of interest and noise measurements on unoccupied frequencies or frequency bands.

Regardless of the cause of the reduced SNR, the link power margin must be increased if the receiving systems are to be left unchanged in the same locations. In other words, the signal strength arriving at the receiving sites will need to be increased. The magnitude of that increase can be estimated from the current measured shortfall and achieved by an increase in radiated power at the transmitting site or relocation of receiving sites, if possible. An increase in radiated power can be realized by either increasing the transmitting power into the antenna or by increasing the antenna gain in the desired directions. Thus, the candidate solutions would typically span a range of transmitter power, antenna system upgrade, and receiver-siting options.

The preferred option is often selected on the basis of cost, acceptable risk, and long-term supportability. For example, an elaborate antenna array driven by relatively low output solid-state power amplifiers may be selected to provide the increased margin. This configuration could offer lower operating costs, lower maintenance requirements, and higher reliability than an alternative approach using a large tube-based power amplifier and a single antenna. The array solution might well have a higher initial cost, but this higher cost would be offset by the lower operating costs (prime power) and lower expected downtime of the tube amplifier solution.

1.4.4 Design by Link Budgets

Link budget design for analog transmission is based on achieving at least a specified minimum SNR at the receiver [21]. For digital systems, however, the key design driver is service delivery. A modern service delivery measure is the number of e-mail messages delivered per hour with at least a certain quality of service (e.g., character error rate). In this application, remote users are "connected" to a network gateway via packet-based digital HF links. For this reason, the delay between user keystrokes and the network return response must be minimized. Consider the simplest class of communication channels conforming to the additive white Gaussian noise (AWGN) channel model, such as fiber optic and satellite channels. With these systems, it is possible to perform a precise link budget analysis and determine the signal SNR required at the receiver for various candidate link configurations to satisfy the design objectives [22,23].

Calculating the available link power budget for a typical HF radio system is far less straightforward than the analogous calculation for satellite or fiber optic communications. As shown in Chapter 2, this complexity arises because HF radio signals suffer propagation distortion of various types and the background atmospheric noise is decidedly non-Gaussian. Moreover, both the signal strength and the noise power vary continuously and are subject to external influences beyond the control of the link designer. Despite these difficulties, it is instructive to attempt HF link design, and hence overall communication system design, by the link budget method. The principal design inputs are the location of the HF terminals and the data transmission requirements. From these inputs, it is possible to derive the transmission distance, propagation characteristics, required antenna take-off angles, and receiving site noise characteristics. The data transmission requirements bound the minimum modem data rate, error-control strategies, and protocol options.

Because many of the link variables are interdependent, it is necessary to fix some of these parameters in order to begin the design process. A useful starting point for this process is to determine the minimum modem data rate and assign a maximum error probability. A minimum 50-byte-per-second transfer rate, for example, implies a data transfer rate of 400 bits in every 500-ms HF transmission burst. The modem must therefore support a minimum transmission rate of 800 bps plus the overhead introduced by the transmission protocols. A figure of 100% overhead is typically assumed, so a 2400-bps modem is indicated for this application because it will provide the necessary throughput with some margin for uncertainties in implementation and performance.

Next, a minimum SNR must be established to support 2400-bps transmission over the assigned HF frequencies at the specified locations. In general, modem performance is highly dependent on the error-control techniques employed. These techniques include (1) forward error correction (FEC), the addition of redundancy bits to enable error bit detection and, if possible, correction at the receiving site; and (2) the reordering of data and coding bits, called *interleaving,* before transmission to space burst errors in the received bit stream in order to improve FEC performance. Note that interleaving can be performed over short or long time intervals depending on the expected duration of error bursts. In an e-mail message communication application, the time-division duplex method of link operation, combined with the need for user keystrokes to elicit a response, typically within 3 sec, limits the use of long interleaving.

To proceed with modem selection, it is necessary to establish the nature of HF channels. Caruna [24] and Priess [25] have shown from simulations and measurements, respectively, that the CCIR *Good* HF channel model [26] is appropriate for Australia. From Chapter 4, a *52-tone modem* is shown to need an SNR as much as 10 dB lower than earlier modems to achieve the same average bit error rate (BER). The SNR/BER modem-performance curves yielding this result assume uniformly distributed bit errors rather than actual burst errors. For this reason, it is not possible to predict accurately the error rate needed to provide the specified quality of service. This result makes the packet transmission error rate much lower than would be expected if the BER were uniform.

Ideally, the curve of "goodput," that is, the error-free packet throughput, would be specified versus SNR for the candidate two modems in the CCIR *Good* channel. For each candidate modem, this figure gives the required SNR values of 15 and 25 dB for the 52-tone and single-tone modems, respectively. These results suggest that the 52-tone modem is the preferable choice for an e-mail-type application.

The next step in the design is the calculation of the necessary ERP defined by the antenna gain pattern and transmitted power needed to achieve a 15-dB SNR. From (1.1), the required ERP to achieve the required SNR value *SNR* is given by

$$\text{ERP} = P_t + G_t - L_t = N_r + \text{SNR} + L_p - G_r + L_r \qquad (1.2)$$

given a specified noise power N_r. The noise power is determined by natural noise power from atmospheric and galactic sources, man-made noise, and interference from other users of the HF spectrum. The first two sources of noise can be predicted from CCIR reports [27,28], whereas data quantifying interference power for any portion of the earth are usually unavailable. Interference data have been collected for Northern Europe and used to generate the Laycock-Gott interference model for that region [29,30].

The signal power arriving at the input of a receiver is determined by the ERP generated by the transmitting station, HF propagation path loss (L_p), the receiving antenna gain, and receiving system losses (L_r). The ERP encompasses the transmitter output power (P_t), the feeder cable losses (L_t), and the transmitting antenna gain (G_t). In practice, the use of high-power amplifiers with output powers, P_t, in excess of 1 to 4 kW is undesirable. This limitation occurs because power levels in excess of these values represents a technological barrier for preferred solid-state amplifier designs. Although output powers above these values can be achieved in practice, they usually entail a parallel architecture or the use of vacuum tube technology. Selection of either of these alternative amplifier technologies would result in higher cost than for the typical solid-state design, including a reduction in reliability and an increase in routine maintenance. For these reasons, increased ERP is best achieved, if possible, with the use of higher-gain antennas rather than higher-power amplifiers.

Once a baseline transmitter power is specified, say, 1 kW, the remaining task in link budget design is to select (or design) suitable HF transmitting and receiving antennas. This baseline value can be reduced if suitable antennas are found to provide the necessary ERP and meet the 15-dB SNR required in the e-mail design example. In fact, the ERP is more accurately known as an effective isotropic radiated power (EIRP). In other words, the EIRP is the ERP in the preferred direction relative to an antenna with a perfect spherical pattern. This "preferred" direction is composed of both the azimuth and elevation "take-off" angles defining the direction of propagation over the earth's Great Circle path between transmitting and receiving stations. Thus, if an EIRP of 10 kW is required, this value can be achieved with (1) a low-gain antenna and a large high-power amplifier, or (2) a smaller amplifier and a high-gain antenna or antenna array. If possible, the selection or design of appropriate HF antennas to achieve the desired EIRP with P_t at or below 1 kW is an important link budget design objective (see Chapter 3).

The most effective approach to obtain an accurate estimate of the propagation path loss, L_p, is to use a computer program designed to predict HF channel performance. These programs are typically based on empirical data describing the ionosphere and other terrestrial and solar phenomena that affect HF propagation. In addition, these programs are used to perform complete link performance analysis including the effects of user-specified antenna gain patterns and predictions of atmospheric and man-made noise. Prediction programs provide an excellent tool for consideration of various theoretical link design options without incurring the cost of a "cut-and-try" approach using actual equipment.

These programs permit the designer to get a feel for the important channel and design sensitivities without incurring the expense of field trials. Several useful prediction programs are discussed in Chapter 2.

Several potential "traps" are inherent in the use of prediction programs for system design. The first such trap is that computer predictions could suggest implementation options that may not be either technically or financially viable. For example, the costs of power amplifiers increases nonlinearly as their output power exceeds about 1 kW. Therefore, high-gain antennas may appear to provide an attractive system solution, but they can become physically large and unwieldy if needed to operate at low frequencies in the HF band. Furthermore, as shown in Chapter 3, achieving high-gain at low elevation angles requires high ground conductivity. This high conductivity is unlikely to be found naturally in arid countries with sandy soils, therefore mandating the use of a large ground plane under the antenna. The latter two facts, coupled with a potential increase in the size of the transmitting site to accommodate a large antenna and expansive ground screen, could render this option unusable. As a result, the design process could shift back toward recommending a transmitter output power in excess of 1 kW. This antenna/amplifier dilemma illustrates the iterative design process best performed by computer prediction of HF link performance. More importantly, it illustrates that the use of predictive programs must be tempered by practical knowledge of the HF channel as well as HF communications equipment.

References

[1] Dunlap, O. E., *Marconi, The Man and His Wireless*, New York: Arno Press and New York Times, 1971, p. 186.
[2] Dunlap, O. E., *Marconi, The Man and His Wireless*, New York: Arno Press and New York Times, 1971, p. 189.
[3] Dunlap, O. E., *Marconi, The Man and His Wireless*, New York: Arno Press and New York Times, 1971, pp. 191–192.
[4] Dunlap, O. E., *Marconi, The Man and His Wireless*, New York: Arno Press and New York Times, 1971, pp. 200–201.
[5] Harrison, G. L., "Functional Analysis of Link Establishment in Automated HF Systems," WP86W00015, McLean, VA: Mitre Corporation, Dec. 1985.
[6] Harrison, G. L., and F. C. Leiner, "Proposed Federal Standard 1045—High Frequency Automatic Link Establishment," WP86W00335, McLean, VA: Mitre Corporation, Sep. 1986.
[7] Harrison, G. L., "HF Radio Link Establishment Systems," *MILCOM'87 Conf. Proc.*, Oct. 1987.
[8] Department of Commerce, *Federal Standard 1045 (FED-STD-1045), Telecommunications: HF Radio Automatic Link Establishment*, Washington, DC: General Services Administration, Office of Information Resources Management, Jan. 1990. Available from GSA Specification Section.
[9] Department of the Army, Information Systems Engineering Command, *MIL-STD-188-141A: Interoperability and Performance Standards for Medium and High Frequency Radio Equip-*

ment, Philadelphia, PA: Naval Publications and Forms Center, Attn. NPODS, Notice 2, Sep. 1993. Available on the Internet at URL ftp://tracebase.nmsu.edu/pub/hf/pubs/mil_std_188_141a.

[10] Department of the Army, Information Systems Engineering Command, *Military Standard 187-721C (MIL-STD-187-721C): Interface and Performance Standard for Automated Control Appliqué for HF Radio,* Philadelphia, PA: Naval Publications and Forms Center, Attn. NPODS, Nov. 1994. Available on the Internet at URL ftp://tracebase.nmsu.edu/pub/hf/pubs/mil_std_187_721c.

[11] Department of Commerce, National Telecommunications and Information Administration, Institute for Telecommunication Sciences, *Proposed Federal Standard 1052 (FED-STD-1052), Telecommunications: HF Radio Modems,* Washington, DC: General Services Administration, Office of Information Resources Management, Jan. 1990. Available from GSA Specification Section.

[12] Johnson, E. E., "HF Radio in the International Information Infrastructure," *Proc. Nordic Shortwave Conf. HF'95,* Faro, Sweden: Nordic Radio Society, Aug. 1985. Reprinted with permission.

[13] American National Standards Association, *ISO Open Systems Interconnection Reference Model, ISO 7498, Information Processing Systems, Open Systems Interconnection, Basic Reference Model,* New York: American National Standards Association, 1984.

[14] *Minutes of HF Radio Federal Standard Development Working Group, Meeting Eight,* Las Cruces, NM: New Mexico State University, Feb. 24–25, 1994.

[15] M'Pherson, P. K., "Systems Engineering: An Approach to Whole System Design," *The Radio and Electronic Engineer,* Vol. 50, No. 11/12, 1980, pp. 545–558.

[16] Defense Systems Management College, *System Engineering Management Guide,* Fort Belvoir, VA: Defense Systems Management College, 1983.

[17] Department of Defense, *Military Standard (MIL-STD-499A): Engineering Management,* 1974.

[18] Hamsher, D. H. (editor), *Communication System Engineering Handbook,* New York: McGraw-Hill Book Company, 1982.

[19] Finkelstein, L., and A. C. W. Finkelstein, "Review of Design Methodology," *Proc. IEE,* Vol. 130, Part A, No. 4, 1983, pp. 213–222.

[20] Defense Systems Management College, *System Engineering Management Guide,* Fort Belvoir, VA: Defense Systems Management College, 1983.

[21] Carlson, A. B., *Communication Systems,* New York: McGraw-Hill Book Company, 1986, pp. 311–341.

[22] Carlson, A. B., *Communication Systems,* New York: McGraw-Hill Book Company, 1986, pp. 512–555.

[23] Proakis, J. G., *Digital Communications,* New York: McGraw-Hill Book Company, 1983.

[24] Caruana, J., *Multipath Propagation Study for the DSTO,* Vols. 1 and 2, West Chatswood, New South Wales, Australia: Ionospheric Prediction Service, Radio and Space Services, Department of Administrative Services, 1994.

[25] Priess, M., "The Measurement and Analysis of Several HF Links Using the DSTO Replay Simulator," *Proc. Fourth Int. Symp. on DSP for Communication Systems,* Co-operative Research Centre for Broadband Telecommunications and Networking, Sep. 1996.

[26] CCIR channel models.

[27] CCIR, "Characteristics and Applications of Atmospheric Radio Noise Data," *Reports to the CCIR, 1983,* Report 322-2, Geneva: International Radio Consultative Committee, 1983.

[28] CCIR, "Manmade Radio Noise," *Reports to the CCIR, 1986,* Report 258-3, Geneva: International Radio Consultative Committee, 1986.

[29] Laycock, P. J., M. Morrell, G. F. Gott, and A. R. Ray, "A Model for HF Spectral Occupancy," *Proc. HF Communication Systems and Technology Conf.,* IEE Publication No. 284, 1988.

[30] Gott, G. F., S. K. Chan, K. Pantijiaros, J. Brown, P. J. Laycock, M. Broms, and S. Boberg, "Recent Work on the Measurement and Analysis of Spectral Occupancy at HF," *Proc. HF Communication Systems and Technology Conf.*, IEE Publication No. 392, 1994.

High-Frequency Sky-Wave Channels 2

2.1 OVERVIEW

The performance of HF sky-wave communication systems is affected by the characteristics of the upper atmosphere, solar emissions, magnetic phenomena, the earth's surface, and the noise and interference environment. This chapter describes the upper atmospheric environment and its effects on HF sky-wave propagation before characterizing the noise and interference environment. In this regard, the chapter provides insight into the physical complexities of HF sky-wave channels in order to justify the automated techniques described in subsequent chapters (see Fig. 2.1).

In addition, some well-known computational algorithms are described that are widely used to predict the performance of HF communication systems. This chapter also includes analyses of measurements from specific propagation paths that provide illustrative examples for comparison of measurements with predictions.

2.2 SKY-WAVE PROPAGATION IN A BENIGN IONOSPHERE

2.2.1 Ionospheric Plasma

The earth's atmosphere is composed of many different gases, and its relative composition varies as a function of altitude, latitude, solar activity, and time. The kinetic temperature of the atmospheric gases is controlled by solar radiation. Short-wavelength solar radiation, such as the extreme ultraviolet (EUV), is sufficiently intense during daylight hours to alter the electronic structure of the ambient gas molecules above altitudes of about 65 km. Above this altitude, the incident solar radiation continuously converts a small percentage of the otherwise electrically neutral gas mixture into an electrically dynamic gas called a *plasma*. Plasma is formed when one or more electrons in a neutral gas molecule

36 Advanced High-Frequency Radio Communications

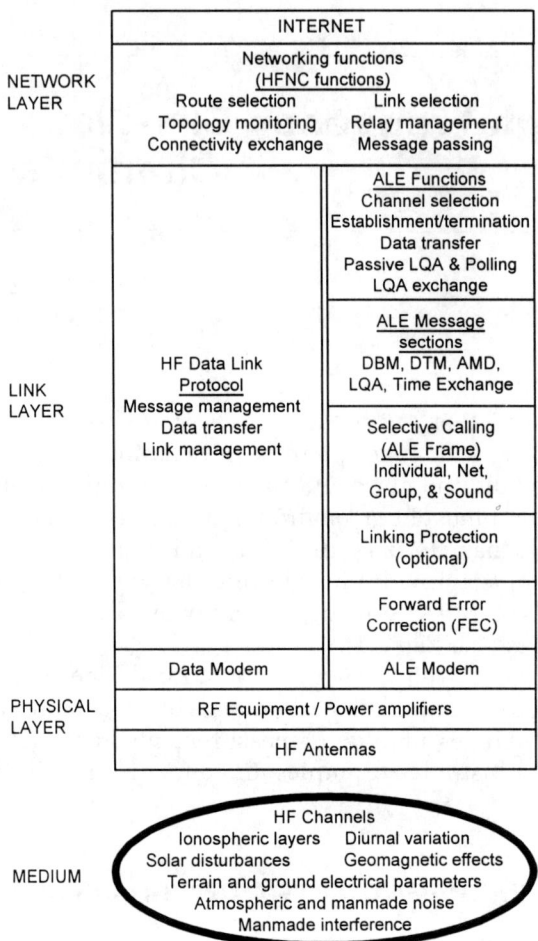

Figure 2.1 The HF channel medium in the hierarchical-layered view of HF communications.

gain sufficient energy from the incident radiation to break out of their nuclear orbits and become free particles. Since an escaped electron has a negative charge, the parent molecule, which now has a net positive charge, is said to be *ionized*. The resulting concentration of ions and free electrons defines the plasma density.

The intensity of solar radiation is correlated with the occurrence of groups of optically visible spots on the sun's surface. The *sunspot number* (SSN) is determined from the number of sunspot groups on the sun's surface that fall within a semicircle about 26 deg in radius centered at the sun's optical center.

The SSN cycle varies from a low near zero to a high of about 200, then back to zero over an approximate 11-year cycle. The current cycle as of this writing, Cycle 22, is shown in Figure 2.2. Two predicted curves are shown as a result of differing hypotheses regarding when the ultimate upturn in SSN values will begin. The SSN is an important solar parameter in determining the long-term characteristics and behavior of the upper atmosphere and its effects on HF sky-wave communication links.

In addition to ionizing a portion of the neutral gas, solar radiation is responsible for breaking down neutral *diatomic* molecules, thereby changing the composition of the upper atmospheric gas. For example, solar radiation at wavelengths less than about 175 nm (including solar Lyman alpha radiation at the 121.6-nm wavelength) breaks diatomic oxygen (O_2) molecules into neutral oxygen (O) through a process known as *photodissociation*. Some of these neutral oxygen molecules are then ionized by solar x-ray and EUV radiation to produce singly charged O^+ ions. Alternatively, O_2 can be ionized directly by incident wavelengths less than about 102.7 nm. Radiation at wavelengths less than about 80 nm can ionize any of the most common neutral molecules (N_2, O_2, O, NO, H_2, He, or H) found at altitudes above about 65 km.

The intensity of ionizing radiation varies with altitude as a function of the angle at which the radiation enters the atmosphere. The production of plasma is therefore dependent on the sun's relative position in the sky, as measured by the solar zenith angle. This production is somewhat offset by loss of plasma due to recombination, diffusion across the day/night *terminator* (the so-called "gray line"), and interhemispheric transport along the earth's magnetic field lines. The gray line demarcates the division between sunlight and darkness in the ionosphere, which is defined here as the region of the upper

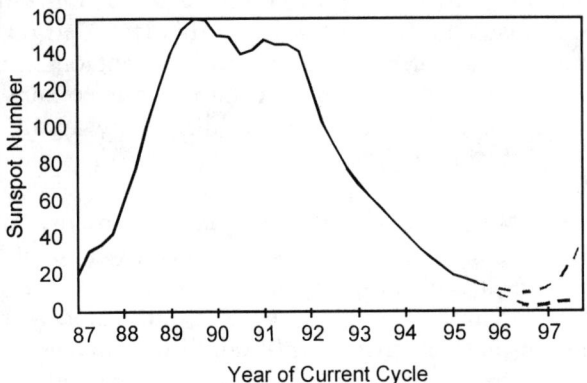

Figure 2.2 Observed and predicted values of sunspot number.

atmosphere where ion densities are greater than about 100 ions per cubic centimeter (ions/cm^3). During daylight hours, the lower boundary of the ionosphere is about 65 to 75 km above the earth's surface, while at night the absence of direct solar radiation causes the lower boundary to move upward to about 100 km. The fact that some ionization exists after sunset is due to the relatively slow recombination of certain ionic species and electrons, diffusion of plasma from the daytime to the nighttime side of the earth, and even reflected solar radiation from the lunar surface.

The different species of gas molecules in the upper atmosphere have different absorption cross sections, and their ability to absorb radiation is a strong function of wavelength. Moreover, the strengths of the bonds between electrons and the molecular nucleus varies from one species to the next. Thus, different radiation wavelengths are responsible for ionizing the various atmospheric gas molecules found at different altitudes. Because the sun emits radiation in a broad spectrum of wavelengths, and because different wavelengths ionize the constituent gases found at different altitudes, ion and free-electron densities exhibit a strong altitude dependence.

2.2.2 Ionospheric Layers

At altitudes below about 80 km, turbulent mixing due to winds and weather patterns creates a homogeneous composition of atmospheric gases known as the *homosphere* [1]. As turbulent mixing diminishes with increasing altitude, *stratification* of the constituent gases becomes more pronounced. At altitudes above about 100 km, the constraints of hydrodynamic equilibrium become apparent and the atmospheric gases begin to form layers. Layers formed from the least massive particles occupy the highest altitude regimes, while the more massive particles form layers at the lower altitudes. The ionized region of the atmosphere, the *ionosphere,* lies primarily in this stratified region. As noted earlier, the different ionospheric gases each have different ionizing wavelengths, recombination times, and collision cross sections, among other characteristics. As a result, the dominant physical processes differ between different altitude regimes within the ionosphere. It is therefore natural to describe the ionosphere as a series of altitude regimes or layers.

When the existence of a conducting region of the atmosphere was first postulated, it was called the *E region* because Sir Edward Appleton used the letter "E" to symbolize the electric vector of the reflected wave [2]. This region was known to exist at an altitude of 100 to 130 km. Later, it was discovered that the ionospheric plasma during sunlit conditions extended to lower altitudes. For reasons of continuity, this lower region was called the *D region.* The ionized region lying at altitudes above the E region was similarly named the *F region* of the ionosphere. Eventually, the F region itself was further subdivided into the F_1 and F_2 layers, with the F_1 layer occupying the lower altitude. The D and F_1 layers were found to exist only during sunlit conditions. In the absence of solar radiation, these layers rapidly disappear due to strong recombination of their constituent particles. The F_1 layer is routinely observed only in middle latitudes, whereas the F_2 layer, observed at all latitudes, produces the dominant characteristic in most altitude profiles of ionospheric plasma density. In fact, the largest plasma density in the earth's mid- and low-latitude ionosphere typically occurs in the F_2 region.

The D, E, and F regions are depicted in Figure 2.3. The altitudes of the various layers and their relative electron densities at any time are a function of latitude, primarily due to the effects of solar zenith angle. For mid-latitudes, the D region typically occupies altitudes between about 65 and 90 km, whereas the E region lies between about 90 and 140 km. The F region occupies the altitude range above 140 km to as high as 1000 km. The peak ionization in the F_2 layer may be found anywhere within altitudes of 200 to 500 km, depending on solar conditions, season, latitude, and local time. In most cases, the approximate heights of the various layers can be easily identified from a plasma density profile, such as the one illustrated in Figure 2.4(a). This profile shows the variability of plasma density with altitude, showing representative values for the various ionospheric regions.

The boundaries between the various ionospheric layers are not distinct. Moreover, the hourly, daily, seasonal, and solar cycle variations in solar activity cause the altitudes of these layers to undergo continual shifting and further

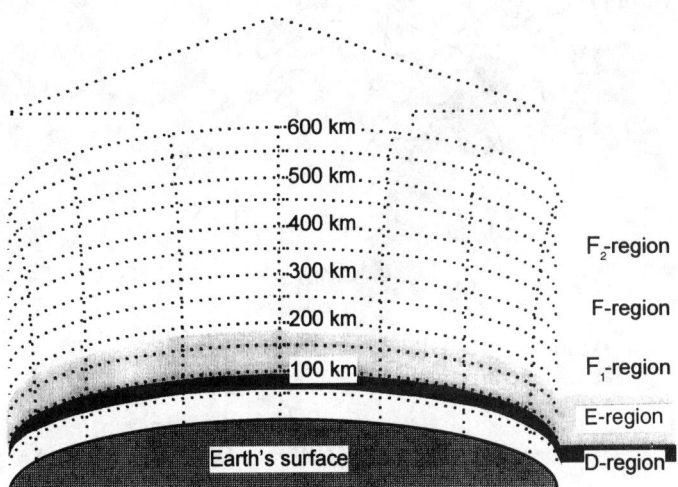

Figure 2.3 The earth's ionosphere and its apparent density layers.

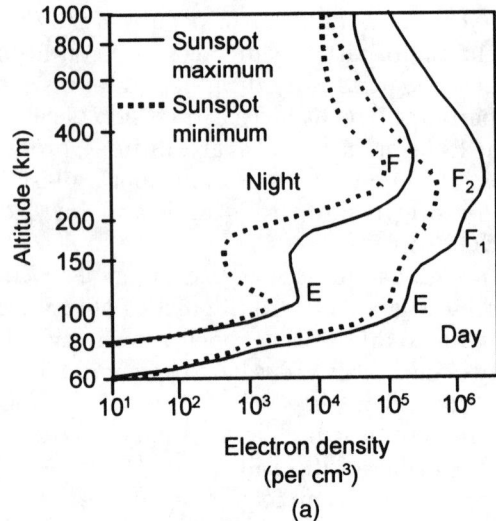

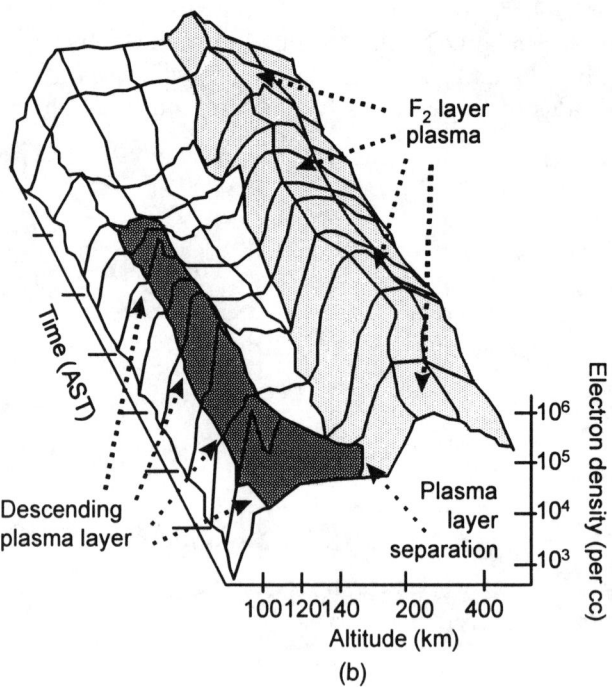

Figure 2.4 Electron density profiles: (a) typical electron density profile and (b) descending layers (*After*: [3]).

substratification. For example, occasional low-density layers forming in the 50- to 70-km altitude regime suggested the existence of a so-called "C" layer below the normal ionization layer in the D region [4]. The E region was found to exhibit occasionally two discernible layers, dubbed E_1 and E_2. The E_1 layer, representing normal E-region behavior, shows a strong solar dependence. The occasional appearance of an E_2 layer was correlated with ionospheric conditions at sunrise and sunset, that is, the gray line. Furthermore, an intermediate $F_{0.5}$ layer has been observed between the E and F_1 layers as well as an $F_{1.5}$ layer between the F_1 and F_2 layers.

In addition to the formation of stable layers, the E and bottomside of the F regions are often populated by so-called "sporadic" layers of enhanced ionization. Despite what their name implies, these layers are quite common. One class of sporadic layers is more commonly known as sporadic E and is identified by the abbreviation "E_s." These layers form primarily at middle and low latitudes near altitudes of 105 km. Sporadic E, or E_s, patches are known to consist of metallic ions produced by meteor ablation in the lower ionosphere. The action of tidal wind systems can form these long-lived metallic ions into thin layers, which are most commonly observed at night when the background densities are low. Layers of sodium (Na) ions produced by similar mechanisms commonly appear in the 90-km altitude range. Finally, intermediate or "descending" layers are routinely observed to drop slowly from the bottom of the F region to the E region in the post-sunset period (see Fig. 2.4[b]). Like E_s, these layers are believed to consist of long-lived metallic ions transported downward by tidal winds. Arguably, these descending layers are observed as $F_{0.5}$ and $F_{1.5}$ layers during ionospheric measurements. As of this writing, the National Aeronautics and Space Administration (NASA) is preparing to investigate these descending layers using ionospheric sounding rockets.

In general, the interactions between ions, free electrons, and background neutral molecules in the ionosphere involve chemical, electrodynamic, and kinetic forces. The existence of charged particles in the ionosphere allows electrical forces to affect the motions of the atmospheric gas. While chemical and collision-related effects occur only between those particles in proximity, electromagnetic forces can act over relatively longer distances. It should therefore be anticipated that the presence of an electrically active gas or plasma, even in small concentrations, will affect the propagation of electromagnetic waves within the ionosphere. It is not apparent, however, that the ionosphere can redirect propagating waves in the HF band to enable long-distance communications.

2.2.3 Sky-Wave Propagation

2.2.3.1 Spreading Loss and Wave Polarization

Consider the path followed by an electromagnetic wave emitted from an antenna in free space, that is, in the absence of the earth and its atmosphere. As the wave propagates, it experiences a decrease in field strength due to the continual outward expansion of the wavefront. Conservation of energy dictates that spreading of the propagating wavefront over this increasing area results in a field strength that decreases in proportion to this increased area. As a result, the radiated wavefronts appear to be continually expanding spherical shells. Since the surface area of a sphere is proportional to the square of its radius R, the power density must decrease in proportion to R^{-2} as the wavefront spreads over continually larger spherical surfaces. For example, the average power density at a distance R (in meters) radiated from an elemental dipole antenna driven by a current of magnitude I is given by [5,6]

$$P_{av} = \frac{1}{2}\frac{|E_\theta|^2}{\eta} = \frac{1}{2\eta}\left[\frac{\eta I\, dl \sin\theta}{2\lambda R}\right]^2 = \eta\left(\frac{I^2\, dl^2 \sin^2\theta}{8\lambda^2}\right)\frac{1}{R^2} \qquad (2.1)$$

where E_θ is the field strength at R in volts per meter, λ is the wavelength, θ is the polar angle in spherical coordinates, dl is the infinitesimal dipole length, and $\eta = 377\Omega$ is the free-space impedance. The R^{-2} spreading loss factor is usually the most significant component of path loss or attenuation, L_p, in radio-wave propagation (see (1.1) in Chapter 1). For this reason, the spreading loss experienced by the transmitted radio wave serves as a lower bound on the amount of path loss expected on an HF sky-wave communications link.

The elemental dipole emits a *linearly polarized* wave such that the electric-field (**E**) vector is parallel to the dipole axis and the magnetic field (**B**) vector is perpendicular, or normal, to its axis. Such a wave is composed of **E** and **B** vectors that maintain these fixed orientations as the wave propagates through free space in the direction given by the wavevector **k**. In the vicinity of the earth, the direct ray can be envisioned as traveling in the *propagation plane* defined by the centers of the transmitting antenna, the propagation path, and the center of the earth as shown in Figure 2.5. If the elemental dipole lies entirely within the propagation plane, then the emitted wave is said to have *parallel polarization*. Strictly speaking, the wave polarization is defined to be the direction of the lines of electric force, that is, the orientation of the electric vector **E**. If the axis of this dipole is also perpendicular to the earth's surface, then the emitted wave and the dipole are said to have *vertical polarization*. Conversely, if the dipole is perpendicular to the propagation plane (out of the page in Fig. 2.5), then the emitted wave has *perpendicular polarization*. In this case, both the antenna and its radiated wave are said to have *horizontal polarization*.

2.2.3.2 Ray-Path Geometry

Although the actual wavefront emitted by a transmitting antenna covers a large area orthogonal to the direction of propagation, it is convenient to envision the

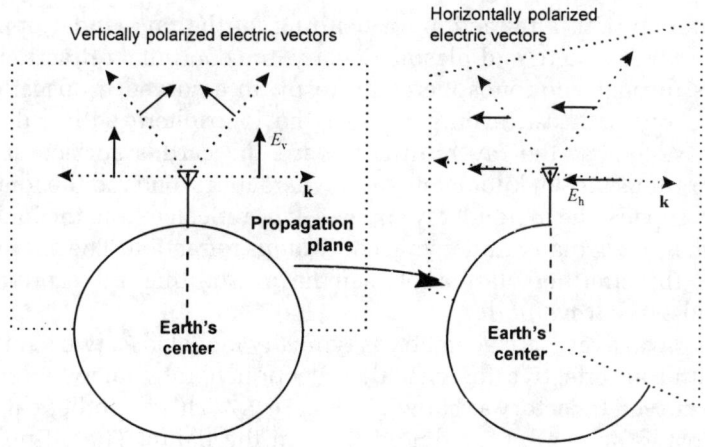

Figure 2.5 Definition of vertical (parallel) and horizontal (perpendicular) polarization.

path followed by the wave as the trajectory of elemental rays as depicted in Figure 2.6. As the figure shows, the ray-path trajectory is essentially a straight line as it traverses the neutral atmospheric region below the ionosphere. Once the ray enters the ionospheric medium, however, it begins to interact with the resident plasma. This interaction results in an increase in the propagation speed

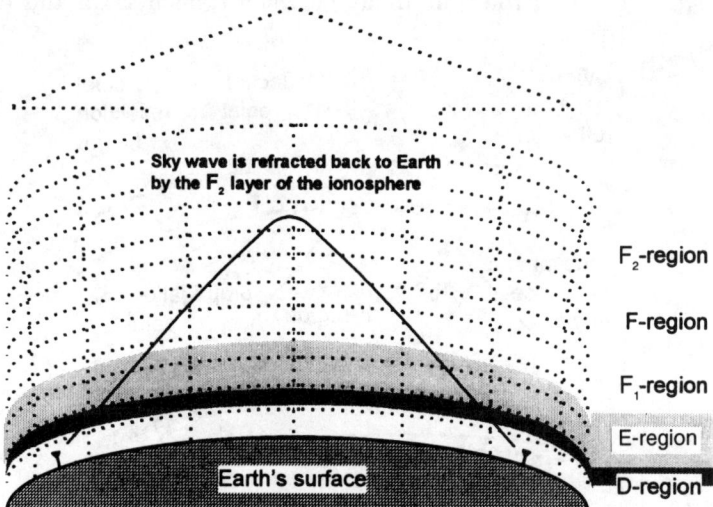

Figure 2.6 Sky-wave path geometry.

of the ray, along with a refraction, or bending, of the ray. Under appropriate conditions of wavelength and plasma density, the continual refraction of a ray as it travels through the ionosphere can result in a curved propagation path. This curved ray path can eventually exit the ionosphere with a downward component such that the ray returns toward the earth's surface at a point hundreds or thousands of kilometers from where it first entered the ionosphere. When this happens, the incident ray appears to be reflected from the ionosphere, even though it has actually undergone continuous refraction. The curved trajectory through this stratified plasma between the transmitting and receiving antennas is called a *sky-wave path*.

The sky-wave ray-path geometry is typically modeled as two straight lines intersecting at an effective (or virtual) reflection point somewhat above the apex of the curved trajectory as shown in Figure 2.7. This hypothetical straight-line trajectory is shown by the dotted lines in the figure. The altitude of the point of intersection of the two dotted lines above the earth's surface (sea level) is called the *virtual height of reflection*. The virtual height is the altitude at which the upward-propagating straight-line ray would appear to be reflected if the ionosphere did not alter both the speed and geometry of the ray path. The figure also shows the ray incidence angle, θ_0, measured as the interior angle between the radio ray incident on the bottomside of the ionosphere and the line formed by the zenith through the reflection point.

The ionosphere always increases the phase velocity of a given ray, so the virtual height is always higher than the actual peak altitude reached by the multiply refracted ray. If the time delay between transmission and reception

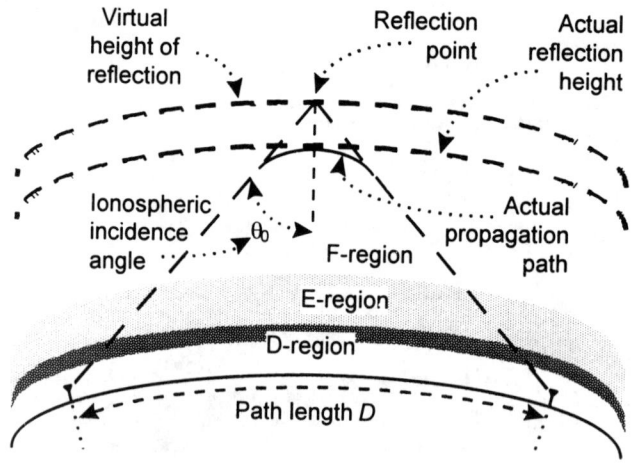

Figure 2.7 Virtual height geometry of sky-wave path.

of the signal is known, the virtual height can always be found by assuming propagation at the speed of light along straight-line trajectories that intersect midway between the transmitter and receiver. Even though HF rays incident on the ionosphere do not actually undergo reflection, the convention followed throughout most HF literature is to refer to the multiply refracted rays as if they were reflected at their corresponding virtual heights. This convention is adopted throughout the remainder of this book.

Long-range HF propagation paths can include reflections from the ground as well as from various ionospheric layers. A single path may therefore include several ground-ionosphere-ground reflections called *hops*. In fact, it is even possible to establish round-the-world (RTW) propagation modes at HF using many hops. Figure 2.8 illustrates several different propagation paths involving various reflection heights and numbers of hops connecting the same two HF stations. Figure 2.8(a) shows a simple one-hop F_2-region reflection symbolized by the conventional abbreviation "$1F_2$." Figure 2.8(b) shows two different ray-path geometries possible at two different frequencies. If the $1F_2$ path occurs at a frequency f_{F_2}, then the two-hop reflection from the E layer, designated "2E," would be possible at a frequency $f_{2E} < f_{F_2}$. As shown in the figure, the 2E path is formed from two independent one-hop E-region reflections and a ground reflection point (see Sec. 2.2.2.1) symbolized by a "-". Figure 2.8(c) shows another possible mode that might be established if a dense sporadic E layer were correctly positioned between the transmitter and receiver, designated "$2F_2$-E_s." Many such single-hop and multihop propagation paths are possible on a single HF link. More than one such path may constitute the HF channel at any one time, and the number and type of specific paths that constitute this channel can change with time.

The ionospheric propagation path followed by a sky wave primarily depends on signal frequency, the incidence angle of the transmitted beam on the ionosphere, and the ionospheric density along the path. The first two variables can often be chosen to optimize link performance, but the ionospheric density cannot generally be controlled except during specialized ionospheric "heating" experiments [7]. The same properties of the ionosphere that permit long-range HF propagation are also responsible for most of its problematic aspects, such as short-wave fading, absorption, spectral broadening, and the propagation of atmospheric noise. In this section, the basic effects of the geophysical domain on HF signals, including effects of the atmosphere, ionosphere, and geomagnetic field, are discussed. For a more complete treatment of these phenomena, the interested reader should investigate a number of the excellent references cited throughout, as well as the more well-known texts on ionospheric propagation [4,8,9]. Subsequent sections address both natural and man-made noise and interference sources.

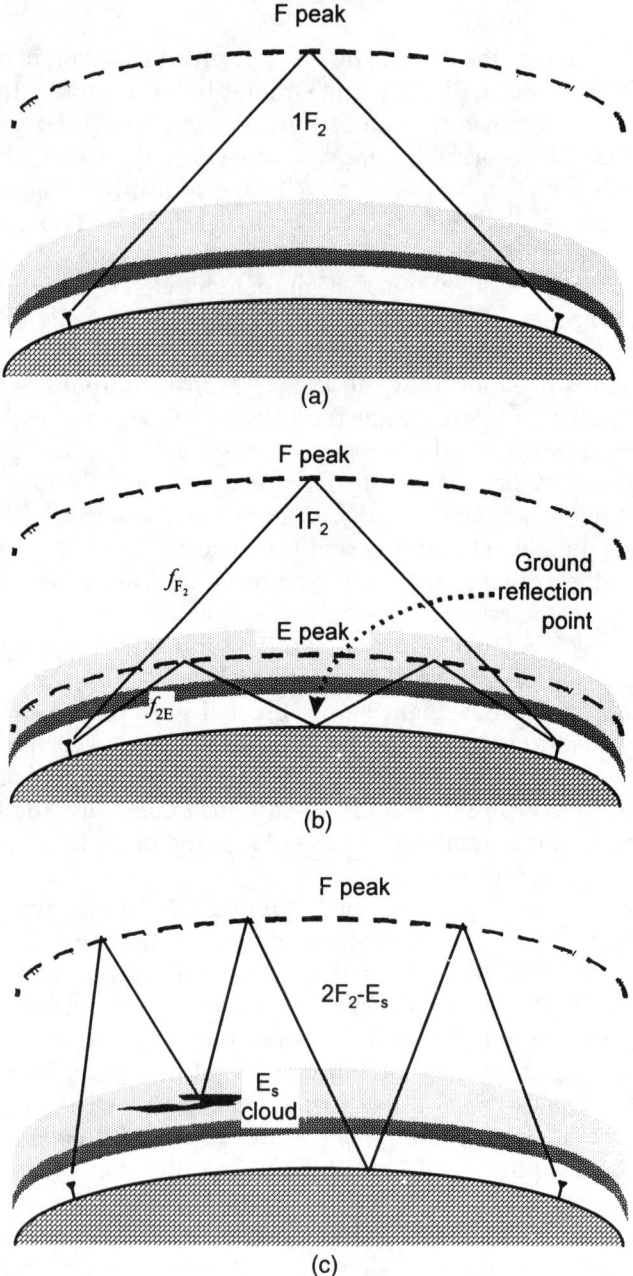

Figure 2.8 Representative sky-wave propagation paths: (a) simple one-hop F-region path, (b) frequency-dependent paths, and (c) two-hop F-region path with intervening E_s reflection.

2.2.3.3 Atmospheric Effects

Below about a 65-km altitude, the neutral atmosphere has very little substantive effect on HF propagation. The sizes of water vapor molecules and other absorptive gases in the medium are insignificant compared to the HF wavelength, so there is negligible interaction between propagating HF waves and the components of the medium. The turbulent mixing of these small molecules throughout the lower atmosphere does, however, create "patchy" variation in the index of refraction for shorter wavelength radiation. This variability results in signal scattering at VHF and higher frequencies.

2.2.3.4 Absorption

Aside from the R^{-2} free-space path loss, or spreading loss, there are significant HF propagation effects in the upper regions of the neutral atmosphere where the ionosphere begins. In this D-region height regime, collision-related ("nondeviative") absorption is usually the largest loss term encountered in HF propagation [10]. The D region exists only under sunlit conditions. Although the *neutral* atmospheric density drops exponentially with increasing altitude throughout the atmosphere to maintain hydrostatic equilibrium, the plasma density is very low in the D region because less than 0.1% of the molecules are ionized. Nevertheless, enough electrons are present at these altitudes to give rise to significant absorption of incident HF signal energy.

In the D region, free electrons in the plasma experience electromagnetic forces due to the oscillatory electromagnetic field of an incident HF wave. As the electrons move under the influence of the applied electric field, they frequently collide with the more abundant neutral particles. Since the mass of an electron is thousands of times less than the mass of a neutral molecule, these collisions result in a significant loss of electron momentum and energy. Since the neutral particles do not oscillate in the applied field, they do not reradiate the energy gained during these collisions. Instead, the bulk temperature of the neutral particles within the illuminated volume is very slightly increased. The net result of this phenomenon is that free electrons in the upper atmosphere effectively *absorb* energy from the wave and disperse it throughout the neutral gas.

The D-region path loss mechanism described above is called *nondeviative absorption*. It occurs where the refractive index of the medium is close to unity, but where collisions between electrons and neutral molecules are frequent. Another mechanism called *deviative absorption* occurs when the wave frequency is close to one of the fundamental oscillatory frequencies of the plasma (defined later in this chapter). Deviative absorption is not a collision-related process; it is better described as a resonance between the HF wave frequency and the fundamental *eigenmodes* of the plasma. While deviative absorption is an important effect, nondeviative absorption typically causes greater attenuation of HF signals as they traverse the ionosphere.

As the preceding description suggests, the absorptive loss term achieves a maximum during local daylight hours when free electrons exist at relatively low altitudes. The ionization efficiency within this altitude range is a strong function of solar zenith angle; hence, the number of electrons available to contribute to the loss process varies with time of day, season, and geographic location. The casual AM listener is well aware of the abundance of worldwide AM radio stations that can be received during nighttime hours. At night, the heavy absorption blanket masking these low frequencies during daylight hours has been temporarily lifted and MF sky-wave propagation is possible.

In general, the nondeviative absorption $A_{dB}(f_1)$ for one pass through an absorbing layer at the frequency f_1 and at an elevation angle θ_z can be approximated as $A_{dB}(f_1) = A_{dB}(f_0)(f_0/f_1)^n \sec(\theta_z)$ dB, where f_0 is a reference frequency lower than f_1, $A_{dB}(f_0)$ is the zenith (vertical path) absorption at f_0, and n is an exponent describing the frequency dependence of the absorption. The exponent n has a value of approximately 2 for low values of absorption.

The $\sec(\theta_z)$ factor describes the ratio between the path at zenith and the path elevated $90 - \theta_z$ degrees. If the earth were flat and the absorbing layer began at the surface, this factor would be equal to the secant of the zenith angle. Due to the earth's curvature, however, the secant factor does not increase above a value of 6 for an absorbing layer. In particular, the significant influence of the secant factor should be noted, because the absorption values expressed in the above formula are in decibels. As is shown later, these parameters are some of the most critical inputs to models of the ionosphere employed to predict HF sky-wave propagation.

Other loss terms become increasingly important at altitudes above about 120 km. At these heights, the ionospheric plasma may be considered collisionless, so nondeviative absorption is negligible. In contrast, resonant interactions (deviative absorption) between the plasma and the HF wave field become increasingly important in this collisionless domain. In particular, when the HF frequency is near one of the fundamental electron oscillation frequencies, the particles can absorb energy from the wave. As we show later, electrons in a *magnetized* plasma such as the earth's ionosphere have several fundamental modes of oscillation. In addition to absorption at resonant frequencies, another resonant damping mechanism called *electron Landau damping* can occur when the incident HF wave field resonates with the thermal speed of the thermal electron population.

Electron Landau damping is a plasma phenomenon in which electrons in the medium absorb energy from an incident wave field that satisfies $n \cdot \omega = \mathbf{k} \bullet \mathbf{V}_e$, where V_e is the velocity of the electrons in the plasma, n is a positive integer, ω is the frequency of the wave, and **k** is the wavevector (|**k**| = $2\pi/\lambda$) [11]. A simple physical analogy to explain Landau damping is a surfer trying to ride a wave at the beach [12]. As the wave approaches, an experienced surfer begins paddling his board toward the beach, so that when the wave eventually reaches him, it will pick up the board and carry it toward the shore. If the surfboard were stationary in the water, the incoming wave would simply pass by, imparting little energy to the board. Good surfers know that the best rides result when the initial speed of their board is only slightly slower than the oncoming wave speed. Similarly, electrons in the plasma moving at speeds close to the HF wave speed can absorb energy from the wave and be accelerated. Kinetic theory calculations show that the magnitude of the Landau damping term is proportional to the slope of the electron distribution function in phase space near the velocity where this resonant condition holds [11]. Similar damping terms from the ion population do not affect HF radiation since the natural oscillations of the ions occur at frequencies well below the HF band. The more sophisticated computer models incorporate these *collisionless* damping factors in their estimates of HF path losses.

2.2.3.5 Refraction

Unlike the lower regions of the neutral atmosphere, the ionosphere strongly influences HF propagation. As previously stated, solar radiation creates the ionosphere by ionizing a small fraction (less than 1%) of the neutral atmosphere at all altitudes above about 65 km. This daytime lower boundary of the ionosphere may be defined as the altitude at which the free-electron density begins to become large enough to allow plasma effects to be discernible. Nondeviative absorption continues to be a strong daytime contributor to HF propagation losses in the lower ionosphere at altitudes up to about 120 km. At higher altitudes, the *neutral* density continues to decrease exponentially, becoming small enough above about 120 km so that collision-related losses can largely be ignored.

In addition to absorption, the presence of ionospheric plasma changes the refractive index of the medium and therefore is largely responsible for the path followed by a propagating HF wave. The complete form of the expression for the refractive index is given in a subsequent section, but a greatly simplified form of the expression useful for illustrative purposes is $n^2 = 1 - f_p^2/f^2$, where n is the refractive index, f is the wave frequency, and f_p is the plasma frequency, which is a function of the local electron density. A convenient approximation for the plasma frequency is given by f_p (Hz) = $9000\sqrt{N_p}$ [13], where N_p is the electron number density expressed in cgs units (number of electrons per cubic centimeter). Note that the refractive index is less than unity. This result implies that the phase velocity of electromagnetic waves propagating in the plasma *will be greater* than c, the speed of light in vacuum. It also dictates that waves entering the plasma from an adjacent free-space medium (where $n = 1$) will be refracted away from the normal to the interface, since Snell's law states that $n_0 \sin \theta_0 = n_1 \sin \theta_1$, where the zero subscripts correspond to free space and the unity subscripts correspond to the plasma medium.

Combining the expression for the refractive index with Snell's law, which describes refraction at a discontinuous planar interface, yields $n_0 \sin(\theta_0) = \sqrt{1 - f_p^2/f^2} \sin(\theta_1)$. Figure 2.9 illustrates the geometry for this Snell's law refraction and defines the angles θ_0 and θ_1. In the figure, n_0 and n_1 are the refractive indices in adjacent stratified layers and θ_0 and θ_1 are the corresponding incident and refracted angles. It is tacitly assumed in the preceding calculation that electron density N_p is essentially homogeneous throughout the ionospheric layer in which the sky-wave signal propagates. This assumption defines the so-called "benign" ionospheric conditions in which these simple calculations provide useful formulas for predicting HF link performance. The Snell's law equations may also be applied to the more realistic case where the plasma density varies with altitude by modeling the medium as a series of thin layers, each having a constant density. Each homogeneous layer contributes a small refraction angle, and the sum of all these angles defines the propagation through the plasma medium. The accuracy of this approximation improves as the thickness of the individual layers is reduced, up to the point where the simple approximation for the refractive index becomes invalid due to plasma scale-size effects.

The expression $n_0 \sin(\theta_0) = \sqrt{1 - f_p^2/f^2} \sin(\theta_1)$ can be used to predict the approximate maximum usable frequency (MUF) for HF sky-wave communications as a function of the ionospheric plasma density. This approximate MUF is known as the standard MUF, and is an approximation to the so-called classical MUF. The classical MUF is the highest frequency propagated exclusively by ionospheric refraction between two points on the earth using the regular ionospheric layers [14]. The classical and standard MUF values depend only on the mode of propagation and path geometry; that is, they are independent of HF radio system parameters and noise. As is shown later, the received signal charac-

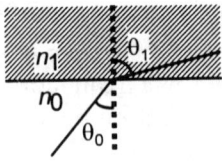

Figure 2.9 Geometry for Snell's law refraction.

teristics are optimal for sky-wave communications when the link operating frequency is chosen at or just below the classical MUF. More recently, it has been recommended that the terms classical and standard MUF be superseded by the term basic MUF, which applies to both parameters and does not differentiate between the actual and estimated maximum propagating frequencies [15]. For this reason, the remainder of this chapter retains the traditional terminology to distinguish between measured and estimated values. However, the expression basic MUF should be considered by the reader to be a recommended term for the system-independent propagating frequency irrespective of how the value is determined.

From a communications perspective, however, the highest frequency that provides communications between two HF stations at a given time and under specific conditions is defined as the operational MUF. The operational MUF accounts for all HF system parameters such as transmitter power, antenna patterns, and receiver sensitivity. In addition, the operational MUF also includes actual environmental conditions such as anomalous propagation from patches of dense ionization and the atmospheric and man-made noise levels at each station. In practice, therefore, the operational MUF value can exceed the basic MUF value.

Recognizing that the refractive index in free space (n_0) is unity, and that refraction in the ionosphere at an angle of 90 deg is the threshold condition for "turning around" an incident HF beam, the Snell's law expression $n_0 \sin(\theta_0) = \sqrt{1 - f_p^2/f^2} \sin(\theta_1)$ can be simplified to $\sin^2 \theta_0 = 1 - f_p^2/f_{MUF}^2$. In this expression, the term f_{MUF} can represent either the standard MUF or classical MUF depending on the availability of historical or real-time ionospheric measurements. Application of a simple trigonometric identity ($\sin^2 \theta_0 + \cos^2 \theta_0 = 1$) and some rearranging of terms yields the well-known secant law for HF propagation $f_{MUF} = f_p \sec \theta_0$, where θ_0 is the incidence angle measured from the normal to the ionosphere and f_p is the plasma frequency. In practice, the classical MUF is typically about 10% higher than the standard MUF owing to approximations used in its derivation [16]. From this point on, the term MUF will refer to the classical MUF, although it is understood that only the standard MUF can be determined from published ionospheric measurements.

Using this knowledge of ionospheric refraction, we can now understand how the MUF varies with incidence angle. As the incidence angle increases, the ground take-off (elevation) angle of the ray decreases and the secant factor increases. For a given ionospheric electron-density profile, therefore, the MUF increases with increasing incidence angle. Alternatively, Figure 2.10 shows how the reflection height of a fixed frequency HF wave increases as the incidence angle decreases. This behavior is observed in practice because the ionospheric density, and hence the plasma frequency, increases more or less

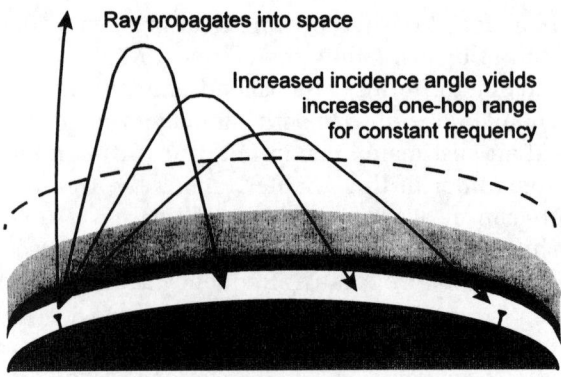

Figure 2.10 Variation of reflection height with incidence angle.

exponentially from about a 90-km altitude to the peak altitude of the ionospheric F_2 region. This peak ionospheric density is typically observed at an altitude of several hundred kilometers above the earth's surface. The simple secant law demonstrates that the plasma frequency in the benign ionosphere is the fundamental parameter controlling the altitude at which an incident HF wave will be reflected. To obtain the maximum range for a single-hop propagation mode, we should therefore use the highest possible frequency and the lowest possible elevation angle that will produce an F-region reflection. As we show in a later section, moving to higher frequencies is also generally beneficial for increasing the propagated signal strength and therefore the SNR of the received signal.

In the preceding derivation of the secant rule, an approximation was used for the refractive index that excluded magnetic field effects and neglected the curvature of both the earth and the ionosphere. As discussed earlier, an HF wave is not simply reflected once at a single critical altitude, instead it is continuously refracted as it traverses the ionosphere. The net effect of these multiple refractions is that the wave changes direction sufficiently to return to the earth's surface. Despite these simplifications, the secant law is extremely useful in HF propagation work, and it forms the basis for many computer prediction models. Armed with this equation and some generic contour maps showing critical frequencies for vertical-incidence reflection [17], approximate operating frequencies and required incidence angles for effective point-to-point communications can be quickly estimated [18].

2.2.3.6 Magnetic Field Effects

The effects of the earth's magnetic field on a propagating HF wave are primarily responsible for the fundamental spectral properties of signals received over HF sky-wave channels. In other words, the magnetic field imposes fundamental limitations on the coherence bandwidth of HF communication links. The coherence bandwidth is the inverse of the overall delay spread, or spread in the time delays experienced by different ray paths between the transmitting and receiving antennas on an HF link. To study these effects from a pure plasma physics perspective, we need a more accurate equation for the refractive index in a plasma. The mathematical development of this equation will reveal that an electromagnetic wave entering a partially ionized and magnetized medium, such as the earth's ionosphere, is effectively split into four different propagating waves. Each of these four waves, or propagation modes, is affected differently as it propagates through the plasma.

At first, distinguishing between the four propagating wave modes may seem unnecessarily complex. Nevertheless, this distinction actually simplifies the mathematical analysis by breaking any electromagnetic plasma wave into its simpler component parts. The technique is somewhat analogous to performing vector analysis by breaking a three-dimensional vector into components lying parallel to the axes of a Cartesian coordinate system. Manipulation of vectors (e.g., sums, differences, dot products, cross products, etc.) is much simpler to perform once the vectors are written in component form. Similarly, calculations relevant to electromagnetic waves propagating in magnetized plasmas are simpler to perform in terms of the four fundamental propagation modes.

Two of the four wave modes, called the *R mode* and *L mode*, are defined only for propagation exactly parallel to the earth's magnetic field (B_E) lines. The other two modes, called the *X mode* and *O mode*, are strictly defined only for propagation exactly perpendicular to the magnetic field. Since all four modes are very special cases (in a mathematical sense), they are rarely observed as distinct waveforms. In practice, the electromagnetic waves that are actually transmitted and received may be considered a superposition of these four individual modes. This physical picture of the resulting multimodal ionospheric propagation will become more clear after a further discussion of wave polarization.

Imagine standing behind a highly directional antenna and looking along the direction of propagation of its transmitted wave. Imagine further that the direction of propagation is parallel to the earth's magnetic field. If it were possible to see the electric (**E**) and magnetic (**B**) field vectors of the wave, they would appear at right angles to each other in the plane perpendicular to the wavevector, **k**, pointing in the direction of wave propagation as depicted in Figure 2.11(a). Note in the figure that the earth's magnetic field vector is directed into the page. For a circularly polarized wave, both the **E** and **B** vectors of the traveling-wave fields rotate together in either a clockwise or counterclockwise direction at a frequency equal to the wave frequency. As they propagate, the magnitudes of these vectors vary sinusoidally, tracing the familiar crests and

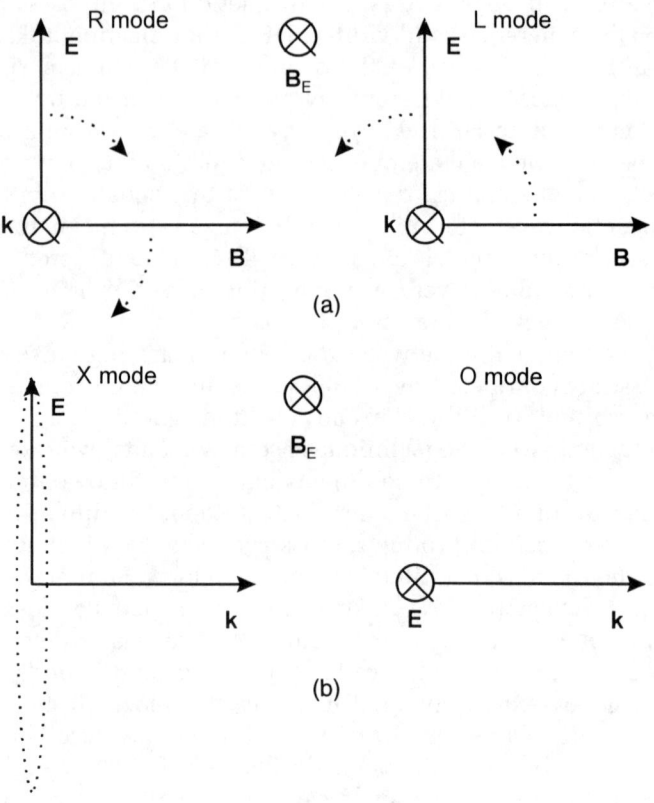

Figure 2.11 Illustrative electric vector diagrams: (a) wave polarization and (b) X and O modes.

troughs of the wave. From our vantage point behind the antenna, clockwise rotation of the **E** and **B** vectors corresponds to the R-mode wave, while counterclockwise rotation corresponds to the L-mode wave. For illustrative purposes, consider that any linearly polarized wave can be described as the vector sum of two circularly polarized waves having identical amplitudes and frequencies, one rotating clockwise and the other rotating counterclockwise. The rotational motion exactly cancels in the vector sum, resulting in linear polarization. Thus any linearly polarized wave can be represented as the superposition of two counterrotating circularly polarized waves of identical magnitudes.

The O-mode wave has its oscillatory, linearly polarized electric field vector parallel to the earth's magnetic field ($\mathbf{B_E}$), while the X-mode wave is elliptically polarized with its electric field vector tracing out an ellipse in the plane perpendicular to $\mathbf{B_E}$. Because this ellipse lies in the same plane as the wavevector **k**, the X-mode wave has a component of its electric field vector parallel to **k**. Figure 2.11(b) illustrates the vector geometries for the X- and O-mode waves. In the figure, the X-mode electric field vector, **E**, traces an ellipse in the plane of the page. Note in the figure that the wave magnetic field vector, **B**, has been omitted for clarity. Each of the four wave modes (R, L, X, and O) propagates differently in a magnetized plasma. For this reason, the composite wave detected by the receiving antenna may have a very different polarization and amplitude from the wave that was initially transmitted into the ionospheric plasma.

In most situations, the wavevector **k** of a transmitted electromagnetic wave is neither exactly parallel nor perpendicular to the earth's magnetic field, but is instead oriented at some arbitrary angle. In this case, the propagation of a linearly polarized wave through the magnetized plasma may be treated mathematically as the superposition of the four distinct modes. Two of these modes propagate exactly parallel to the earth's field (the circularly polarized R and L modes) and the other two modes propagate exactly perpendicular to the earth's field (the elliptically polarized X mode and linearly polarized O mode). The time-varying vector sum of the respective **E**, **B**, and **k** vectors of the four modes gives the instantaneous **E**, **B**, and **k** vectors for the complete sky-wave propagating mode. The use of these wave modes to describe propagation in a plasma is a very helpful mathematical artifice. Caution is warranted, however, since it is misleading to think that the transmitted wave actually propagates in separate directions simultaneously. As discussed earlier, describing the wave propagation in terms of the four modes is loosely analogous to describing an arbitrary vector in terms of its Cartesian coordinate unit vectors.

Before introducing a more general form of the expression for the refractive index, it is helpful to introduce a new characteristic oscillation frequency that exists only in magnetized plasmas. This characteristic frequency, called the gyrofrequency, quantifies the natural oscillation experienced by charged particles in a magnetic field. These oscillations are caused by the Lorentz force **F** = q**v** × **B** arising from random thermal motions with velocity **v** of particles with charge q in a magnetic field **B**. In real plasmas, the mass difference between ions and electrons is so large that ion motion is negligible relative to the electrons for many practical calculations. For time scales relevant to HF propagation frequencies, only electron motions need be considered. The electron gyrofrequency is given by $f_g = eB/2\pi m_e$, where e is the fundamental unit of electronic charge, B is the magnetic field strength, and m_e is the mass of an electron. For most of the earth's ionosphere, the gyrofrequency falls in the range of 1.0 to 2.0 MHz.

If we define and $X = (f_p/f)^2$ and $Y = (f_g/f)$, where f is the transmitter frequency, then a more complete expression for the refractive index in a collisionless, magnetized, cold plasma can be written as

$$n^2 = 1 - 2X(1 - X)/[2(1 - X) - Y_T^2 \pm \sqrt{Y_T^4 + (2 - 2X)^2 Y_L^2}] \quad (2.2)$$

where $Y_T = Y \sin \theta$ and $Y_L = Y \cos \theta$ and θ is the angle between the wavevector **k** and the earth's magnetic field. This expression is called the collisionless Appleton-Hartree equation. Similar forms of this equation, including a more complicated version that includes collision-related effects, can be found in plasma physics textbooks such as the excellent introductory text by Krall and Trivelpiece [11] or other references for this chapter [12,18,19–21].

The various electromagnetic propagation modes discussed earlier are all embodied in (2.2) [22]. For instance, the "±" sign in the denominator gives rise to the ordinary, or O mode ("+" sign), and the extraordinary or X mode ("–" sign). Similarly, the R mode results from using the minus sign in the denominator when $\theta = 0$, while the L mode is derived from employing the plus sign when $\theta = 0$. Because both the R and X modes result from assuming the same numeric sign in the denominator, the R mode is sometimes considered to be a special case of the X mode for parallel propagation. Similarly, the L mode is considered a special case of the O mode. This simplification is *incorrect* from the strict plasma physics perspective, since all four modes are distinct with different phase velocities and dispersion relations. If the plasma is not magnetized, the Appleton-Hartree equation (2.2) can be simplified because the gyrofrequency ceases to exist, so $Y_T = Y_L = 0$. In this case, the R, L, and X modes of propagation are no longer defined, so only the O mode exists. The plasma was not assumed to be magnetized in the case assumed earlier along with Snell's law for deriving the secant rule.

The effect of a propagating wave's oscillating electric field vector is to perturb electron motions in the plasma. Since electrons in a plasma can move more easily along a magnetic field than across it, the O-mode wave (with **E** parallel to **B**$_E$) is unaffected by the magnetic field in the plasma. Electrons responding to the O-mode oscillation simply race back and forth along the field line, but are not impeded by it. In contrast, the X-mode electrons experience an oscillatory **E**-field force that attempts to move them across the geomagnetic field, but the magnetic moment set up by the moving electrons does not allow such motion to occur. It is this effect that makes the X-mode wave elliptically polarized, with a component of its electric field vector in the wave's direction of motion (but still perpendicular to **B**$_E$).

The phase speeds of the various modes are different, and they depend on the characteristics of the plasma encountered along the propagation path as well as on the wave characteristics themselves. As a result, the propagating modes in the received signal may be out of phase with one another as compared to the original transmitted signal. Because of the dynamic behavior of the ionosphere, the relative phase relationships between propagation modes change in time. Because the received signal is the vector sum of the signals represented by the four modes, the result is destructive interference that produces the received signal level fading and limited-bandwidth characteristic of HF channels.

Plasmas are electromagnetically active, so the various cutoffs and resonances between the four propagating modes provide a very rich and mathematically complex field of study. For example, lightning storms generate large broadband electromagnetic pulses that can travel long distances in the ionosphere. The superposition of these overlapping pulses is exhibited as static in HF receivers and is called *atmospheric noise*. The wideband R-mode waves generated by lightning travel easily between the Northern and Southern Hemispheres of the earth, but the different frequencies generated by the lightning pulse travel at different speeds. Since HF R-mode waves have higher phase velocities than lower frequencies, the result is a descending whistle-like tone, since the HF portions of the oscillation are the first to arrive at the receiver. The most commonly heard "whistlers" have frequencies below a few hundred kilohertz, because higher frequencies are strongly absorbed by resonant ionospheric electrons. AM band radio operators are undoubtedly familiar with such whistler mode waves and their characteristic descending tones.

The Faraday rotation effect [12] is another interesting aspect of the different wave modes. Essentially, this effect is caused by different phase velocities of the various modes in the ionospheric plasma. A plane-polarized electromagnetic wave transmitted into the magnetized ionospheric plasma will emerge with its plane of polarization tilted relative to the source. Utilizing our modal breakdown of the wave, it can be understood that this polarization rotation occurs because the R and L modes have different phase velocities, so their E-field vectors undergo varying degrees of rotation in traversing the same distance. The degree of rotation is related to the magnetic field strength, so this effect can be used to measure the magnetic field in a plasma-rich environment.

2.2.3.7 Ionospheric Path Loss

As previously stated, free-space spreading loss (R^{-2}) usually causes the greatest attenuation of the sky-wave propagated radio wave. However, the relationship between signal frequency, incidence angle, and ionospheric plasma density can lead to sky-wave effects that range from benign (small additional losses) to severe (no signal refracted back to earth). In the former case, the ionosphere causes the signal to refract while spending little time in the lower ionosphere, where nondeviative absorption is substantial. In the latter case, either the frequency is too high or the incidence angle is too small, so the transmitted signal passes completely through the ionosphere and continues into space. Between these two extremes lies a very wide range of possibilities and corresponding path loss values, including many permutations of ground range, attenuation, reflection heights, and hop numbers. Figures 2.12(a) through (d) show four examples that illustrate some of these possibilities.

Figure 2.12(a) shows a daytime ionospheric plasma density profile for a mid-latitude site on the west coast of the United States. This profile was predicted by an ionospheric model for the given day number, sunspot number, time, and location. The peak plasma density in the F region is nearly 10^6 electrons/cm^{-3} at about a 300-km altitude. Figure 2.12(b) shows the family of propagation paths for a broad-beamwidth antenna operating at 8.7 MHz. The maximum single-hop ground distance for this link is about 1250 km for a single-

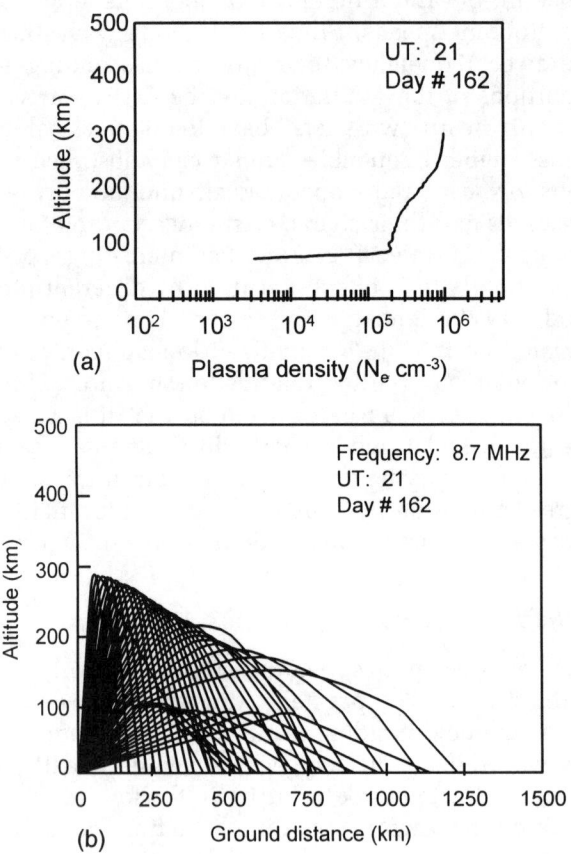

Figure 2.12 Ray propagation versus frequency: (a) normal daytime plasma density for SSN = 140, (b) ray paths at 8.7 MHz, (c) ray paths at 13.6 MHz, and (d) ray paths at 19.0 MHz.

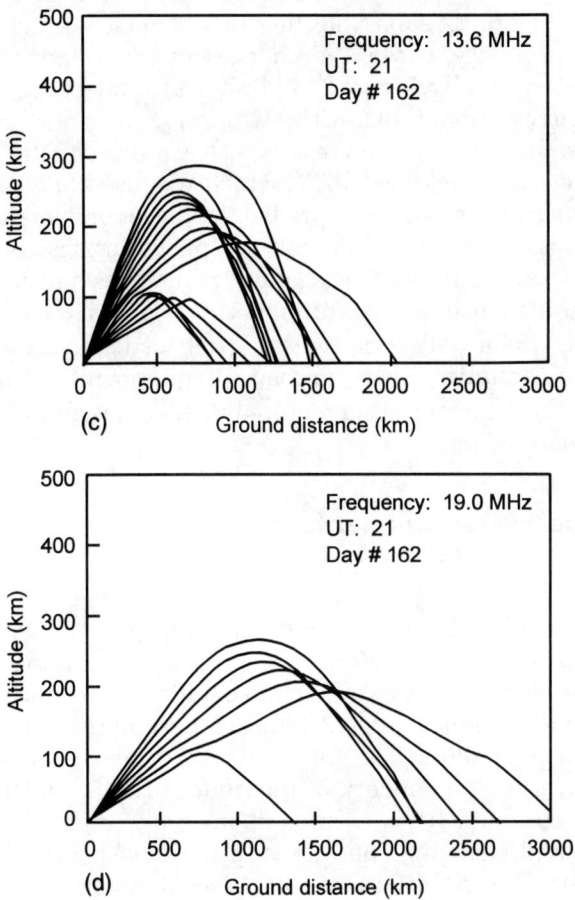

Figure 2.12 (continued)

hop F mode, or a slightly shorter path length for a single-hop E mode. Because the E-mode rays refract in the lower ionosphere, they travel greater distances at low altitudes where the high neutral particle density causes more absorption. The F-mode rays generally travel a shorter distance through the lower ionosphere, so they are typically able to deposit more power at longer ranges. Although the path length traveled at given altitude cannot be directly determined from these plots, it can generally be assumed that the ray travels more slowly while it is undergoing refraction.

Figure 2.12(c) shows somewhat different behavior for a higher frequency transmission at 13.6 MHz through the same ionosphere. The number of propagating rays is smaller, but the maximum achievable range is nearly twice the range shown in Figure 2.12(a). Note that this behavior is due in part to the spherical earth geometry as well as changes in the antenna pattern as the frequency increases. Finally, Figure 2.12(d) shows the effect of a further increase in signal frequency to about 19.0 MHz, which provides a further increase in path length from the Figure 2.12(c) results. Later in this chapter we discuss the computer prediction model IONORAY, which was used to generate these figures. We show that such models can predict the power delivered to the receiver for each of the possible propagation modes depicted in these figures, allowing the HF user to choose optimal propagation frequencies and incidence angles over a wide range of ionospheric conditions. This selection is based on achieving the maximum allowable path loss, L_p^*, (see (1.1)) needed to achieve the desired SNR and meet communication performance requirements. If the actual path loss exceeds L_p^*, then the minimum required SNR value cannot be achieved for a fixed link power budget.

2.2.4 Ionospheric Measurements

2.2.4.1 Ionograms

To compute the standard MUF for a particular propagation path, it is necessary to know the ionospheric density along the path as it changes over time. Measurements of the plasma frequency as a function of altitude are routinely performed with vertical-incidence ionosondes. A typical ionosonde station at the midpoint of a propagation path through a normal daytime ionosphere is depicted in Figure 2.13. A worldwide network of ionosondes continuously monitors frequencies and virtual heights at sufficient intervals, typically every 15 min or hourly, to record changing conditions in the ionosphere. The ionosondes typically operate by transmitting a train of pulses at increasing frequencies over some nominal range in the HF band (e.g., 2 to 20 MHz in 50-kHz steps). A receiving system records the reflected signal from the ionosphere, and processing electronics determine the time delay between transmission and reception for each fixed-frequency pulse. This delay is multiplied by the speed of light to determine the virtual height of the ionospheric reflection. Since a pulse at a fixed frequency will be reflected where $f = f_p$, the result can be plotted as an *ionogram*, which is a graph showing the reflection frequency (related to ionospheric plasma density) versus virtual height.

The frequency at which the ionosonde signal no longer reflects from a given layer is called the *critical frequency* for that layer. Once the ionosonde frequency exceeds the critical frequency, the HF wave penetrates the ionospheric layer and no return signal is detected from that layer. Multiple time-staggered returns often reveal the presence of different layers at different alti-

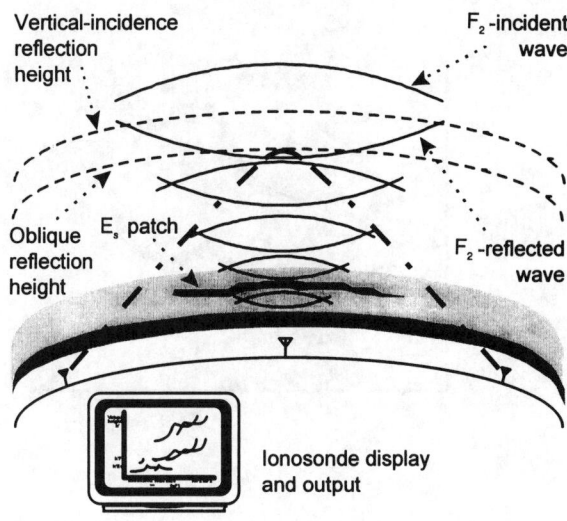

Figure 2.13 Vertical-incidence ionosonde.

tudes as depicted in Figure 2.14(a). Ultimately, the ionosonde operating frequency becomes sufficiently high that no significant ionospheric reflection is possible, so the signal traverses the ionosphere and enters outer space. When this occurs, the ionogram exhibits a *cusp* at the frequency corresponding to the maximum density in the layer, as was shown in Figure 2.14(a). HF users can interpret ionograms from stations worldwide to help choose appropriate frequency bands, standard MUFs, and viable layers for reflecting their transmitted signals.

Since only two traces are typically seen in an ionogram, they have been (unfortunately) dubbed the O and X modes for the "ordinary" and "extraordinary" modes, respectively. However, the O and X modes seen in an ionogram are *different* from the rigorously defined electromagnetic wave modes discussed earlier. In the context of an ionogram, the O-mode trace is simply the linearly polarized portion of the reflected signal displayed on the ionogram, that is, the portion that would be reflected from the same height whether or not the ionospheric plasma was magnetized. Similarly, the X-mode trace on an ionogram is the portion of the returned signal that is not linearly polarized and whose reflection height is affected by the earth's magnetic field. As a result, the X-mode trace exhibits a slightly higher critical frequency than the O-mode trace on an ionogram.

Since the O mode always exists for $f > f_p$, it is the O-mode trace on an ionogram that is commonly used to characterize the ionospheric propagation

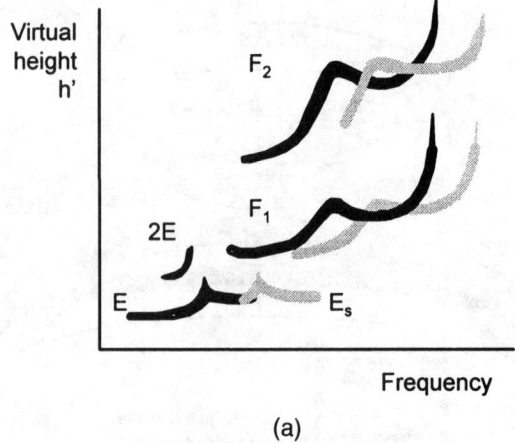

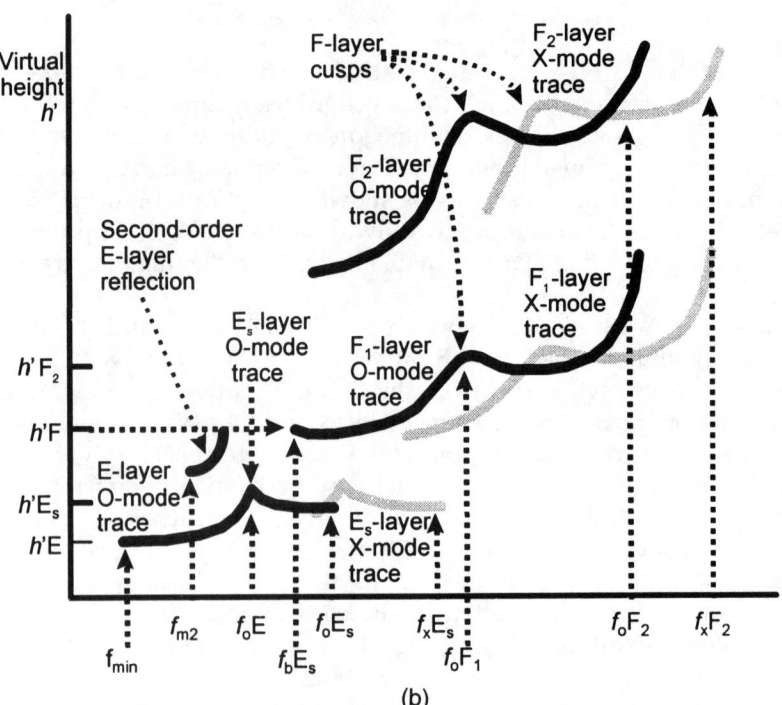

Figure 2.14 Ionosonde measurements: (a) layer heights and (b) critical frequencies and heights.

medium. For example, the term "f_oF_2" is used to define the O-mode frequency corresponding to the plasma frequency of the peak density stratification in the F_2 layer. Worldwide networks of vertical-incidence ionosondes routinely gather data on f_oF_2 and the peak height of the layer, known as "h_mF_2," around the clock. The data collected by these ionosondes are used to provide near real-time ionospheric conditions for HF users, to collect empirical data needed to determine ionospheric plasma density distributions, and to accumulate a database essential for the development and maintenance of computer-based HF link performance-prediction tools.

The interpretation of ionograms requires the identification of critical frequencies associated with reflection from ionospheric layers. In addition to critical frequencies, ionogram traces also exhibit what are defined as *blanketing* and *top* frequencies. The blanketing frequency is the lowest frequency for which a discernible ionogram trace is present, and the top frequency defines the upper bound at which a distinct trace is observed. In addition to f_oF_2, several other frequencies and virtual heights are routinely used to characterize ionospheric conditions, including the following [23].

- f_{min}, the lowest frequency showing evidence of a trace on the ionogram, that is, at which the lowest frequency trace begins to become transparent;
- f_{m2}, the minimum frequency at which ionosphere-ground-ionosphere reflections (second-order reflections) are observed;
- f_oE, the ordinary wave critical frequency which causes a cusp in the height of the lowest E-layer trace;
- f_bE_s, the lowest ordinary wave critical frequency at which an E_s trace is discernible;
- f_oE_s, the ordinary wave top frequency at which a continuous E_s trace is observed;
- f_xE_s, the extraordinary wave top frequency at which a continuous E_s trace is observed;
- f_oF_1, the ordinary wave critical frequency of a discernible F_1 layer;
- f_xF_2, the extraordinary wave critical frequency of the F_2 layer.

In addition to h_mF_2, the important heights associated with an ionogram include the following [24].

- $h'F$, the minimum virtual height of the entire F-layer trace;
- $h'F_2$, the minimum virtual height of the ordinary wave trace for the highest stable stratification in the F region;
- $h'E$, the minimum virtual height of the entire E-region trace;
- $h'E_s$, the minimum height of the trace corresponding to f_oE_s.

Figure 2.14(b) is a representative ionogram showing several of these standard critical frequency and virtual height parameters derived from the ionosonde output plot shown in Figure 2.14(a). As another example, Figures 2.15(a) and (b) show plots of a representative plasma density and the associated ionogram trace, respectively. Note that the cusps in the ionogram trace of Figure 2.15(b) correspond directly to the peak plasma-density values in Figure 2.15(a).

Figure 2.16 shows two global data sets for predicted f_oF_2 values in megahertz at 1200 universal time (UT) during a period of high sunspot activity, that

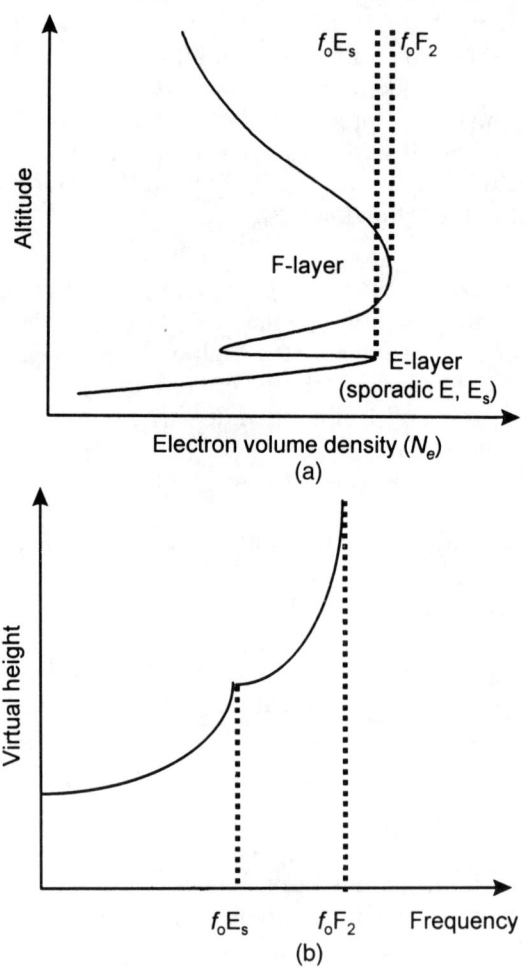

Figure 2.15 Representative plasma densities and displays: (a) F- and E_s-layer plasma densities and (b) idealized ionogram.

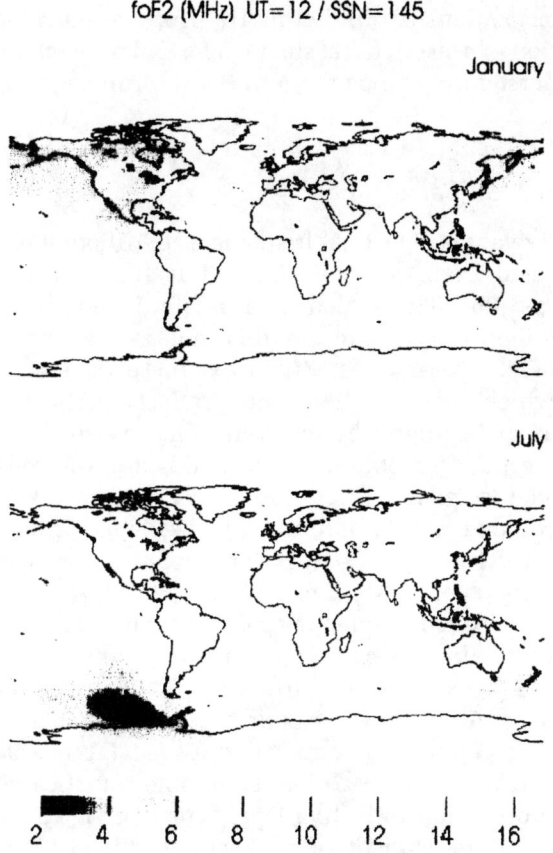

Figure 2.16 Global maps of predicted f_oF_2 contours at 1200 UT with SSN = 145.

is, periods of maximum solar radiation. The top figure corresponds to a solar zenith angle appropriate for January, whereas the bottom figure represents conditions in July. The lightly shaded regions have relatively high f_oF_2 values, indicating high plasma densities; the darker areas indicate low f_oF_2 values corresponding to lower densities. The grayscale color bar at the bottom of the figure relates the shading patterns to f_oF_2 values in megahertz. A comparison of the January and July figures shows the wide swings in f_oF_2 values observed from season to season, even when all other geophysical parameters are held constant. For example, the figure shows an f_oF_2 value of about 5.5 MHz over the central United States at 1200 UT in January, while in July the corresponding value approaches 10 MHz. Since the critical frequency varies as the square root of the plasma density, this seasonal difference represents a more than 300%

change in the plasma density. Predicting HF propagation conditions and variability as a function of season, time, solar cycle, and other geophysical parameters involves understanding ionospheric variability, an ongoing area of research [25].

2.2.4.2 Standard MUF Calculations

In practice, hourly values of critical frequencies of different ionospheric layers are manually scaled from the basic ionosonde traces using internationally adopted standards [26]. The scaled data are formatted according to a data structure convention consisting of monthly tables of hourly values for scaled frequency and height parameters. Primarily, these data include the critical frequency f_v for which the ionosphere would reflect a vertically launched radio wave from one or more ionospheric layers. The measured values of critical frequency from vertical (or oblique) ionosondes are referred to as values of maximum observed frequency (MOF). Ideally, the virtual reflection height h' would also be provided or, at a minimum, the scale factor for determining the standard MUF value from MOF values for single-hop propagation to a fixed distance. For the F_2 layer, a MUF factor $M_{3000}F_2$ (MF_2) is often provided to convert the F_2-layer critical frequency f_oF_2 MOF values into the corresponding standard MUF values for a single-hop range of 3,000 km. This MUF factor could then be used to estimate the virtual height, if it were not available, for computation of standard MUF values at ranges other than 3,000 km.

A FORTRAN computer program called *IONOSTATS* was developed [27] to input UAG-23-formatted data [28] and compute statistics of the critical frequencies and the dominant reflecting-layer type (e.g., F_2, F_1, E, or E_s). IONOSTATS determines these statistics for measured critical frequencies, such as the mean, standard deviation, and cumulative distribution, versus time of day for any designated period of hours, days, months, or years. By using the secant rule described in an earlier section and ionospheric curvature corrections, these vertical-incidence MOF measurements were used to compute standard MUF values versus ground range. To address forward propagation, therefore, the IONSTATS program was modified to permit extrapolation of IONSTATS vertical-incidence results to oblique incidence for a one-hop F-region or one- and two-hop E-region propagation modes. The algorithms used to perform this extrapolation were implemented in the IONOLINK computer program.

Solar UV radiation dominates the production of the ionospheric electron densities that determine the classical MUF values on an hourly basis. For this reason, the use of vertical ionosonde data from a polar location, such as Scott Base in Antarctica, assures wide-ranging variation in the ionospheric conditions. In particular, the sun is below the horizon (zenith angle $\chi > 90$ deg) for all of June and above the horizon ($\chi < 90$ deg) for all of December at Scott Base

(78°S latitude) with the expected diurnal variation (see Fig. 2.17). For this analysis, the half-year period from April to September will be called a period of *low visible sun*, while the combined periods from January to March and October to December will be called *high visible sun*.

Ionosonde data measured at Scott Base in Antarctica from the years 1970 through 1983 were available for this study. As a result, years with minimum and maximum daily-average SSNs during this period could be combined with the periods of minimum and maximum visible sun to span the full range of solar conditions observed at Scott Base from 1970 through 1983. During this period, the low values of SSN occurred during 1975, in which SSNs averaged 19 during the high visible sun period and SSNs averaged 24 during the low visible sun period. The highest values of SSNs during this 14-year period occurred in 1979, for which the maximum-sun and minimum-sun SSN values averaged 172 and 133, respectively.

Figures 2.18(a) and (b) show the cumulative standard MUF and mode distributions, respectively, for minimum SSN (1975) and low visible sun at Scott Base on a 50-km path with Scott Base at the midpoint. The 50-km range corresponds to the well-known near-vertical-incidence sky-wave (NVIS) case, whose results are approximately the same as the ionosonde results for vertical-incidence reflection, that is, a zero range. In other words, the NVIS case is merely HF sky-wave communication at very short ranges, which is well predicted by the standard formula. Clearly, only the F_2 mode is propagating and the standard MUF values are primarily limited between 2 and 6 MHz. An infrequent occurrence of 6 to 10 MHz is also apparent that disappears from 1200 to 1800 UT, that is, local nighttime (no sun) hours.

Figures 2.19(a) and (b) show the corresponding low visible sun for the 50-km link, but for the maximum SSN year of 1979. These figures show that

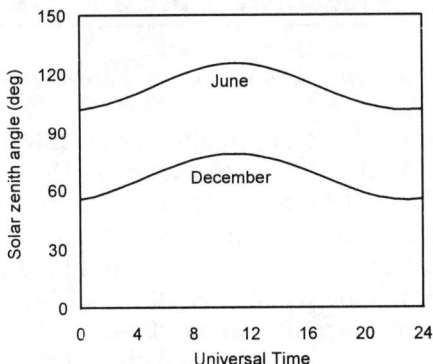

Figure 2.17 Solar zenith angle at Scott Base, Antarctica.

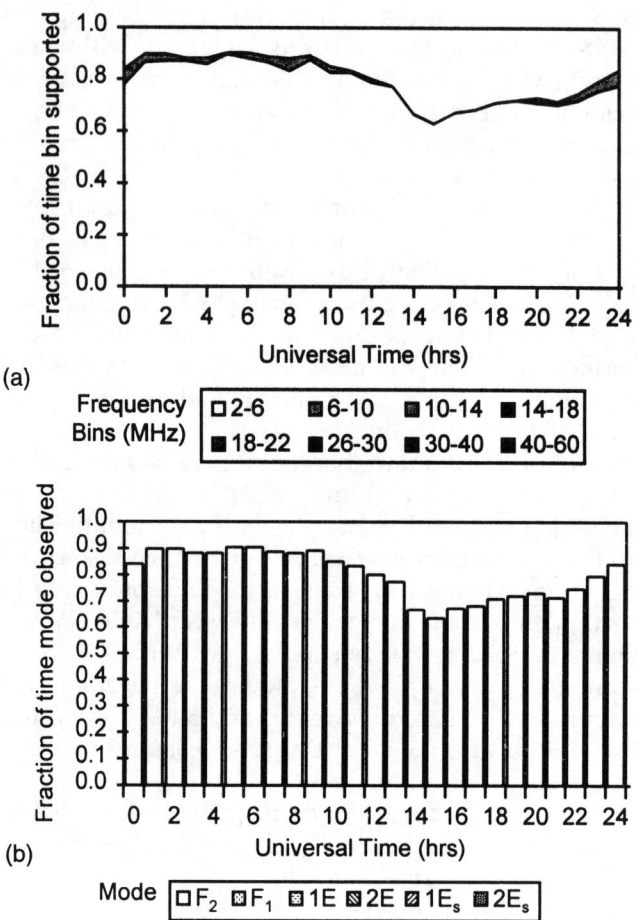

Figure 2.18 Minimum SSN, low sun, 50-km path: (a) standard MUF and (b) sky-wave modes.

the increased ionization indicated by the higher SSN value has significantly increased the standard MUF values and added reflecting layers other than F_2, such as sporadic E layers (E_s). In this case, the 2- to 6-MHz frequency range provided the dominant standard MUF values during the nighttime hours, whereas the 6- to 10-MHz band dominated during the periods of visible sun. The figure also shows some use of the 10- to 14-MHz band, reaching most probable occurrence between 0000 and 1200 UT. As expected, the increased ionization during the sunlit period raised the critical frequencies of the F_2 layer, increased the significance of sporadic E and, therefore, increased the corresponding MUF values.

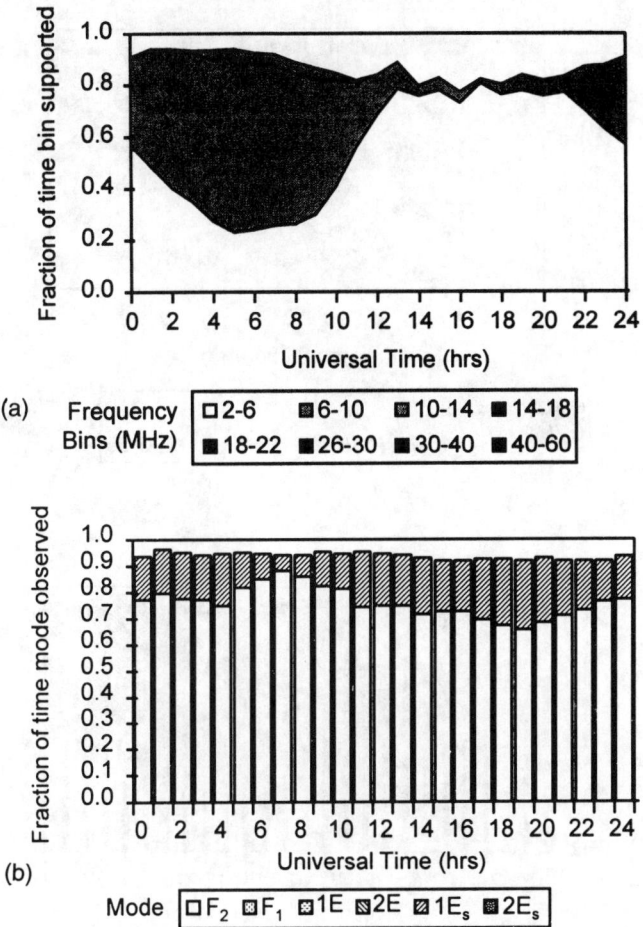

Figure 2.19 Maximum SSN, low sun, 50-km path: (a) standard MUF and (b) sky-wave modes.

Figures 2.20 and 2.21 show the maximum SSN, low sun standard MUF and mode statistics for 1,000- and 2,000-km ground ranges, respectively, assuming the Scott Base midpoint. A comparison of these figures clearly shows an increase in standard MUF values with a corresponding increase in the sporadic E contribution, all resulting from increased ground range. This result was expected because the increased path length increases the $\sec(\theta_0)$ factor, thus producing higher standard MUF values. The sporadic E layers become relatively more useful than the F_2 layer with increasing range due to their lower altitude. In other words, the MUF values contributed by sporadic E increase faster with

70 Advanced High-Frequency Radio Communications

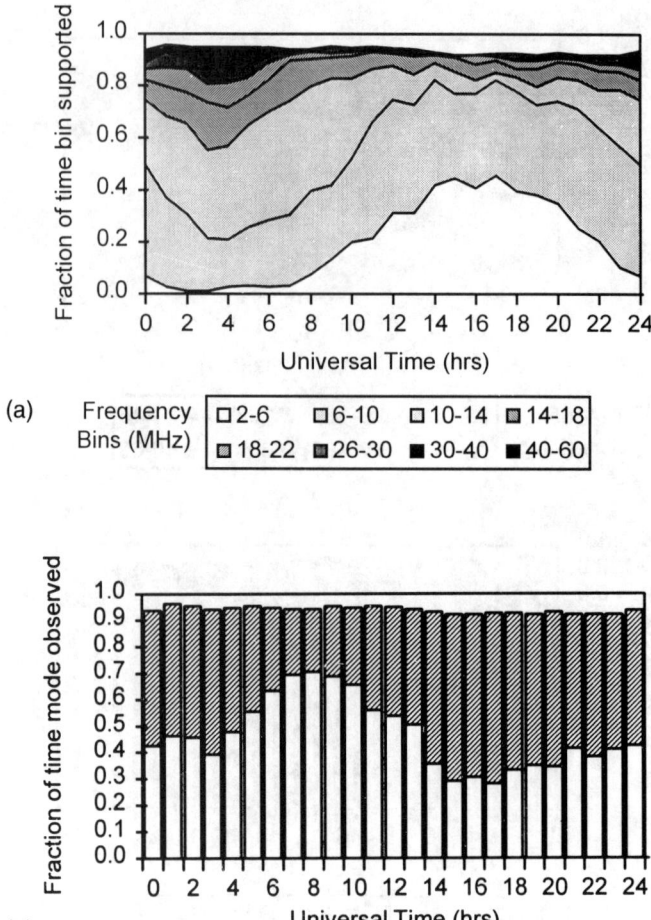

Figure 2.20 Maximum SSN, low sun, 1,000-km path: (a) standard MUF and (b) sky-wave modes.

ground distance than the F_2 layer contribution. This result is most pronounced when comparing Figures 2.19 (50-km path) and 2.21 (2,000-km path). Finally, Figures 2.22(a) and (b) and Figures 2.23(a) and (b) show the 1,000- and 2,000-km path length results, respectively, for maximum SSN (1979) and the high visible sun period. The high sun results show F_1-layer contributions not observed in the corresponding low sun results.

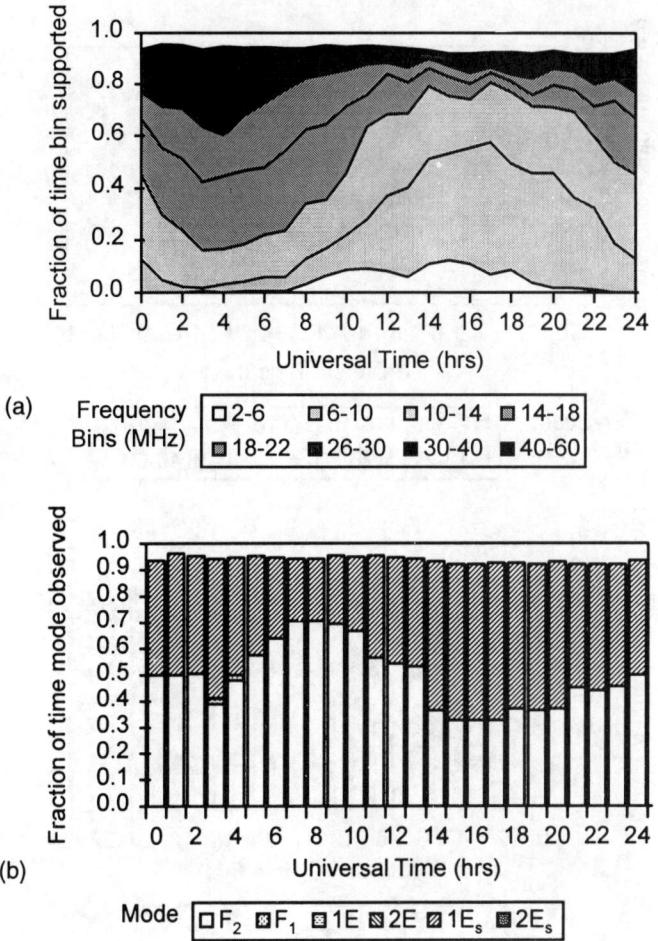

Figure 2.21 Maximum SSN, low sun, 2,000-km path: (a) standard MUF and (b) sky-wave modes.

2.2.5 Link Performance Prediction

2.2.5.1 Background

As HF communication systems find increasingly widespread applications, the need to predict their performance accurately increased correspondingly. Specifically, it is desirable for the HF system operator to know in advance the frequencies and take-off angles that will provide optimal performance for a given set of geophysical conditions. Early attempts to quantify this information involved

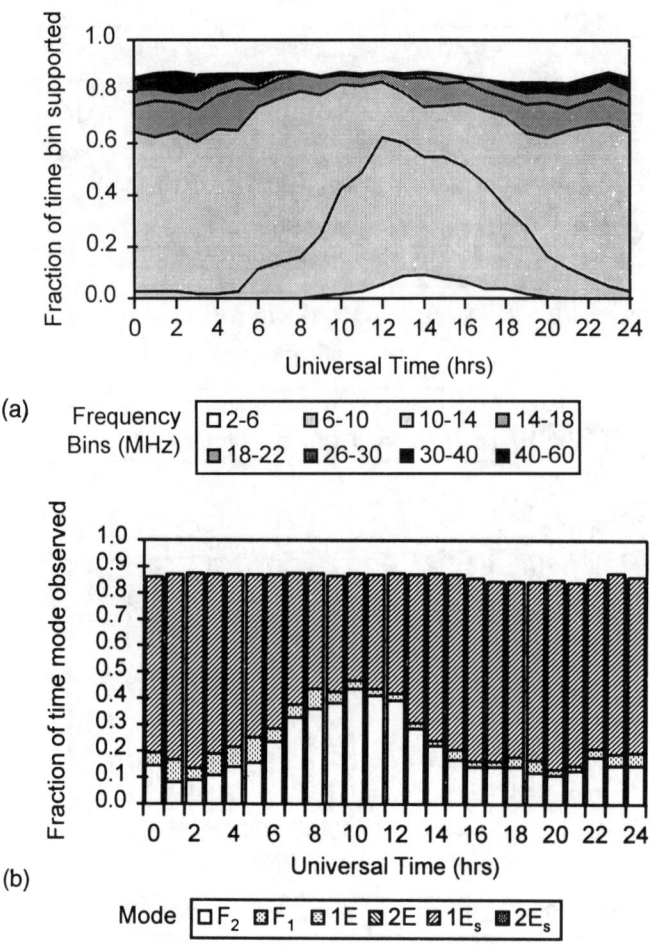

Figure 2.22 Maximum SSN, high sun, 1,000-km path: (a) standard MUF and (b) sky-wave modes.

look-up tables containing recommended operating frequencies as a function of time, season, and other geophysical parameters. As computational systems have improved and system memory has increased, ever more sophisticated modeling schemes have been developed.

Several computer models have been developed to predict plasma behavior and dynamics [28], including the performance of HF communication systems. In general, these models predict the best HF link operating frequencies and antenna take-off (elevation) angles versus the geographic location of transmit-

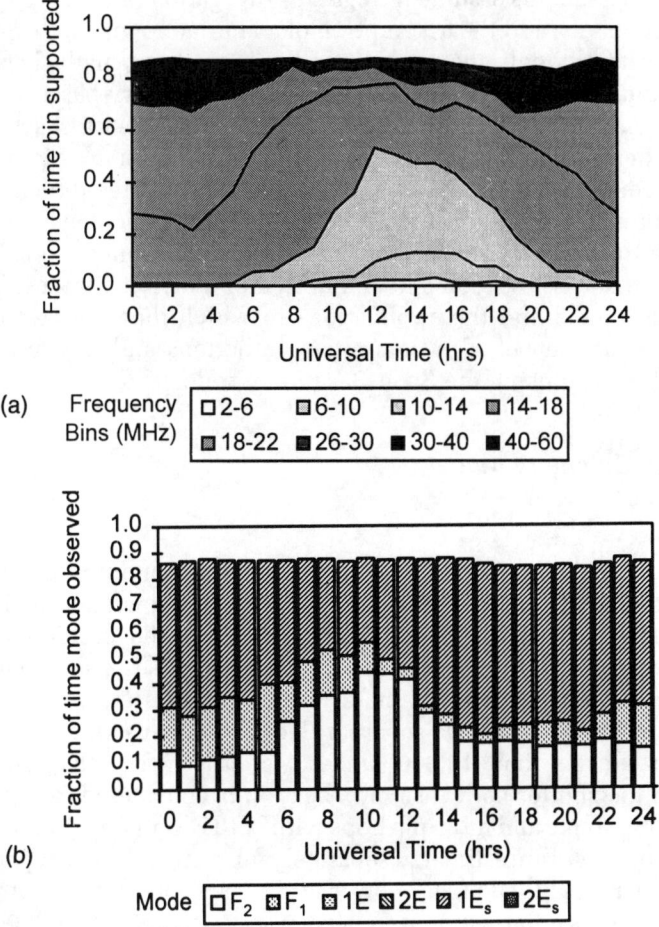

Figure 2.23 Maximum SSN, high sun, 2,000-km path: (a) standard MUF and (b) sky-wave modes.

ting and receiving antennas, antenna gain patterns, and link power budget, among other factors. In this regard, a value of path loss, L_p (see (1.1)), is computed for each viable sky-wave path between the transmitting and receiving antennas. Clearly, the "best" operating frequency and corresponding propagation path will yield an $L_p = L_p^*$ value meeting minimum link power budget requirements, that is, providing adequate SNR to the modem (see Chapter 4) for acceptable voice or data transmission. Some models also determine the likelihood of multipath interference from simultaneous propagation paths providing similar signal levels at the receiver input.

The simplest models assume triangular propagation paths and use Snell's law, flat-earth models, and the assumption of a laminar ionosphere. Somewhat more sophisticated models analyze all feasible O-mode ray paths between the transmitting and receiving antennas. These virtual geometry-based models are in widespread use and usually yield acceptable accuracy for most system planning efforts when benign ionospheric conditions are dominant. More sophisticated models generally incorporate some type of two- or three-dimensional ray tracing in conjunction with empirically derived ionospheric density models to determine the trajectory of HF rays as they traverse the ionosphere. The next section briefly discusses several such models, along with the trade-offs associated with each model and the applications to which they are best suited. It presents a comparison between computer predictions and measurements of standard MUF values using the Scott Base ionosonde.

2.2.5.2 Virtual Geometry Models

IONCAP

Of the available HF prediction models, the Ionospheric Communications Analysis and Predictions (IONCAP) model is the most well known and widely used. The IONCAP model was developed as a computer-based predictive tool for HF propagation through the ionosphere [16]. IONCAP uses a semi-empirically derived table of values to approximate HF propagation losses between two points specified by the user. Two separate algorithms are used within IONCAP, one for paths greater than 10,000 km and another for paths shorter than 10,000 km. In the shorter path case, propagation is considered for all possible modes including single and multiple hops with both E- and F-region reflections. The propagation losses are estimated separately for each path, and superposition is used to estimate the total signal power at the receiver from all paths. In the long-path case, the same treatment is used over domains within 2,000 km of both the transmitter and receiver. Losses experienced by the signals propagating more than 2,000 km from both the transmitter and receiver are estimated using a fixed loss-per-kilometer formula.

The IONCAP approach treats the ionosphere as an unmagnetized, horizontally stratified plasma medium over a flat earth. Simple algebraic approximations are used to simulate the ionospheric density profile using linear, parabolic, and exponential height dependencies based on a set of control (reflection) points garnered from a semi-empirical model. The ionospheric parameters used in the model are linearly scaled from values of the ionospheric layer densities and peak heights relevant to the midpoint between user-specified transmitter and receiver locations at sunspot minimum and maximum. Snell's law is then used with the resultant density profile to calculate a table of critical frequencies for various oblique incidence angles. In this quasivirtual-geometry approach, these incidence angles are determined from all viable ray paths between specific transmitter and receiver coordinates. The estimated losses for each path are then evaluated as described earlier and compared to determine which path delivers the maximum signal power to the receiver. Correction terms are used in an attempt to offset the effects of a spherical earth and ionospheric boundary. The ionospheric model used by IONCAP is simple, but it provides reasonable accuracy and executes quickly on modern personal computers [16].

The IONCAP algorithm is constructed to use propagation preferentially above the MUF as discussed by Wheeler [29]. Empirical evidence indicates that losses for above-the-MUF propagation exceed the losses predicted by simple models. In some cases, however, this result is offset by the longer ranges achieved by transmitting at higher frequencies. Separate subroutines within IONCAP estimate losses, E- and F-layer parameters, and environmental noise effects. IONCAP uses table look-up procedures in order to speed execution and binary input files for antenna patterns, ionospheric density profiles, and CCIR atmospheric noise data [30] to minimize memory requirements.

The empirical basis of IONCAP signal and noise predictions permits link performance to be specified as the percentage of time for which a given SNR will be achieved, called *time availability*, for a specified frequency. IONCAP also computes the change in link power budget in decibels needed to achieve a specified circuit "reliability," that is, the percentage of days within the given month that the minimum required SNR will be achieved in a specified hour. Re-executing IONCAP after altering the link power budget to the recommended value results in the desired SNR achieved for the required time availability at the given frequency. At this point, IONCAP indicates that the link performance is achieved with a 50% service "probability." In other words, exactly half of the HF systems using the required link power budget will achieve the minimum required SNR for the required time availability, whereas the other half of these systems will not achieve this performance. IONCAP computes a standard deviation in decibels of link power budget for the Gaussian service probability distribution, which can be used theoretically to achieve any desired service probability value.

IONCAP was developed by the Institute of Telecommunication Services (ITS) beginning in the early 1970s to support U.S. government HF circuit planning and operations. More recently, a PC version was developed and distributed. Since that time, IONCAP has gone through several revisions both by the original developers, ITS, and other U.S. government organizations. For example, the Voice of America Coverage Analysis Program (VOACAP) [31] was developed by the U.S. Information Agency both to correct bugs in the IONCAP program and to provide an enhanced PC-based graphical user interface (GUI) for broadcast coverage prediction. The VOACAP GUI was developed not only to provide a

DOS Windows input capability, but also to tailor VOACAP outputs and batch execution control so that broadcast coverage maps can be generated and plotted. VOACAP has been further upgraded since its creation and was made available on the Internet [31].

Recently, another corrected version of the IONCAP program, called *ICEPAC* [32] was developed. ICEPAC added an improved high-latitude model and made additional improvements to the circuit reliability calculations and the noise model. A new DOS-based GUI developed by Los Alamos National Laboratory, called *PROPVUE*, was also added to ICEPAC. An MS Windows GUI, called *IONWIN* [33], has also been developed for IONCAP and is compatible with both VOACAP and ICEPAC. IONWIN provides a variety of standard Windows input/output features, including map-based selection of transmitting and receiving locations and SNR contour plots.

ASAPS

The model employed by the Australian Defence Science and Technology Organization (DSTO) in the Austral-Asian region is embodied in a program called the Advanced Stand-Alone Prediction System (ASAPS) created by the Ionospheric Prediction Service (IPS) of DSTO [34]. This program is based on propagation and noise models published by the CCIR (see Sec. 2.5) and the results of research performed by the IPS. The program can be used in a number of ways, but a typical use of the model begins with the entry of input data quantifying the chosen HF links and direct ASAPS to estimate the expected SNR for several specified frequencies of operation. Experiments have confirmed that ASAPS accurately predicts the workable frequency band and can select usable frequencies within this band. The SNR predictions have also proven accurate for link design purposes, providing an accurate estimate or measurement of noise power is entered.

Similar to IONCAP, ASAPS requires the following input: (1) location of transmitting and receiving stations, (2) antenna types (the program then extracts the necessary antenna characteristics from a library or further user input), (3) time of day and day in year for the prediction, (4) transmitter power, (5) available frequency set, (6) man-made noise level or CCIR model, (e.g., rural, business, industrial, etc.; see Sec. 2.5), and (7) T index, a parameter related to the sunspot activity quantified by SSN (IONCAP uses SSN). Given these inputs, ASAPS can then generate the following outputs: (1) HF path distance and great circle path bearings from site to site, (2) best usable frequency (BUF) for each hour and the corresponding ray-path elevation angle, (3) signal power, total noise power, SNR, and (4) probability that each of the likely propagation modes is present, provided for each hour of the day at each of the nominated frequencies. Note that if a noise category rather than a noise power is input, ASAPS will employ the upper decile of that noise category as the noise power rather than the median or mean value. This convention accounts almost exactly for the discrepancies noted between the SNR predicted using ASAPS and experimental results. It is then necessary to integrate ASAPS output over time, frequency usage, and geographic locations to provide availability contours for alternative system configurations.

Comparison With Measurement

A comparison of standard MUF values with predictions from IONCAP and ASAPS was performed to illustrate solar effects on propagating frequencies. (The results shown should not be used to draw conclusions about the relative accuracy of these two programs.) The IONOLINK program was used to extrapolate standard MUF values from critical frequencies measured by the Scott Base ionosonde. These MUF values were compared with the corresponding IONCAP and ASAPS predictions to illustrate the effects of ground range, sunspot number, time of year, and time of day on MUF values.

IONOLINK was used to process Scott Base ionosonde data from June and December in 1975 and 1979. These months were chosen because they represented the smallest and largest monthly average SSN values, respectively, during this 14-year period. The 1975 monthly average values of SSN were 24 in June and 17 in December, while in 1979, these values jumped to 151 and 200, respectively. A measurement-prediction comparison was performed for MUF values, which were not exceeded for 10%, 50%, and 90% of the hours in each 1-hr time slot during the day for each day in a month. In other words, about 30 hourly measurements contributed to computing these percentages for each hour of a typical day in each month. Figures 2.24(a), (b), and (c) show the measurements and predictions for ranges of 50, 1000, and 2000 km, respectively, with minimum sun (June) and minimum SSN of 24.

The IONCAP and ASAPS-predicted MUF values are typically 1 to 2 MHz above the corresponding measured values for the 10% not-exceeded value. For the 50% not-exceeded value, this difference increases to about 2 to 3 MHz, and for the 90% case, the predicted MUF values are about 2 to 5 MHz greater than the corresponding measurements. The distribution of predicted MUF values is spread over a wider range and exhibits more diurnal variation than the extrapolated ionosonde measurements. Figures 2.25(a), (b), and (c) are plots of the measurement-prediction comparison for 50, 1,000, and 2,000 km, respectively, for maximum visible sun (December) and maximum SSN (200). As expected, the increased SSN yielded an increase in both measured and predicted MUF values, but increased prediction error is also apparent. This increased prediction error is arguably due to the increased occurrence of sporadic E propagation,

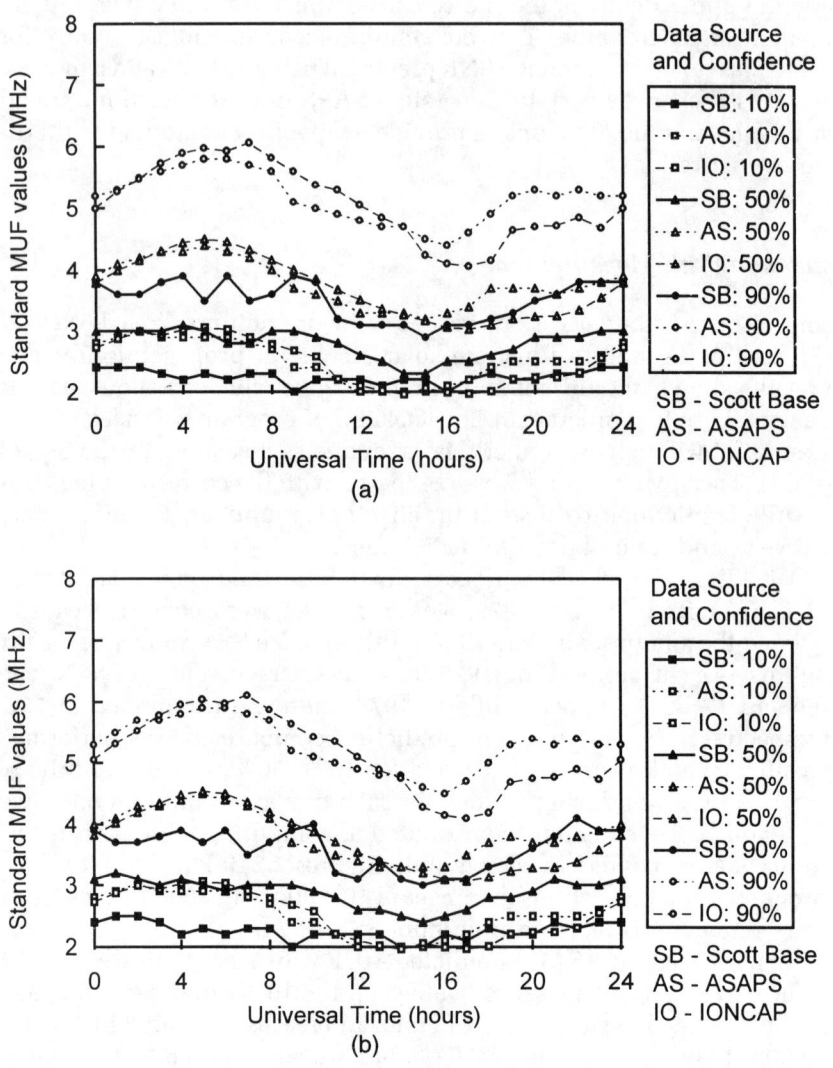

Figure 2.24 Standard MUF for minimum SSN, lowest sun month: (a) 50-km path, (b) 1,000-km path and (c) 2,000-km path.

which is not modeled by either IONCAP or ASAPS. Sporadic E propagation modes are apparent in the plots shown in Figures 2.19 through 2.23. Another reason for the fluctuation in prediction error is the small number of measured values, about 30, used in determining the numerical MUF distribution for each measurement hour.

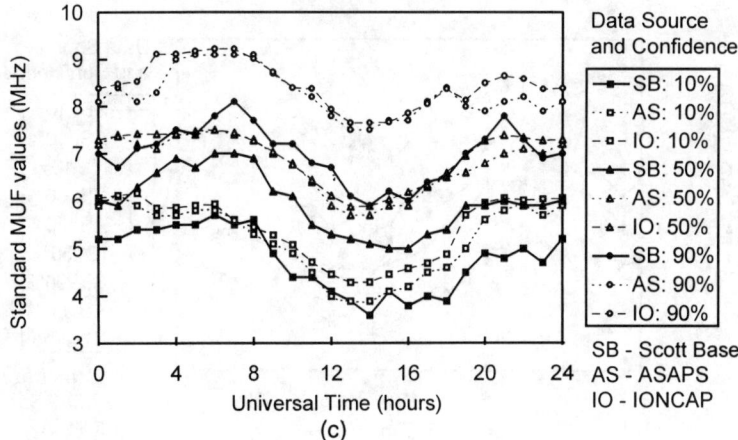

Figure 2.24 (continued)

2.2.5.3 First Principles Models

AMBCOM

In addition to the simple models described in the previous section, the continuing development of ever faster and more powerful computers is now at a point where fairly sophisticated ray tracing can be performed using complex models run on personal computers or workstations. The AMBient COMmunications (AMBCOM) program, developed by SRI International, is a two-dimensional ray-tracing code using the geometric optics solution to the wave equation and Fermat's principle to treat significant ionospheric *tilts*. These tilts are exhibited at high latitudes, near the geomagnetic dip equator, and near the daylight/darkness plane in the ionosphere, the so-called "solar terminator." AMBCOM will predict asymmetrical hop geometries and unconventional propagation modes inherent in the ionospheric model. These geometries and modes can result in significant (factor of two or more) differences in the ray take-off and arrival angles not predicted by models based on virtual geometry. AMBCOM provides both point-to-point HF link performance predictions and point-to-area (e.g., broadcast) predictions for over-the-horizon (OTH) radar performance prediction. Since AMBCOM is a two-dimensional ray-tracing model, however, it cannot employ ionospheric profiles with irregularities transverse to the direction of propagation.

IONORAY

The complexity of electromagnetic wave interactions with the plasma medium suggests that a functional predictor of HF propagation in the ionospheric plasma

80 Advanced High-Frequency Radio Communications

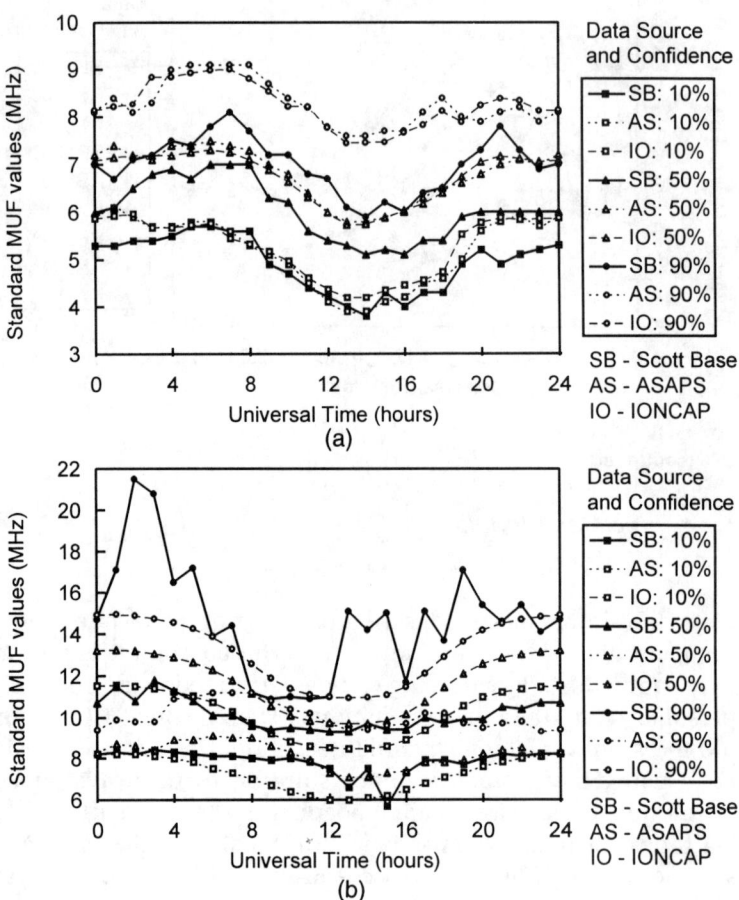

Figure 2.25 Standard MUF for maximum SSN, highest sun month: (a) 50-km path, (b) 1,000-km path and (c) 2,000-km path.

should be fully three dimensional, since the interactions of the various wave modes require full vector mathematical treatments. Ionospheric scientists at Science Applications International Corporation (SAIC) have developed a three-dimensional ray-tracing model called *IONORAY*. IONORAY overcomes many of the shortcomings of the virtual-geometry models by using a more sophisticated iterative integration routine in conjunction with well-known empirical models of the propagation environment. Figure 2.26 shows a block diagram of the various software modules that constitute the IONORAY algorithm. The blocks labeled "Controls" and "Environmental Specification" in the diagram list code

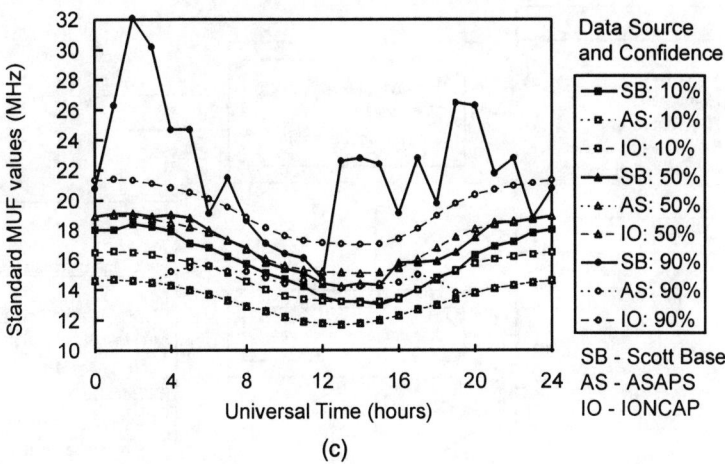

Figure 2.25 (continued)

inputs and show which geophysical models are affected by each input. Five main models are used to specify the geophysical environment within IONORAY:

1. *International Reference Ionosphere (IRI):* An empirically based model that specifies the altitude, plasma density, composition, and temperatures for each ionospheric layer.
2. *Mass Spectrometer and Incoherent Scatter (MSIS) atmospheric model:* An empirically derived code that specifies neutral atmospheric temperatures and composition as a function of solar output, time, and season, used to determine collision frequencies needed to determine nondeviative absorption losses.
3. *International Geophysical Reference Field (IGRF) magnetic field model:* A widely used model specifying the orientation and magnitude of the geomagnetic field. Model coordinates are updated every five years using data from a worldwide magnetometer network.
4. *Auroral Oval module:* A subroutine that uses one of three auroral models to specify the location of the auroral oval. This information can be interpreted by the user to provide a relative confidence level for propagation paths near the polar regions.
5. *Layers module:* A subroutine designed to insert plasma irregularities into the background density provided by the IRI model. This module can be used to assess the importance of sporadic E layers or other plasma irregularities to the propagation mode under consideration.

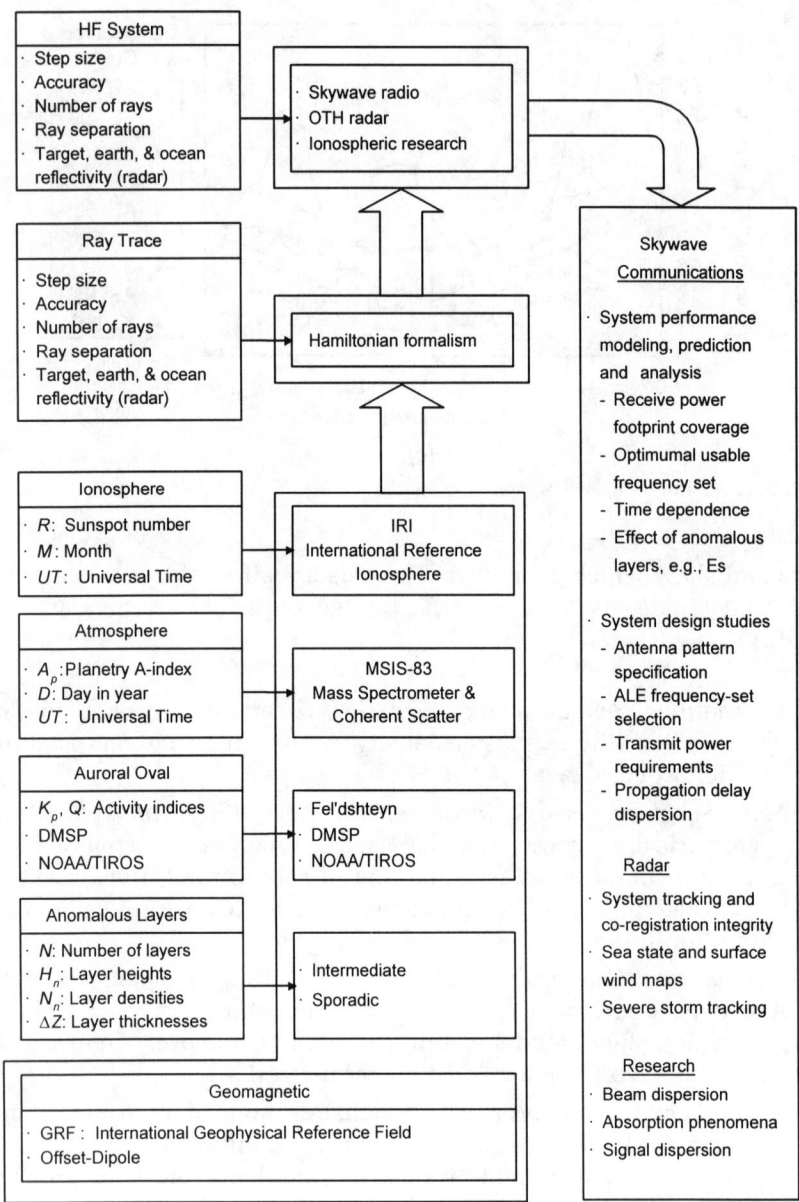

Figure 2.26 IONORAY flowchart.

The IONORAY code merges the five main routines outlined into a single package capable of specifying the global ionospheric conditions at any time, location, altitude, season, and sunspot number. In this respect it is a very powerful tool for testing system capabilities or simulating the effects of ionospheric variability (such as sporadic E layers) on HF system performance.

In addition to the user inputs required to run these five models, specific information about the system must be provided to produce accurate predictions. This information includes transmitter and receiver antenna patterns, operating frequency, coordinates of the transmitter and receiver, and transmitter power. User input is also needed to specify the number of individual rays to be simulated (traced) and their relative spatial density. For instance, the user may specify 5-deg increments in elevation between test rays to obtain a first look at various propagation paths, and then rerun the program specifying 0.5-deg elevation increments centered around a particular elevation (such as the peak of the main antenna pattern lobe) to improve the resolved view of the path to be used.

The ray-tracing module within IONORAY iteratively solves the three-dimensional dispersion relation for high-frequency electromagnetic waves in a cold plasma. At each time step, the module calls the various model subroutines and obtains new values for the plasma density, temperature, and magnetic field. These values are then used to recalculate the plasma frequency, collision frequencies, and the gyrofrequency. Finally, the code solves the ordinary differential equations describing wave propagation and obtains a new set of wave conditions that includes direction of propagation and field magnitudes in each of the three spatial dimensions. These new values are used as the initial conditions in the next time step, and the location of the phase front is modified according to the new direction of propagation.

The accuracy of this technique is limited by the size of the time step taken between successive calculations and the approximations introduced by the linearized equations describing the *cold plasma dispersion relation*. The code typically takes several hundred time steps as it traces the path of propagation through the ionosphere for each ray. To prevent overly large steps through the region where refraction is strong, one of the user inputs determines the maximum time step allowed. In addition, the code self-regulates its step size based on the magnitude of the changes calculated at each step. For example, if the values of the field magnitudes or wavevector orientation undergo a particularly large variation at a given time step, the algorithm disregards the new values, reduces the step size, and recalculates. This procedure prevents the code from stepping through a region of large gradients with an inappropriately large step size, which could introduce substantial errors.

The outputs of the IONORAY code include the field magnitudes at the location where the traced ray intercepts the earth's surface. At this point, the program employs a variety of procedures depending on the application. For example, in an over-the-horizon radar application, the appropriate procedure is to determine the total power reflected and delivered back to the receiver. This is easily accomplished by doubling the path loss from the one-way path. Alternatively, the user may wish to simulate a multiple-hop communication channel. In this case the location of the ray's footprint is simply used as the starting location for an isotropic radiator with a power level equal to the power deposited by the first hop. The IONORAY predictive code has been used in both these scenarios with good results. Although the IONORAY program is mathematically complex and slow to execute in comparison to simpler models, in some cases it is more useful because it can more accurately simulate system performance in the presence of irregularities.

SKYCOM

SKYCOM is an HF and refractive VHF ionospheric ray-tracing model developed by the Ionospheric Support Branch of the Solar Terrestrial Dispatch. SKYCOM employs either the CCIR or URSI ionospheric models to produce critical F_2-layer frequencies in agreement with those of the IRI model and employs spherical ray-tracing algorithms to predict all signal paths. SKYCOM provides user-controlled modeling of various disruptive effects on HF communications, including the ionospheric density changes with changing geomagnetic activity, sporadic E layers, solar flares, absorption, and D-region absorption. SKYCOM provides the capability to (1) sweep through various transmitted ray elevation angles with frequency and time of day fixed, (2) sweep through frequencies while fixing elevation angle and time of day, and (3) sweep through time of day while holding frequency and ray elevation angle fixed. SKYCOM provides broadcast signal coverage maps of received signal quality, propagation delay, ionospheric focusing/defocusing, and multipath effects across the broadcast coverage area. In contrast with IONCAP and ASAPS-type programs, SKYCOM is not a communication-system performance model, it is a high-fidelity modeling tool for studying the propagation of sky-wave propagated HF (and lower VHF) radio signals through the ionosphere.

2.2.5.4 Model Trade-Offs and Applications

One of the most important engineering trade-offs between the various types of computational HF propagation models is speed of execution versus accuracy. For instance, the IONCAP model discussed earlier executes very quickly on a high-end (486 or better) PC, but it sacrifices some accuracy in the interest of processing speed. For example, IONCAP introduces errors by executing a fixed-point integration routine for all paths, rather than a self-adjusting routine based on the inherent accuracy at each time step. For complex paths through an irregular medium these errors may be substantial. In addition, the flat earth and laminar ionosphere assumptions implicit in virtual geometry models relinquish some accuracy but speed processing considerably. These simplifications lead to easily solved algebraic equations for the ray paths instead of the more rigorous time-stepping integration routines used by more complex algorithms.

In contrast, the more rigorous models execute more slowly on a computer than the virtual geometry programs but produce more accurate coordinate registration. This registration is useful for transmitter antennas with narrow patterns, such as those sometimes used in military applications. The IONORAY model was developed under contract with the U.S. Air Force (USAF) to help predict the performance of a powerful over-the-horizon radar system. By systematically varying input parameters such as time of day, season, and sunspot number, the performance of the system could be predicted months or even years into the future (e.g., to predict performance over a complete solar cycle). This capability allows the USAF to determine under what conditions the radar has a high probability of detecting incoming missiles or planes, or more importantly, under what geophysical conditions the system is less reliable. Furthermore, it shows how the *barrier location* (footprint of the HF rays on the earth's surface) varies as a function of frequency, time, season, and sunspot number. In the over-the-horizon radar application, the location of the barrier determines the amount of advance warning of intrusion, so precise coordinate registration is of major importance.

The IONCAP model is often used for frequency selection in point-to-point communications applications. In such cases, it is simple enough to vary the frequency using a trial-and-error approach until a link can be established. While first principles approaches would work for this application, the additional execution time required might be unacceptable. Because each HF radio-wave propagation problem is unique, the selection of appropriate numerical tools for investigation of the various sky-wave modes should be subject to the problem constraints.

2.3 SKY-WAVE PROPAGATION IN A DISTURBED IONOSPHERE

2.3.1 Global Phenomena

2.3.1.1 Overview

The ionosphere is created by solar x-ray and extreme ultraviolet radiation incident on neutral atmospheric particles above about a 65-km altitude. Although the well-known solar cycle produces changes in solar radiation output over an approximate 11-year period, these changes are sufficiently slow that they are associated with the benign ionosphere (see Fig. 2.2). By contrast, abrupt changes in solar radiation at the ionizing wavelengths influence not only the peak plasma density of the ionosphere, but the spatial distribution as well. By directly perturbing the ionospheric plasma density, these disturbances can significantly affect HF propagation modes and disrupt communication systems reliant on them.

The subject of this book is HF communications and not ionospheric physics, so readers desiring an in-depth study of these phenomena are encouraged to consult the text by Kelley [35]. The impact of these phenomena on HF radiowave propagation is measured by signal attenuation produced by absorption and signal fading and distortion resulting from time dispersion, or *delay spread*, and frequency dispersion, or *Doppler spread* [36] due to time-varying multipath propagation. Delay and Doppler spreads are the standard measures of signal distortion when dealing with signals propagated via a dynamic dispersive medium and are expressed as the twice the root mean square (rms) of delay or Doppler shift, respectively, about their mean values.

Sky-wave signals propagated over mildly disturbed ionospheric paths exhibit minimal Doppler spread, on the order of 1 Hz, and moderate time dispersion, on the order of 100 ms. Signals propagated over moderately to severely disturbed channels are characterized by time-varying multipath and scatter from moving irregularities that are spatially distributed about the great circle path between transmitter and receiver. These irregularities can induce delay spreads of 300 ms to 1 ms and Doppler spreads of 5 to 20 Hz [37]. The larger spread values are associated with the most disturbed ionospheric conditions, in which weak irregularities dominate the ambient ionization and for which signal scattering is the primary propagation mechanism. The received signal level from such a scatter propagation path has been observed to be as much as 30 dB below the level of signals propagated by reflection from the ambient background ionosphere [38,39].

2.3.1.2 Day/Night Terminator

Normal day-to-night and night-to-day transitions create sharp density irregularities in the ionosphere and form the so-called "day/night terminator." As the sun rises at ionospheric altitudes, the production of plasma from the ambient neutral population increases dramatically. In conjunction with abrupt increases in absorption, this phenomenon can create relatively sharp horizontal plasma gradients that refract HF waves in azimuth and elevation. Path lengths for propagation at a given frequency will often change dramatically as the plasma density and layer heights of the ionosphere are modified by sunrise. Diffusion of plasma gradually smooths and disperses these terminator gradients and reduces their impact on propagation. For periods on the order of one hour near sunrise, however, HF communication can be adversely affected. Although the effect of the day/night terminator can be significant, its predictable daily occurrence combined with its brief duration reduces its impact on well-engineered HF systems.

Terminator effects are not always detrimental to HF communications. In particular, many HF users have noted the existence of very stable, relatively undistorted propagation modes for mid-latitude paths. These propagation paths are parallel to the terminator and are typically observed shortly after sunset. This so-called "gray-line propagation" is associated with quiescent ionospheric conditions that immediately follow sunset at mid-latitudes. Once sunlit conditions cease, the plasma densities in the ionospheric E and F_1 regions begin to decay through recombination of electrons and ions, while the D region (where most collision-related absorption occurs) disappears completely. These effects are also observed at high latitudes where the daytime F_1 layer is frequently as strong, or stronger, than the F_2 layer, leading to enhanced dispersive effects for F_2-layer reflection. These relatively quiescent conditions can persist for about one hour after sunset and provide ideal conditions for HF sky-wave communication.

Sporadic E and Intermediate Layers

Sporadic E, as the name implies, refers to E-layer ionization effects that are anomalous in the sense that they are not directly related to incident solar radiation. Different forms of sporadic E, having different characteristics and production mechanisms, are found at low equatorial latitudes, mid-latitudes, and high latitudes. They share the common characteristics that they are all E-layer phenomena, their occurrence is not predictable, and they all have a disruptive effect on HF communications.

Traveling Ionospheric Disturbances

Traveling ionospheric disturbances (TIDs) are the generic name for a class of perturbations whose signature is a change in the ionospheric plasma density profile that appears to propagate from the magnetic polar regions to the equatorial regions [10]. TIDs are sometimes associated with solar flares and magnetic disturbances, but at other times they seem independent of such events. In any case, the induced fluctuations in plasma density and/or layer heights can have noticeable effects on the effective range and received signal strength of HF signals passing through the affected portion of the ionosphere.

Solar Flares

Solar flares are explosive releases of solar energy in the form of radiation at wavelengths from x-rays to visible light and mass ejecta ranging from alpha particles to very hot plasma. The radiation reaches the earth in a few minutes, but relativistic alpha particles and protons can take from 0.5 hr to several hours to reach earth depending on their energy. Hot plasma will usually arrive on the order of 36 hr after the flare [40]. Although they can occur throughout the solar cycle, solar flares are most frequent during solar maximum, the active phase of the 11-year solar cycle when the SSN achieves its peak values (see Fig. 2.2). The source of the energy released during solar flares is resident in huge magnetic-field structures that protrude into the solar corona (see Fig. 2.27) from the hot fluid comprising the solar surface in which they are embedded.

Solar flares are relatively short-lived events, but the associated increase in ionizing x-ray and other high-energy radiation can lead to dramatic increases in the ionospheric electron content in the 65- to 90-km altitude range over the entire sunlit hemisphere. This altitude regime is called the ionospheric D region [41] and, as already stated, it is the region in which most HF absorption occurs. During a flare, the electron density in the D region may increase as much as an order of magnitude, leading to a sudden and drastic increase of nondeviative absorption. Solar flares and the increased radiation fluxes associated with them also affect the E and F layers of the ionosphere. Because these higher altitude regions are essentially collisionless, and therefore do not contribute to nondevi-

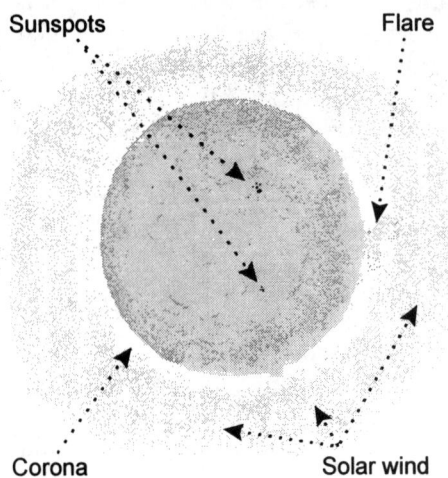

Figure 2.27 The sun and its emissions.

ative absorption, their response to solar flares is less dramatic. Nevertheless, enhanced ionization at these altitudes persists for longer time periods than at D-layer altitudes because of the decreasing electron-ion recombination rate with height. Propagation effects associated with the immediate effects of strong solar flares include short-wave fadeout (SWF), also known as *HF blackout*, as well as sudden phase anomalies (SPAs) [10]. SWF is caused by dramatically increased absorption due to enhanced electron density in the D region. SPA events are associated with abrupt changes in wave reflection height generating an observable phase advance in received HF signals.

Coronal Mass Ejection

In recent years, *coronal mass ejection* (CME) events have been correlated with sudden changes in magnetospheric structure and subsequent ionospheric disturbances. CMEs are associated with active but relatively stable regions on the sun (sometimes called coronal holes) that emit a supersonic stream of hot plasma. If the solar latitude and longitude of the region is such that the stream is directed at the earth, the effect is similar to a solar flare. CMEs differ from solar flares in that they (1) emit no photon radiation, (2) recur at the rotation period of the sun, approximately 28 days, and (3) are most prevalent around solar minimum, the quiet period of the solar cycle. The lifetime of an identifiable coronal hole is variable, but they usually last for several solar rotations. Once a CME has occurred, therefore, the next occurrence of a CME is predictable to within a period of several hours because the solar rotation period is known.

Magnetic Storms and Substorms

Solar flares and CMEs are responsible for a sudden enhancement in the density and velocity of the supersonic hot plasma streaming outward from the sun past the earth, the so-called "solar wind" depicted in Figure 2.27. Through the mechanism of the enhanced solar wind, solar flares and CMEs are indirectly responsible for two major ionospheric disturbance phenomena, magnetic storms and magnetospheric substorms. Both magnetic storms and magnetospheric substorms result from the impact of the enhanced solar wind on the earth's magnetosphere, which causes changes in its geometric configuration as depicted in

Figure 2.28.[1] Magnetic storms affect primarily the equatorial and mid-latitude ionospheres, whereas magnetospheric substorms are confined to the high-latitude regions of the earth's ionosphere.

The storms can cause changes in the movement of charged particles in the polar cap, strengthening of electric currents in the ionosphere, equatorward expansion of the auroral region, and increased ionospheric density in all layers. These effects have a profound impact on the motions and distribution of plasma in the ionosphere and corresponding effects on HF communication systems

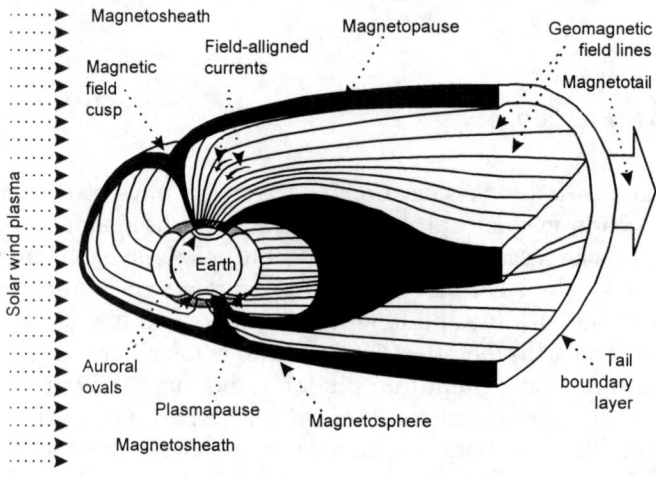

Figure 2.28 The earth and its magnetic field in the solar wind (*After*: [10], Fig. 5.1.).

---

[1]The action of the solar wind plasma impacting the earth's geomagnetic field is (1) to compress the field lines on the sunward side of the earth and (2) to sweep the earth's magnetic field lines emanating from the north and south polar regions into a very elongated region on the night side of the earth. The effect of the solar wind is to form a cavity within which the geomagnetic field lines and the terrestrial ionization are contained, and outside of which the solar wind plasma and sun's interplanetary magnetic field (IMF) dominate. The cavity within which the earth's geomagnetic field, the most distant terrestrial plasma structures, the Van Allen radiation belts, and the ionosphere are contained is called the *magnetosphere*. The magnetosphere extends to roughly 10 earth radii (10 R_e) on the sunward side of the earth and hundreds of earth radii on the nighttime side. The boundary between the magnetosphere and the interplanetary medium is called the *magnetopause*. The *magnetotail* is that region of the nighttime side magnetosphere that lies beyond the region (i.e., greater than 4 to 6 R_e) of high plasma density. The *plasmapause* is the boundary between the region of relatively dense terrestrial plasma and a region of substantially reduced plasma in more outlying regions of the magnetosphere. It is also regarded to be the boundary between regions of plasma that co-rotate with the earth and regions that do not. The region is essentially contained within geomagnetic field lines that extend to about four earth radii from the earth at the equator, but this boundary is variable depending on magnetic storm conditions.

that use this medium. These phenomena and their impact on HF communication links are described in later sections for each geographic region of the earth.

2.3.2 Latitude-Dependent Phenomena

Sky-wave communication in equatorial, auroral, and polar regions is generally acknowledged to be more problematic than in mid-latitude regions. Plasma density irregularities that significantly impact HF radio performance are an important, and sometimes dominant, feature of the ionosphere in these regions. In particular, sky-wave signals received over these paths exhibit enhanced delay and Doppler spreads that result in significant distortion of received signals. The severity of this signal distortion is determined by the rms fluctuation of electron density. This fluctuation is due to the magnitude, motion, and spatial distribution of irregularities in the electron density compared with the background ionization density. Even weak irregularities embedded within an otherwise benign ionosphere can cause small but detectable signal distortion. On the other hand, prominent signal distortions are observed in cases where the irregularities are the dominant feature of the ionospheric electron density.

Recent work by Wagner et al. [38,42,43], Wagner and Goldstein [39], and Basler et al. [44,45] provides the most comprehensive set of high-latitude HF sky-wave channel measurements yet available. These measurements are currently being supplemented by measurements from *Project Damson* [46], which is attempting to develop a statistical database for the parameters of the high-latitude HF sky-wave channel. A more limited set of measurements of irregularity phenomena is available for the equatorial region [44,45,47]. These measurements, in conjunction with research performed at the U.S. Air Force Phillips Laboratory [48], have produced a rich literature on the geophysics of irregularity formation and their effects on coincident HF and VHF radio links.

The following sections discuss ionospheric disturbances characteristic of different geographic regions of the earth. In this context, the low latitudes refer to latitudes within several degrees of the Equator. The mid-latitude region is assumed to extend from low latitudes to the region of the ionosphere impacted by auroral phenomena. The high-latitude region consists of latitudes above the auroral boundary and has been further subdivided into auroral and polar regions to better explain the associated phenomena unique to these regions.

2.3.2.1 The Low-Latitude (Equatorial) Region

Equatorial spread F (ESF) [49-51] is a phenomenon more commonly known as equatorial clutter by HF communication system operators. From the standpoint of ionospheric physics, ESF is an upwelling of low electron density globules, or *depletion* regions, traversing from the bottomside to the topside of the F-region ionosphere. This phenomenon is caused by the well-known Rayleigh-

Taylor fluid instability, which occurs when a dense fluid is suspended above a less dense fluid [52]. The presence of the geomagnetic field and the conductivity of the lower ionosphere affect the instability in a plasma, so that ESF occurs shortly after sunset at low (equatorial) latitudes, where the magnetic field is nearly parallel to the earth's surface. ESF formation is linked to the rapid disappearance of the lower part of the F region (F_1 layer) as the sunlight fades. For this reason, the ESF phenomena are usually observed between 1900 to 0300 local meridian time.

From an HF communications perspective, ESF is particularly damaging because the *scale* sizes (the typical dimensions of the irregularities) of the upwelling density "bubbles" range from centimeters to hundreds of meters. In fact, the largest bubbles can measure 1,000 km from north to south and 100 km from east to west. These bubbles always follow the sunset, moving in an easterly direction at speeds of 100 to 200 m/s. The wide distribution of bubble sizes, coupled with their movement, suggests that ESF can interfere with radio communication at frequencies from HF through the microwave band. At HF, ESF produces multipath from bubble and normal ionospheric reflection, Doppler shift from the moving plasma forming the irregularities, and dispersion caused by refraction of the incident wave from the nonlaminar interfaces between the bubble and the background plasma. In fact, even microwave satellite communication channels experience scintillation due to ESF.

Few direct measurements of the communication properties of the equatorial channel have been performed. However, HF channel-probe measurements at 10.96 MHz were reported for a 2,160-km east-west path along the geomagnetic equator during a period of low solar activity in the summer of 1986 [44,45]. These conditions are known to be ideal for the formation of spread F. The *scattering* function [53] obtained from these measurements is depicted in Figure 2.29. The figure plots contours of SNR below peak value on the delay-Doppler plane, with the delay axis spanning values between 7.49 and 8.49 ms and the Doppler axis spanning values between ±40 Hz. The figure shows delay spreads on the order of 300 to 400 ms and Doppler spreads on the order of 2 to 4 Hz, both measured at the −10-dB (from peak) SNR contour.

Larger spread-parameter values caused by ESF are anticipated on north-south propagation paths than east-west paths. Measurements on a 1,600-km north-south transequatorial path in South America were made in 1994 [47,54]. Tentative results of these measurements indicate delay spreads of up to 1 ms and Doppler spreads between 2 and 10 Hz. Both of these values correspond to spreads measured at the −10-dB SNR contour of the scattering function.

2.3.2.2 The Mid-Latitude Region

HF propagation at mid-latitudes can be adversely affected by a number of factors, including the day/night terminator, TIDs, sporadic E, solar flares, spread

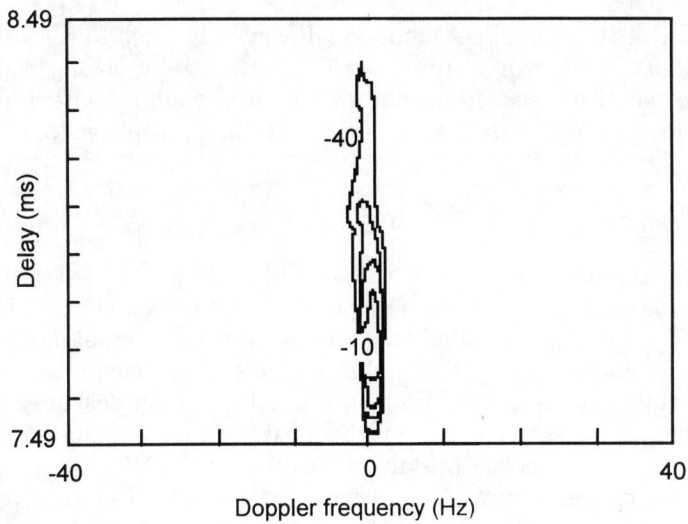

Figure 2.29 SNR contour plot versus delay time and Doppler frequency.

F, and magnetic storms. In particular, the manner in which magnetic storms and mid-latitude spread F affect HF sky-wave propagation at these middle latitudes requires further discussion. The spread F phenomena is all too familiar to users of the mid-latitude HF channel.

Sporadic E and Intermediate Layers

Mid-latitude sporadic E is an irregularity most familiar to HF communicators. Mid-latitude sporadic E layers are thin patches of long-lived ions (primarily metallic) that are believed to be ablated from meteors as they pass through the earth's atmosphere [3,55]. Because the recombination rates of metallic ions are extremely low in the ionosphere [56], these thin layers can persist for many hours before being neutralized by recombination and dispersed by diffusion.

The action of neutral wind shears in the lower ionosphere, in conjunction with the action of the geomagnetic field, can create convergence zones where these ions coalesce (bunch up) to form thin, high-density layers. Sporadic E layers typically form at altitudes between 105 and 110 km and persist for many hours (see Fig. 2.3). Intermediate layers are initially created at higher altitudes and then descend over a period of several hours to the 105- to 110-km altitude range as was shown in Figure 2.4(b). Little is known about the spatial extent of either sporadic E or intermediate layers, but it is known that their peak densities can sometimes exceed that of the higher altitude F region. When this occurs, these layers can reflect incident HF waves at much lower altitudes and prevent (blanket) reflections from the F layer, thereby greatly reducing the expected range of transmission. In addition, strong empirical evidence exists that these *Es clouds* also drift laterally within the ionosphere [57].

Magnetic Storms

The major result of a magnetic storm at middle latitudes is its effect on the shape and peak electron density of the ionospheric F layer [13,58]. The characteristics and effects of individual magnetic storms differ widely. Depending on the time a magnetic storm begins, the peak electron density can rise or fall sharply from its quiet-day equilibrium value. At mid-latitudes, however, a drop in the F-layer peak electron density is the most common outcome. The decrease in F-layer density can pose difficulties for users of the HF channel because of concomitant changes in the MUF for a given path. The MUF can change abruptly and unpredictably, making the choice of an optimum HF link operating frequency more difficult, particularly for manual operators. Furthermore, the changes in layer shape can lead to an apparent change in the height of the F layer. In communication terms, this height change is equivalent to a change in the delay of the received signal. In the case of a reduction in electron density, the band of frequencies available to HF users shrinks, sometimes drastically, and often results in a very high density of users with the concomitant interference problems.

Mid-Latitude Spread F

Normally, the height profile of ionospheric electron density is vertically stratified and smoothly varying with height. As in the low latitudes, spread F is a phenomenon in which irregularities of electron density embedded within the otherwise benign ionosphere cause a distortion of signals propagated via the medium. The name *spread F* is derived from the appearance of the traces on a standard vertical ionogram. Instead of a sharply defined trace, the trace appears to be either spread in frequency (frequency-spread spread F), or in delay time (range-spread spread F). Frequency-spread spread F suggests multipath reflection, whereas range-spread spread F suggests scattering from a spatially diffuse irregular medium.

At middle latitudes, spread F is a sporadically occurring nighttime phenomenon whose production mechanism is not well understood. In this context, the delay spread of the received signal is usually related to the spatial spread of the irregularities affecting signal contributions, whereas the Doppler spread is related to their motions. The resulting signal amplitude will depend on whether the received signals are multipath reflected or scattered from the irregularities. In the case of mid-latitude spread F, delay spreads of up to 1 ms have been observed, but Doppler spreads are usually much less than 1 Hz [59].

2.3.2.3 Auroral Region

The auroral ionosphere is the region of greatest particle precipitation from the earth's outer radiation belt, the magnetosphere, and the interplanetary environment. It is usually considered to include the region of observed optical emissions, the auroral oval, and the region most affected by auroral absorption, the diffuse auroral absorption region. In addition, it includes the subauroral region of diminished nighttime F-layer electron density, the so-called nighttime F-layer "trough" region. The auroral oval and the diffuse aurora are regions in which particle precipitation (primarily electrons) plays a significant, at times dominant, role in ionization production. For this reason, the diverse phenomena associated with the auroral region are inherently irregular in both space and time.

As in the polar cap, sunlight fosters an intense ambient plasma density that masks many possible irregularities in ionospheric density. However, in the absence of ultraviolet ionization at night in the fall, winter, and early spring, the available ionization is wholly dependent on particle precipitation as well as ion movement from the sunlit ionosphere to the nighttime side. This movement of charged particles results in the so-called "two-cell convection pattern",[2] as depicted in Figure 2.30. During these periods, a transauroral HF channel is subject to many forms of density irregularities and therefore exhibits effects from the greatest range of disturbances.

The detailed geometry of the auroral region depends on the degree of magnetospheric disturbance measured by the *Q index*. The Q index takes on values from zero to seven, with values greater than four corresponding to magnetospheric substorm conditions. Figure 2.31 is a sketch of the geometry for the northern high-latitude region with a Q index of three (i.e., $Q = 3$). The diagram sketches the nominal location of the auroral oval, the "diffuse" auroral absorption region, and the F-layer trough region. The radial coordinate of the sketch is geomagnetic latitude measured from the magnetic pole, while the azimuth coordinate is the corrected geomagnetic local time (CGLT), with the sun's meridian at 1200 CGLT.

---

[2]Within the polar cap, plasma typically flows antisunward in a strip directly over the pole, and sunward along the auroral oval. This flow naturally divides the polar cap ionosphere into two kidney-shaped regions, forming a two-cell convection pattern. Other convection patterns sometimes form as a result of magnetic storms and substorms. Figure 2.30 shows the standard two-cell convection pattern.

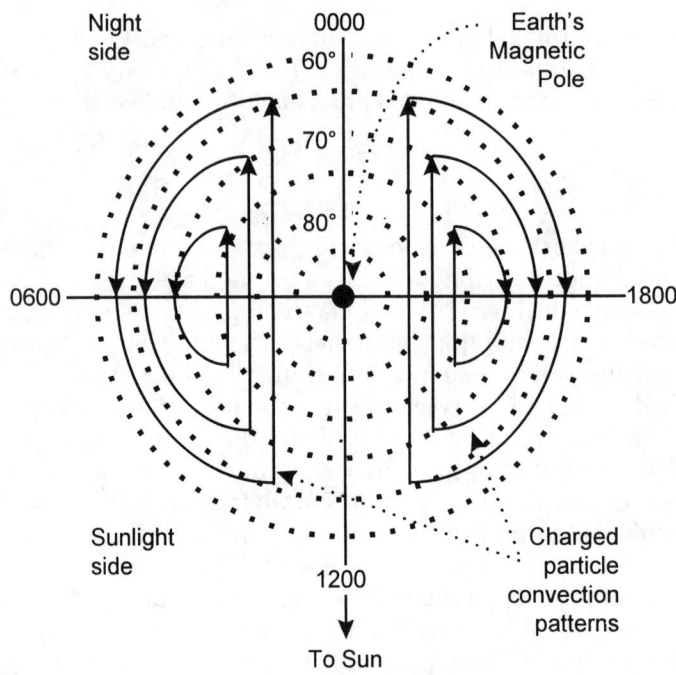

Figure 2.30 Two-cell ionospheric plasma convection pattern in the F layer.

As shown in Figure 2.31, the auroral oval [60,61] is asymmetrically distributed about the magnetic pole. The noon sector of the oval forms a narrow latitudinal strip centered on a geomagnetic latitude of approximately 75 deg, while the midnight sector is generally much wider and centered at a lower magnetic latitude (between 65 deg and 70 deg). The shape and the latitudinal dispersion of the oval is described by the disturbance index Q. During magnetically disturbed conditions, the oval expands equatorward, with the major expansion occurring on the nightside equatorward edge of the oval.

Auroral effects are classified as *discrete* or *diffuse* based on their temporal and spatial characteristics [62,63]. These characteristics take into account all observed effects including optical emissions, electron precipitation events, geomagnetic perturbations, and absorption effects. Discrete auroral effects are observed to occur most frequently at night with a broad maximum centered about 2200 CGLT, whereas diffuse effects are fairly widespread in magnetic local time but are primarily a morning phenomenon with a broad peak centered on 0800 CGLT [62]. The discrete aurora is generally structured, dynamic, and optically bright; the diffuse aurora is spatially diffuse, slowly varying, and optically faint. Presumably, the discrete aurora is associated with electron pre-

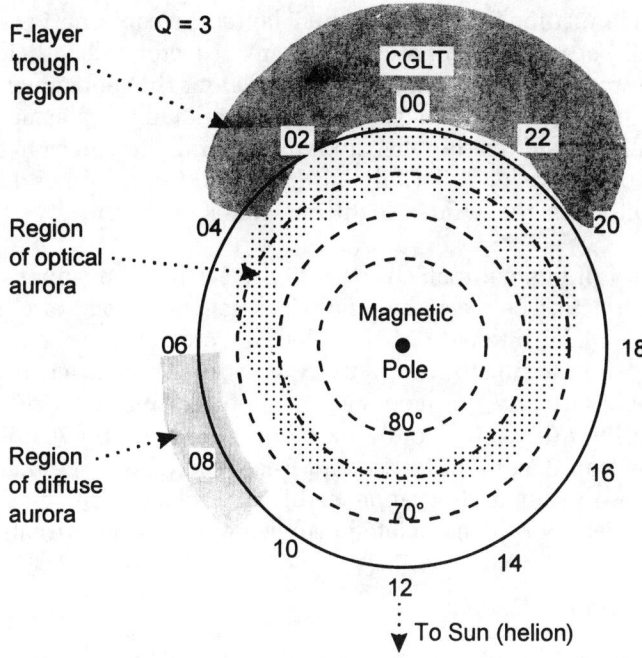

Figure 2.31 The northern high-latitude region (*After*: [39], Fig. 1.).

cipitation that is a prompt response to chaotic events in the storm-time solar wind. Occurrence of the diffuse aurora is related to precipitation events that have been buffered in the earth's magnetosphere and dispersed spatially and temporally in the process.

The diffuse aurora is present during quiet magnetic conditions as well as disturbed conditions, whereas the discrete aurora is primarily a disturbance phenomenon. Both types of aurora intensify during magnetospheric substorms. During severe substorms, auroral absorption can be strong enough to cause extended periods of radio blackout, provided the radio propagation path pierces the D layer in regions of enhanced D-layer ionization. As indicated in Figure 2.31, the diffuse auroral absorption region is located in a circular annulus centered about the magnetic pole, between approximate magnetic latitudes of 60 and 67 degrees. It lies well south of the auroral oval on the day side sector and overlaps the oval on the nighttime side. The degree of overlap is described by the Q index. The F-layer trough region lies adjacent to the equatorward-edge of the nighttime side oval, between the meridian of the day/night terminator in the dusk sector and about 0200 CGLT in the morning sector.

Auroral ionization in both the E and F layers is a combination of photoionization from direct solar illumination in the sunlit sector and collision-produced ionization from electrons streaming earthward along the geomagnetic field lines. In addition, the nighttime F layer is also populated by plasma blobs, large regions of enhanced ionization that enter the auroral F region from the midnight sector of the polar cap [51,64]. Collision-produced ionization and blobs are the dominant factors in the winter nighttime ionosphere and give it its unique character.

The entry of energetic electrons into the earth's atmosphere depends on a complex interaction between the solar wind magnetoplasma and the terrestrial magnetosphere. The intensity of this electron precipitation source increases with magnetospheric substorm activity. Magnetic activity is a general indicator of substorm activity and, in many cases, is a direct indicator of precipitation activity. Familiar effects of electron precipitation are (1) the visible aurora, (2) increased electron density at D-, E-, and F-layer heights, (3) enhanced auroral radio-wave absorption, and (4) enhanced E-layer electric currents, such as the auroral electrojet, and the associated geomagnetic field perturbations.

Electron Precipitation Effects

The deposition height and density of collision-produced ionization depends on the highly variable energy and flux of precipitating electrons and therefore is related to conditions in the local interplanetary medium and ultimately on the sun. This variability, along with variability in the direction of the IMF, gives the auroral ionosphere its unpredictable and rapidly changing character. Precipitating electron energies vary depending on their source region [63]. The three energy groups (~0.1 keV, 1 to 10 keV, and >40 keV) are represented in both the discrete and the diffuse aurora, but the latitude-time sectors affected by them are different [63]. The effects of the low-energy electrons (~0.1 keV) for both discrete and diffuse aurora are confined primarily to the auroral oval F layer. For the 1- to 10-keV electrons, which are primarily responsible for E-layer effects, the area affected by the discrete aurora is confined to the auroral oval. The region affected by the diffuse aurora overlaps the auroral oval on the nighttime side but lies just equatorward of the oval on the day side. In the high-energy domain (>40 keV), the discrete aurora is confined to the nighttime side oval. In this case, the diffuse aurora is found in an annular region about the magnetic pole between approximately 60 deg and 67 deg geomagnetic latitude as depicted in Figure 2.31. The high-energy diffuse aurora overlaps the auroral oval on the nighttime side but is located approximately 10 deg below the equatorward-edge of the oval on the day side. The high-energy electrons of the diffuse aurora affect primarily the ionospheric D layer and are responsible for enhanced auroral absorption.

Auroral Irregularities and Blobs

The auroral zone has long been recognized as a region characterized by very intense field-aligned irregularities. Early backscatter measurements at VHF [65,66] revealed the presence of aspect-sensitive scattering from irregularities of electron density in the auroral E layer, thereby confirming the existence of field-aligned irregularities [67]. In addition, HF backscatter measurements by Bates [68] established the existence of auroral echoes from field-aligned irregularities at F-layer heights. In regions such as the winter or nighttime auroral ionosphere, ionization production is almost wholly dependent on energetic particle precipitation. Furthermore, the energy spectrum of the entering particles, which is variable, determines the height of deposition of collision-initiated ionization. Therefore, energetic particle precipitation into the auroral atmosphere is an inherently irregular process in both space and time.

Polar Patches

Polar patches are an additional source of enhanced ionization in the auroral F layer. Once formed, patches (1) convect across the polar cap from the noon to the midnight meridian, (2) enter the nighttime side auroral F layer at the midnight meridian, and then (3) return to the noon meridian by way of the dawn and dusk sectors of the auroral region. Co-rotation of the neutral atmosphere, along with the velocity shears inherent in the return path of the two-cell convection pattern (Fig. 2.30), result in a distortion of the patches into relatively thin elongated regions of irregular electron density. These elongated irregularities have been given the rather inelegant title of blobs [51,64]. These blobs are generally aligned with the equatorward edge of the auroral oval, adjacent to the high-latitude boundary of the nighttime side F-layer trough region (see Fig. 2.31).

The high-latitude boundary of the F-layer trough is also a region of strong irregularity development owing to the steep horizontal gradients of electron density that exist in this region. The combination of the blobs and the F-layer trough boundary make this region a highly irregular portion of the auroral ionosphere. In addition to the auroral boundary blobs, smaller patches of irregularities are found within the auroral oval. These patches are related to isolated regions of particle precipitation. Finally, additional F-layer blobs are found on the poleward boundary of the auroral oval [64]. All of these factors combine to make the auroral ionosphere a particularly formidable sky-wave communications medium, especially during the late fall, winter, and early spring seasons when the background ionospheric density is minimum.

Auroral Sporadic E

Collision-generated E-layer ionization is inherently sporadic since it does not follow the usual diurnal pattern of ionization generated by solar radiation. During magnetically quiet sunlit periods, it is most likely that solar photon radiation will be the dominant source of ionization production at E-layer heights. During late autumn, winter, and early spring, however, collision-generated ionization will be the dominant production mechanism. At these times, E-layer ionization will be quite variable and unpredictable, and sporadic E-layer formation becomes a common phenomenon. Auroral sporadic E differs from mid-latitude sporadic E in that the irregularities tend to be aligned with the magnetic field. In this sense, auroral sporadic E bears some resemblance to equatorial sporadic E in which the irregularities are also field aligned.

During magnetic storms, the 1- to 10-MeV electron precipitation is greatly enhanced. This enhancement results in a sharp increase in E-layer electron density and in the E-layer electric current system. Under these circumstances, very intense E_s layers develop, usually with a critical reflection frequency that far exceeds that of any higher layers. Once these layers are formed, F layers are no longer accessible to ground-launched HF radio waves because of E_s-layer blanketing. Any communication link into or out of the auroral ionosphere must then propagate via the E layer, at least on the initial or final hop of the path. Generally speaking, the E layer is so irregular and variable during existence of strong sporadic E plasma densities that signals propagated via that mode will be subject to significant delay and Doppler spread effects [59].

Auroral Absorption

As previously discussed, the ionosphere attenuates radio waves when significant ionization is present in the dense upper atmosphere and lower ionosphere, the D layer. D-layer ionization is a naturally occurring daytime phenomenon at all sunlit latitudes. It is caused by the hydrogen Lyman alpha UV and the 1- to 8-Å (where 1 Å = 10^{-10} m) x-ray radiation bands of the quiet sun. During an intense solar flare, the 1- to 8-Å band is intensified by two to three orders of magnitude and the x-ray emission spectrum is extended to shorter wavelengths, resulting in an extension of D-layer ionization to lower heights. The net effect of these changes is a sudden and severe increase in nondeviative D-region absorption and the concomitant HF radio blackout, which usually persists for the duration of the flare event.

At auroral latitudes during late autumn, winter, or early spring, the primary production mechanism for D-layer ionization is the electron precipitation flux with energies greater than 40 keV. As previously mentioned, this precipitation occurs mainly in the night oval for the discrete aurora and in the diffuse auroral absorption region for the diffuse aurora. The diffuse auroral absorption zone is the more troublesome of the two phenomena. Although the diffuse absorption usually peaks around 0800 CGLT, it is a very broad peak that extends essentially around the clock. During 40-keV electron precipitation events, propagation paths that pierce the D region in the nighttime auroral oval or the diffuse auroral absorption band will likely experience serious disruption causing HF link outages. During active magnetic disturbance conditions, the 40-keV flux appears to be sporadic, judging by the manner in which absorption effects appear to wax and wane. During major magnetic storms, communication links that intercept the D layer in the diffuse auroral absorption zone may be unavailable 24 hr a day for as long as the storm lasts. Since these storms could continue for several days, depending on the degree of the disturbance, HF communications could be unavailable for extended periods.

Nighttime F-Layer Trough

The nighttime, mid-latitude, ionospheric trough is a region of depressed F-layer electron density occupying several degrees of latitude just equatorward of the auroral oval [69-72] as shown in Figure 2.31. This trough extends from roughly the location of the solar terminator to approximately 2–4 hr past the midnight magnetic meridian. Aside from a diminished MUF, the trough region is characterized by a very steep electron density gradient at its boundary with the auroral oval [70,72]. This gradient can be significant in terms of fostering instabilities leading to plasma irregularities and associated radio-wave scattering phenomena. The major factors contributing to the formation of the F-layer trough are the opposing effects of (1) the eastward atmospheric co-rotation with the earth and (2) the westward polar cap plasma convection on the nightside of the earth [71,72].

Communication Characteristics of the Auroral Channel

The ionization resulting from particle precipitation in the auroral zone creates field-aligned irregularities in both the E and F regions. These irregularities are sufficiently dense to scatter radio waves in the HF and VHF bands. The scattering is increasingly aspect sensitive as the frequency is increased, with the greatest scattered signals resulting from radio waves normally incident on the magnetic field-aligned irregularities. *Side-scattering* from these auroral irregularities in the E region north of the great circle path between transmitter and receiver are well known at VHF [73]. For auroral-scatter paths at E-region altitudes below 120 km, link geometry for specular side-scatter requires that the transmitter and receiver be located below about 62 deg of latitude. F-region reflection at altitudes above 250 km requires that the two stations be located below 40 deg latitude.

Channel probe measurements [42,43] indicate that spread conditions on the nighttime transauroral channel do not depend strongly on magnetic conditions. By contrast, spread conditions on the daytime auroral channel are quite sensitive to disturbed magnetic conditions. Evidence of enhanced daytime spread conditions exists [42,74] for values of $K_p = 3$, where K_p is a quasilogarithmic three-hourly index of planetary magnetic activity. K_p is quantified on a scale from zero to nine, in which zero indicates no activity.

Examples of nighttime conditions on a transauroral channel are shown in Figures 2.32(a) through (e). These measurements were made in March 1992 on a 1,300-km path between Sondrestrom, Greenland, and Keflavik, Iceland, shortly after midnight UT. Magnetic conditions were mildly unsettled, $K_p = 2$, and the midpoint of the path lay near the polar boundary of the nighttime oval. Figure 2.32(a) shows an oblique path ionogram depicting ionospheric conditions at the end of the measurement period. The ionogram shows a complex structure of individual traces, all diffuse, but with strong returns at their leading edges. These traces suggest a complicated ionospheric spatial environment consisting of a number of strong irregular regions distributed about the propagation path. Typical slant-F scatter behavior, characterized by a monotonically increasing minimum delay with frequency [75], is observed at frequencies above approximately 10 MHz. Above 14 MHz, the ionogram trace is totally diffuse, implying propagation based purely on scatter mechanisms. This result supports the view of a generally irregular ionosphere with embedded field-aligned irregularities.

Two scattering functions, one measured at 7.5 MHz and the other at 14.5 MHz, have been selected for examination. The 7.5-MHz frequency was chosen because it lies in a frequency range where individual traces and strong signals are visible on the ionogram, whereas the 14.5-MHz frequency was selected because it lies in the purely slant-F scatter region of the ionogram. The scattering function of Figure 2.32(b) is a three-dimensional representation of the distribution of received signal energy on the delay-Doppler plane for the 7.5-MHz return. It shows a relatively weak sporadic E return at a delay of approximately 350 ms. This return is followed by two nearly contiguous returns corresponding to signals from F-layer irregularity regions. The peak amplitude of the scattering function in the region of the F-layer return (~94 dB) is comparable with that obtained from a daytime specularly reflected signal during magnetically quiet times. This observation suggests that the measured signal was reflected rather than scattered.

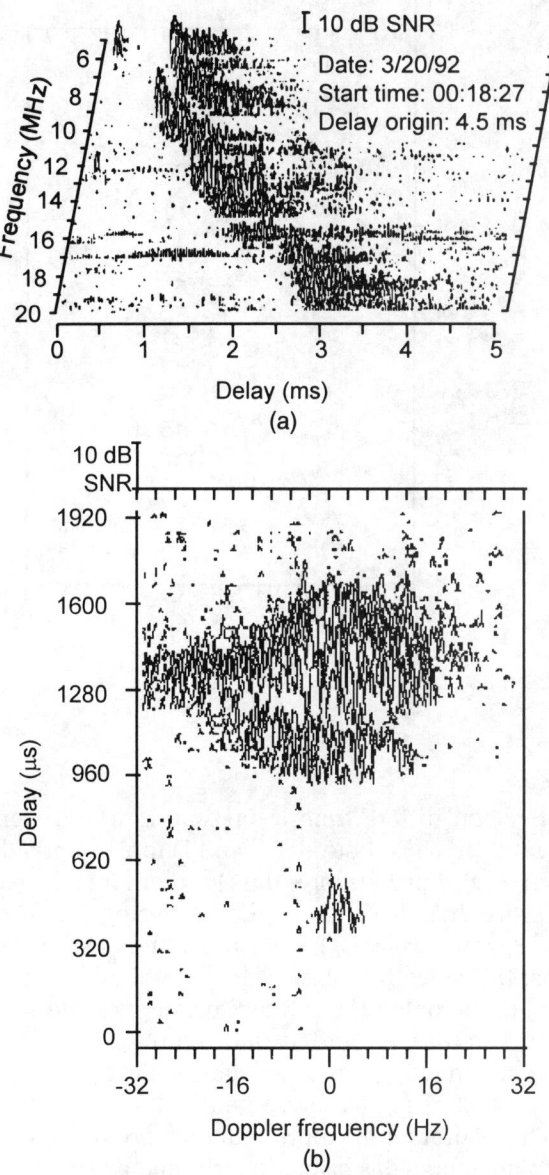

Figure 2.32 Channel probe measurement results from a transauroral channel on a 1,300-km path in Greenland: (a) oblique ionogram, (b) scattering function for 7.5 MHz, (c) contour plot of 7.5-MHz scattering function, (d) scattering function for 14.5 MHz, and (e) contour plot of 14.5-MHz scattering function.

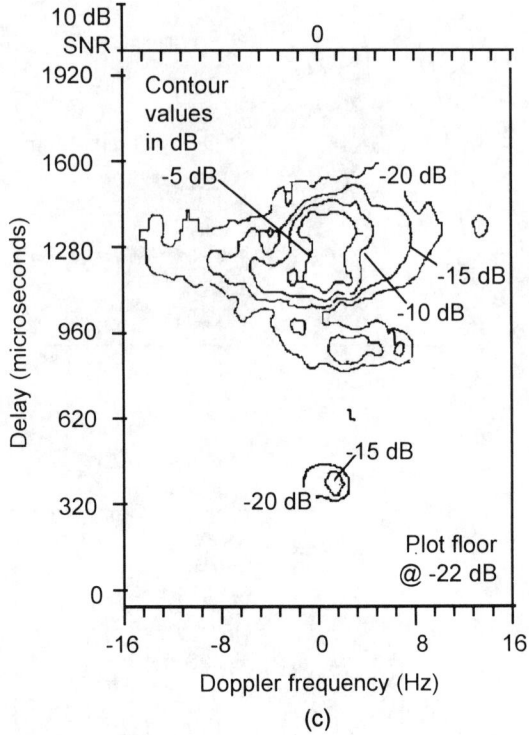

Figure 2.32 (continued)

The F-layer return differs from a daytime reflected signal, however, in that it is extensively spread in both delay and Doppler (spread values ~200 ms and 6 Hz, respectively). Furthermore, the Doppler spreads exhibited by many of the stronger individual features in the scattering function are unusually broad. This broad spread suggests that the reflecting medium was continuously changing, and that the reflecting surfaces had a limited lifetime compared with the duration (~4s) of the data sample used to calculate the scattering function. These facts, taken together, suggested that the observed multipath reflection condition was from a medium in a constant state of change. The scattering function in Figure 2.32(b) is also presented in the form of a contour plot in Figure 2.32(c). The contour plot representation provides a clearer picture of the manner in which energy has been redistributed on the delay-Doppler plane as a result of passage through the channel. Each contour represents an amplitude level in decibels relative to the peak spectral density of the scattering function. Contour levels of −5 to −20 dB in steps of -5 dB are shown on the plot with the innermost contour representing −5 dB.

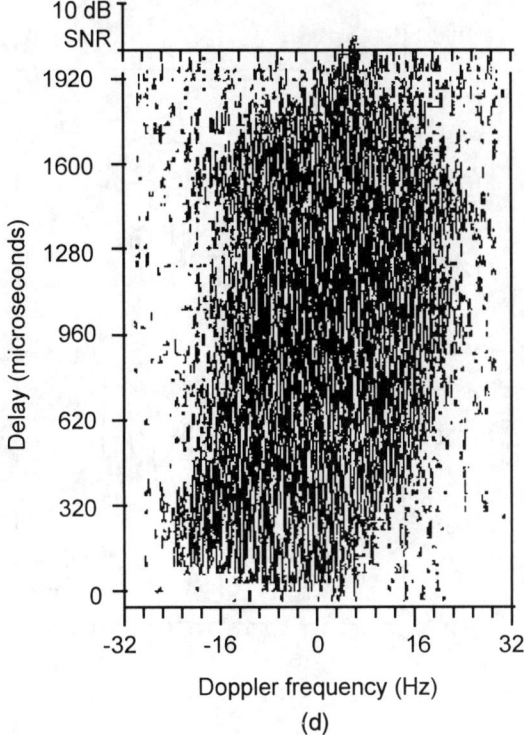

Figure 2.32 (continued)

The peak spectral amplitude for the 14.5-MHz scattering function (see Figs. 2.32[d] and [e]) is about 62 dB, which is about 30 dB below that of the peak signal measured at 7.5 MHz. The delay and Doppler spreads of the signal at 14.5-MHz are 750 ms and 15 Hz, respectively, as compared with values of 200 ms and 6 Hz at 7.5 MHz. All of these factors support a hypothesis of a scatter mechanism being responsible for the received signal at 14.5 MHz, as opposed to a specular multipath reflection mechanism as at 7.5 MHz. This hypothesis was supported by the analysis of the ionosonde results presented in Figure 2.32(a).

Results based on measurements of transauroral propagation [39,42] indicate that three types of signals are observed on the channel depending on ionospheric conditions. During magnetically stable daytime conditions, signal characteristics comparable to those of a quiescent mid-latitude channel are observed. During disturbed conditions, multipath signals are observed that are comparable in amplitude to those of the quiet daytime channel, but with considerably more spread in delay time and Doppler frequency. Finally, under disturbed conditions, one may also observe scatter-type signals, which are about

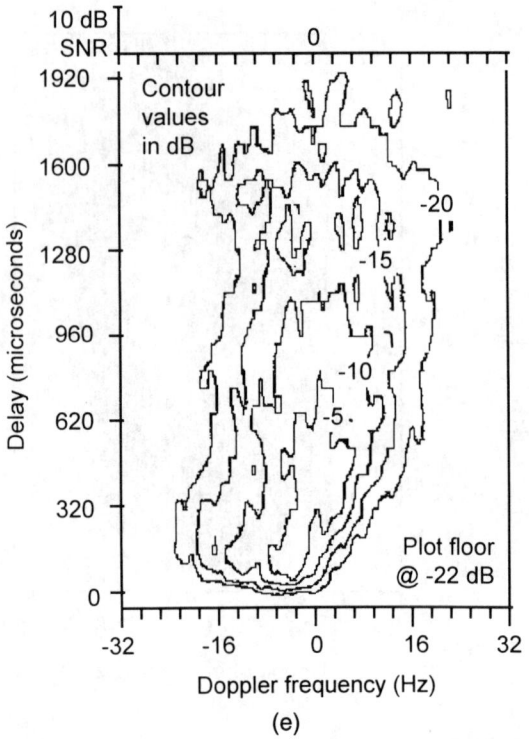

Figure 2.32 (continued)

30 dB smaller in amplitude than the specular-multipath signals albeit with somewhat more spread. These results can be concisely expressed in a *channel performance map*, which is a scatter plot of measured points plotted on a graph whose ordinate is signal amplitude, in decibels. The abscissa of the channel performance map is the channel spread factor [76], where the spread factor is expressed as the dimensionless product of the measured delay and Doppler spread values.

An illustration of a channel performance map is given in Figure 2.33 for a case involving two experiments conducted by the Naval Research Laboratory (NRL) between Sondrestrom, Greenland, and Keflavik, Iceland [42]. One daytime measurement was performed at a time of mildly disturbed conditions, and a second measurement was performed at night during relatively quiet magnetic conditions. In spite of the presence of some weak scatter returns, the daytime experiment showed clear evidence of specularly reflected signals at all frequencies probed. The nighttime measurement showed evidence of specular multipath and scatter returns at frequencies below about 8.5 MHz, but only scatter returns at frequencies above 11.5 MHz.

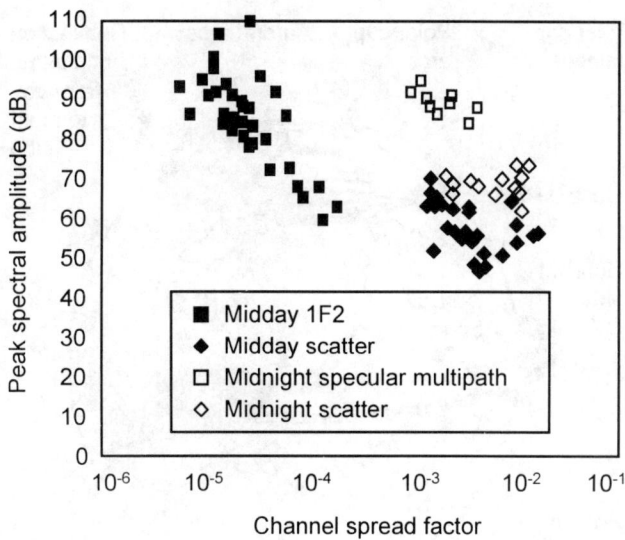

Figure 2.33 Channel performance map.

2.3.2.4 Polar Region

Polar Arcs and Patches

The polar region (or polar cap) is defined as the region within the poleward boundary of the auroral oval. The primary ionospheric irregularities observed in the polar region are called polar *arcs* and *patches* and are depicted in Figure 2.34. These irregularities are the dominant ionization features in the polar winter ionosphere, when the absence of solar radiation results in a weak ambient ionosphere. Arcs and patches are known to move quite rapidly, and it might therefore be expected that HF propagation via these irregularities would show significant delay and Doppler spread values. This result is quite important for HF users in Antarctica, who must often rely on HF communication in the absence of continuous satellite communications. During polar summer, the ionospheric irregularities such as arcs and patches are masked by the dense ionospheric plasma produced by the continuous exposure to solar radiation. Based on a private communication from Basler, Wagner reports [37] that channel measurements during polar summer are characterized by much smaller delay and Doppler spreads than those found during the polar winter.

Polar arcs are narrow, elongated irregularities of enhanced plasma density that can reach up to 1,000 km in length by about 100 km in width. These arcs are aligned with the earth-sun line and are believed to be formed by "soft"

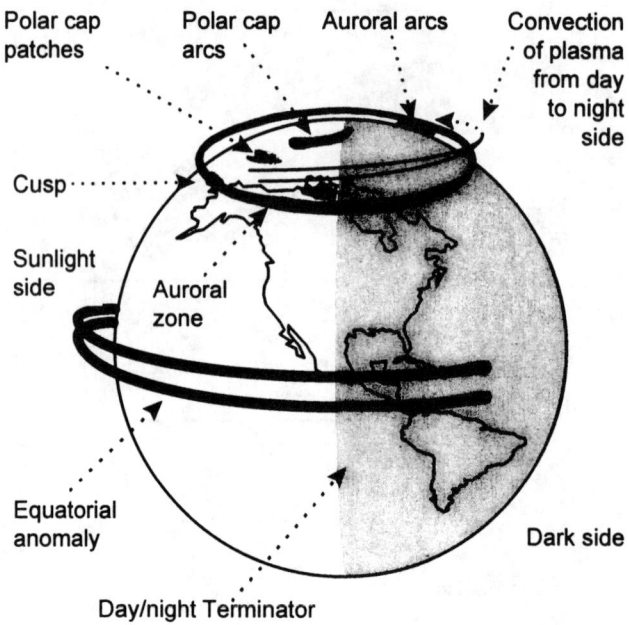

Figure 2.34 Polar arc, patches, and plasma convection.

electrons (energies less than 500 eV). Polar arcs have been observed with optical imagers and ionosondes, which have shown that arcs occur most frequently in "quiet" magnetic conditions during solar maximum. Arcs are not considered significant contributors to the polar ionosphere during solar minimum. Arcs produce scintillation on satellite links and spread traces on vertical-incidence ionograms. No diurnal variation in the polar arcs has been reported, although they are known to drift from the dawn to dusk meridians at speeds between 100–250 km/s.

Polar patches are large amorphous plasma irregularities with diameters of approximately 1,000 km. Patches are believed to originate in a sunlit portion of the subauroral ionosphere where they are subject to the high-latitude two-cell convection pattern (see Fig. 2.30). In other words, patches are conjectured to originate at the point of maximum solar illumination (noon meridian) in the subauroral zone. Since the auroral oval is centered about the magnetic pole, but its orientation is fixed relative to the earth-sun line, the earth's rotation changes the geographic latitude of the auroral subsolar point in universal time. In this context, the subsolar point corresponds to the point in the auroral zone with the greatest solar zenith angle (angle between the local horizon and the sun) and therefore the greatest ionizing radiation. This geographic modulation of the solar zenith angle of the auroral subsolar point results in an apparent diurnal variation in UT of the occurrence rate of polar patches.

The formation rate of patches is believed to be independent of the solar cycle, although as expected, the plasma density within a patch is enhanced during periods of solar maximum. The precise mechanism by which patches become detached from the background ionosphere is not firmly established. Once detached, however, they move across the polar cap (noon to midnight meridian) with the regular plasma convection pattern at speeds of about 1,000 m/s. Because of the low background ionization in the polar winter ionosphere, large density gradients exist at the boundaries of patches. These large gradients are believed responsible for instability processes [64] that lead to the development of a broad spectrum of irregularity sizes, thus impacting a broad spectrum of communication systems.

Polar Cap Absorption

An important disturbance feature of the polar ionosphere is induced by a solar proton event (SPE), which is associated with the most intense solar flares. During these events, relativistic solar protons are ejected from the sun and arrive in the vicinity of the earth within approximately 30 min. Although these protons are deflected by the geomagnetic field at lower latitudes, they nonetheless gain entry to the earth's atmosphere via the polar region. Protons having energies >10 MeV penetrate to D-region heights where they produce ionization in the dense neutral atmosphere, thus creating the conditions for a strongly absorbent medium. These events are called polar cap absorption events, or PCAs.

PCA events are caused by high-energy protons (>10 MeV) arriving at the earth during a solar proton event (SPE), which is associated with the most intense solar flares. These protons create ionization at altitudes between 40 and 90 km above the earth's surface, which includes the D region. The resulting ionospheric absorption during a PCA event usually occurs in the altitude range of 45 to 80 km. Because of the decreasing electron-neutral collision frequency with increasing height, most of the absorption occurs in the height range of 40 to 60 km. The absorption decreases with increasing frequency and increases with increasing propagation path length through the absorbing region. The increased ionization associated with an SPE attenuates, at least to some degree, all radio waves passing through the D region of the polar ionosphere.

Communication Characteristics of the Polar Channel

Channel probe measurements of the polar channel were performed by Basler [44,45] on a path between Narssarsuaq and Thule, Greenland, in October 1984.

Figure 2.35(a) shows a depiction of the spread of signal energy (represented by contours of SNR) in frequency (Doppler spread) and time (delay spread) observed during one of the 5-min sample periods on October 3. Large variations in both delay and Doppler spread were observed between the results recorded on subsequent 5-min sample periods. The ionogram associated with the sample

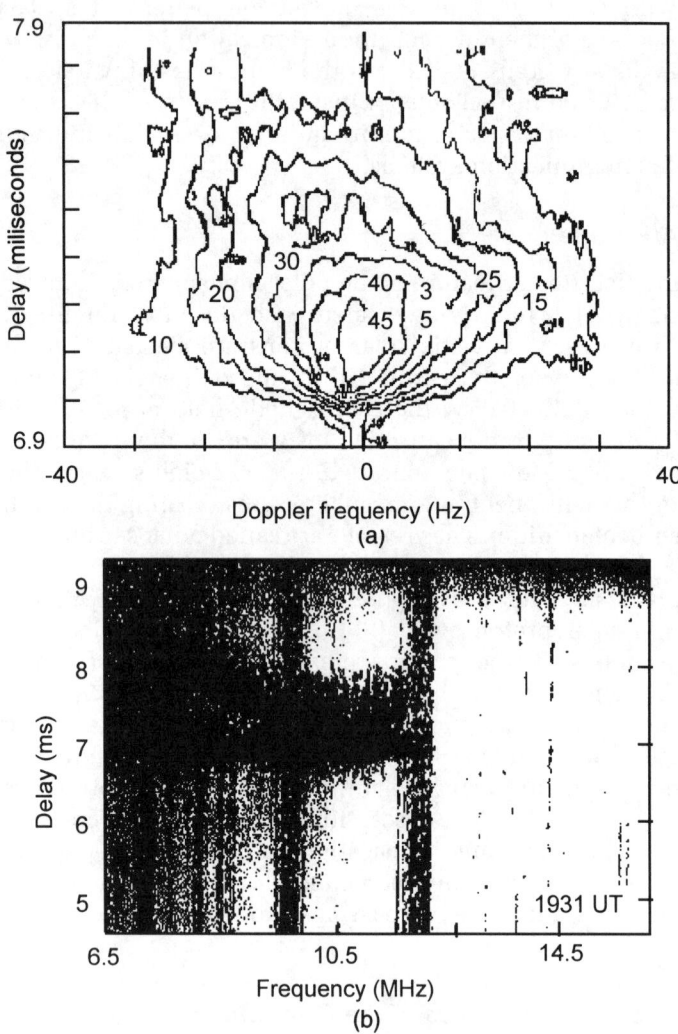

Figure 2.35 Channel probe measurement results from polar channel: (a) SNR contours of delay spread and Doppler frequency and (b) ionogram showing diffuse one-hop F_2 mode.

depicted in Figure 2.35(b) shows that these spreads were obtained from a diffuse one-hop F_2 mode.

Average Doppler and delay spreads have been derived for this path over an equinoctial period using data from March and October. The computed values for Doppler spread varied from 12 to 15 Hz, whereas the average delay spread, computed at the −20-dB contour at zero Doppler value, was found to be 0.25 ms averaged over the whole day. We should point out that these measurements were made in a period of low solar activity, judging by the monthly mean sunspot numbers for the period. Given what is known about the solar cycle dependence of arcs and patches, it is most likely that these propagation conditions were due to the presence of polar patches.

Statistics derived from riometer measurements in Thule [77] showed that significant PCA events occurred about 2% of the time in the period from 1962 to 1972. During such events, HF communication using paths through the affected region could be interrupted for periods of up to several days, depending on the severity of the event. The impact of PCA-induced absorption on HF communications was observed during a propagation experiment conducted by the Naval Undersea Warfare Center (NUWC) and the Phillips Laboratory in May 1991 [78]. As part of this experiment, an HF channel probe was operated between Sondrestrom AFB and Thule AFB. The probe transmitted a pair of spaced tones according to a predetermined frequency-use schedule. On May 13, a significant PCA was observed by the 30-MHz Thule riometer as shown in the plot of one-way zenith absorption (A_z) of Figure 2.36. If the ionospheric conditions above midpath on the Sondrestrom-Thule link were similar to those

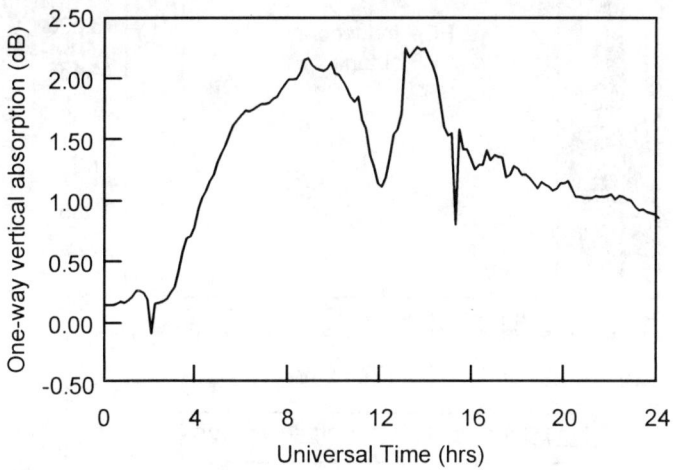

Figure 2.36 PCA-induced attenuation above Thule, Greenland, on May 13.

above the Thule riometer, then the time line of the riometer values plotted in Figure 2.36 would, with appropriate scaling, be representative of the time history of absorption on the HF probe link.

The channel availability measured during the May 13 HF probe measurements is presented in Figure 2.37. These results show an apparent correlation between the absence of usable probe frequencies and the time interval for which the Thule riometer recorded one-way 30-MHz vertical absorption in excess of 1 dB. The 1-dB absorption value was exceeded at about 0430 UT at Thule while the HF outage began at about 0500 UT. Similarly, the ZA value dropped below 1 dB between 2000 and 2200 UT with the subsequent restoration of probe connectivity in the upper HF band at about 2030 UT. The channel availability of frequencies in the upper HF band evident in the figure immediately upon restoration of HF probe connectivity is characteristic of sporadic E layer formation following PCA events [79].

2.3.3 Summary

Sky-wave HF communication is widely regarded as the most challenging radio communication medium. A useful guide to the most significant phenomena in

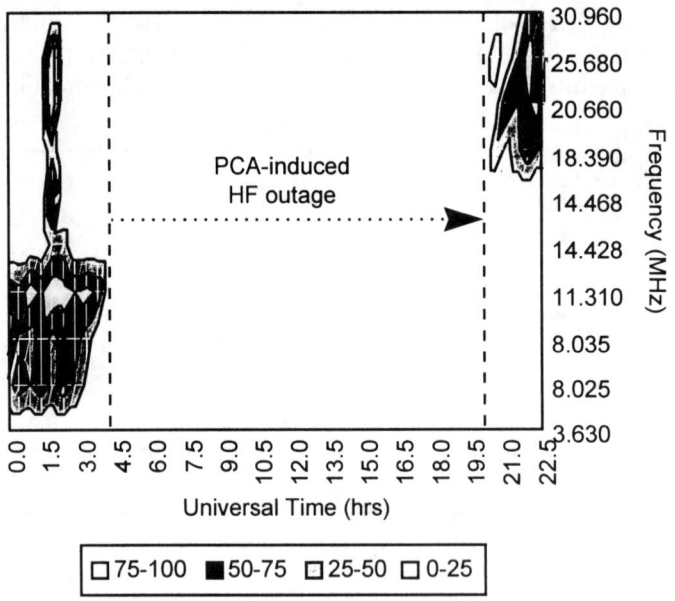

Figure 2.37 HF probe frequency availability on May 13.

HF sky-wave radio is given by Maslin [8]. Some of the phenomena are summarized below to help explain the relative merits of the various HF system technologies discussed throughout the remainder of this book. Table 2.1 summarizes the various sources of ionospheric disturbances and ionospheric irregularities discussed in this chapter and the corresponding impact on HF sky-wave signals. The durations shown for the various propagation events, as well as their geographical locations, are typical of observed conditions and do not indicate absolute limits. Except for spread F and sporadic E events, all of the phenomena listed in the table have been correlated with solar-cycle variations. The various irregularities fostered by these ionospheric phenomena exhibit propagation characteristics dependent on their size and shape relative to the signal wavelength. For this reason, HF systems that can adapt their operating frequencies to changing environmental conditions offer the best available technique for reliable sky-wave communications, particularly at high latitudes. In addition,

Table 2.1
Summary of Principal Ionospheric Disturbances and Their Effects

Event	Duration	Location	Effect
Terminator	Minutes	Global along day/night and night/day boundaries	Frequency deviation and phase shifts
Sporadic E	Few hours	Global, but more common at low and mid-latitudes	Decreased range from F-region paths, multipath fading, time and frequency dispersion
Solar flare	20-30 minutes	Global, with significant polar effects	Increased absorption from SWF and SPA events
TIDs	Minutes to hours	Global	Localized absorption increases with changes in reflection heights
Spread F	Minutes to hours	Equatorial	Multipath fading, random phase shifts, time and frequency dispersion
Magnetic storms	Few days	Global; severe near poles	Increased absorption
Substorms	Hours	High latitudes	Increased absorption, multipath fading, time and frequency dispersion
CME	Hours	High latitudes	Increased absorption, multipath fading, time and frequency dispersion
Polar patches and arcs	Hours	Polar latitudes	Multipath fading, time and frequency dispersion
PCA	Hours to days	Polar latitudes	Increased absorption

these systems must accommodate the received signal level fading, time, and frequency dispersion expected for such disturbed channels.

As has been shown, the ionosphere is a dispersive medium that spreads the signal in both time and frequency. In addition, it is usual to receive a radio signal that has been reflected from more than one ionospheric layer, a phenomenon known as *multipath propagation*. The enormous variability of the ionosphere makes it difficult to model the HF channel adequately; however, the CCIR [80,81] has proposed three standard simulation channels, termed "Good," "Moderate," and "Poor," that have been widely adopted as a basis for comparison of HF modems (see Table 2.2). These channel models consist of two equal-amplitude, independent Rayleigh-fading sky-wave paths representing reflections from different layers in the ionosphere. Reviewing the delay and Doppler spread values observed for disturbed conditions, it is evident that even the CCIR Poor channel can often serve only as a lower bound on the signal distortion expected under these conditions.

Table 2.2
CCIR Channel Characteristics

Effect	Good	Moderate	Poor
Time delay	0.5 ms	1 ms	2 ms
Fade rate	0.1 Hz	0.5 Hz	2 Hz

2.4 NOISE

2.4.1 Background

As discussed in Chapter 1, the performance of a radio link is determined by the signal strength of the desired signal and any other signals present in the channel. The transmitter power, the path loss, and the gain of the transmitting and receiving antennas define the level of the desired signal. Other signals present in the channel are a combination of galactic noise from celestial radio sources, atmospheric noise from terrestrial lightning, noise generated within the receiving equipment, and finally man-made noise and interference. This last component includes signals and noise from other users of the radio spectrum propagated via ionospheric sky waves, via ground waves, or otherwise generated in proximity to the receiving antenna. The importance of the individual noise and interference sources will vary with time of day, season, geographic location, and frequency. Although one source may be dominant for any given set of conditions, the contribution from all potential sources must be evaluated during the design of an HF station needing receive capability.

The ultimate bound on any communications system is the noise power n_t that arises from the thermal agitation of electrons in conductors and is given by $n_t = kT_0B$, where k is Boltzmann's constant (1.38×10^{-23} JK^{-1}), T_0 is the system *noise* temperature, usually taken to be 290 K, and B is the receiver bandwidth in hertz. The noise floor in an HF radio system is rarely set by thermal noise, but this level forms a useful reference point for the more energetic forms of noise. Thus, total received noise power is conventionally expressed in terms of the effective antenna noise factor F_a in decibels above kT_0B for a lossless antenna. Figure 2.38, adapted from Maslin [8], shows typical values of F_a for a variety of noise sources at three frequencies in the HF band.

2.4.2 Atmospheric and Man-Made Noise

Atmospheric noise originates from lightning strokes, with the largest contribution generated by the tropical rain forest regions in Asia, Africa, and South America. The wideband noise from individual strikes in these regions is propagated by the ionosphere over large regions of the earth. Far from these lightning strokes, the noise level is the superposition of lightning-emitted radiation from many sources. The combination of these many impulses yields the background "static" normally heard on HF and MF receivers. The atmospheric noise contri-

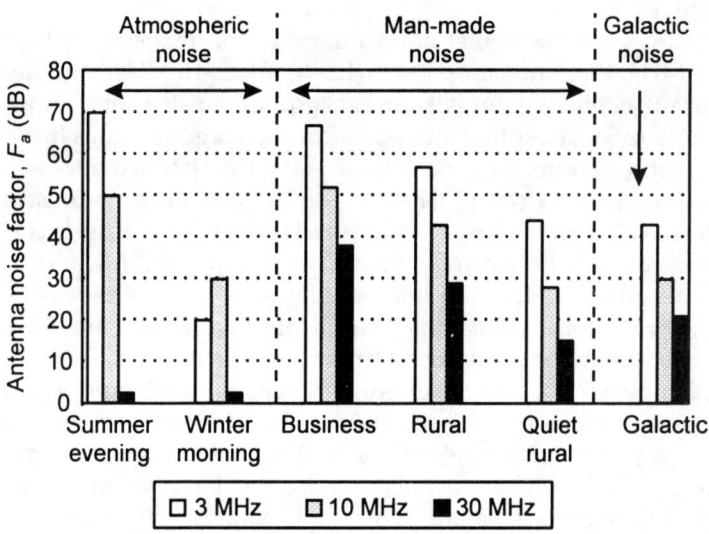

Figure 2.38 Effective antenna noise factors for a range of noise sources. (*After*: [8].)

bution within ground-wave range of lightning discharges will appear as significant noise bursts overlaid on an otherwise homogeneous background level generated from distant sources. This background noise level will decrease significantly at high latitudes and the many contributions will combine to form a very stable noise signal. Even at high latitudes, however, occasional discharges between clouds will generate discernible noise bursts.

Atmospheric noise is characterized by short pulses with random recurrence superimposed on a background of random thermal noise. Average atmospheric noise levels tend to remain constant over periods as long as an hour. Because these levels are well correlated over wide areas, it has been possible to produce worldwide contour maps for each season and illustrate its frequency dependence. Worldwide maps of atmospheric noise were published by CCIR in Report 322-3 [30]. Previous editions of this report overestimated the noise level at high latitudes and should therefore not be considered indicative of these locations. Report 322-3 presents atmospheric noise values on world maps of the excess noise in decibels (F_a) above kT_0 at 1 MHz available at the terminals of a short lossless vertically polarized antenna. World maps are available for four seasons referenced to the Northern Hemisphere, that is, winter (December, January, February), spring (March, April, May), summer (June, July, August) and autumn (September, October, November). The diurnal variation is covered through six maps, each covering a 4-hr period in local time for each season. Each map has an associated set of scaling curves for the frequency range from 10 kHz to 30 MHz.

The CCIR maps show that atmospheric noise decreases with frequency up to 2 to 3 MHz, then increases somewhat to about 15 MHz, and then sharply decreases again, achieving the galactic background noise level at about 25 MHz. Similarly, Maslin [8] states that atmospheric noise predominates in electromagnetically quiet locations (no man-made noise) at frequencies below about 20 MHz. The expected level of galactic noise and man-made noise at a quiet receiving site is also shown on these graphs. Data on the variability of the noise level, such as the ratio between the median and upper and lower decile values of noise amplitude as well as the standard deviation for these results, are given in a separate set of curves. Note that these values do not account for lightning-generated noise from local thunderstorms [30].

The parameters F_{am}, the mean noise factor above kT_0, and V_d, the voltage deviation, are defined as the difference in decibels between the rms and average noise envelope voltages. Parameter V_d is used to predict the amplitude probability distribution (APD) of atmospheric noise. A V_d value of 1.05 indicates Gaussian-distributed noise values, whereas V_d values larger than 1.05 indicate the presence of increasing levels of non-Gaussian impulsive noise. CCIR predictions of atmospheric noise level statistics are quite accurate in the evening and night hours when this noise contribution exceeds potential levels of man-made noise. During morning and daylight hours near built-up areas, the noise observed can be predominantly man-made noise, with levels below those reported in Report 322-3 for a "quiet receive site." Higher values of V_d correlate with higher F_{am} values, indicating the presence of man-made impulsive noise.

Lauber [82] has presented measurements of atmospheric and man-made noise at a number of quiet and urban sites in eastern Canada. The frequencies free from interference close to 2.5, 5, 10, 15, 20, and 25 MHz were examined through four seasons as used in Report 322-3. The survey found levels of man-made noise exceeding the CCIR predictions for a quiet site [83] by 20 to 30 dB at urban and industrial environments. These findings agree well with the predictions of man-made noise by Spaulding and Disney [84].

Atmospheric noise levels below the predicted levels of man-made noise can be obtained in isolated locations such as those found at high latitudes, provided efforts are undertaken to eliminate local noise sources such as ignition systems, electrical generators, and computer equipment. Numerical methods to predict the noise environment at a given site have appeared within the last 10 years. A numerical version of Report 322-3 is presented by Spaulding and Washburn [85], and a numerical method to evaluate atmospheric, man-made, and galactic noise is presented by Spaulding and Stewart [86].

An important concern in the use of communications performance-prediction tools is their reliance on many stochastic processes in the determination of the SNR provided by an HF channel at a particular frequency. Some of these processes, such as the generation of man-made noise, are treated as deterministic quantities for ease of analysis. For example, some models depict man-made noise power as a constant value as a function of frequency, typically 3 MHz, with a linear variation with frequency defined by a slope of −27.7 dB per decade. This empirical model can be appropriate for some receiving sites, such as a ship at sea, but less applicable for more dynamic conditions. More specifically, Spalding and Disney [84] show that man-made noise power is strongly time dependent. Rush hour traffic and industrial and office plant activation during the working day were shown to yield changes in excess of 20 dB. In fact, the upper decile of the noise in all noise environment categories is about 15 dB above the corresponding median value. Ideally, man-made noise would be considered a nonstationary stochastic process with the expected diurnal variations as well as account taken for the differences between weekend and work days. As a result, link performance predictions would show that less EIRP would be required to achieve a specified SNR value (see Chapter 1) during the early morning hours when the HF channel is often least effective, but man-made noise levels are also at their lowest values at these times.

Man-made noise arises from all manner of sources, such as electrical machines, power lines, and ignition systems. It decreases with increasing frequency and varies considerably with location, particularly in built-up areas.

Figure 2.39, also adapted from Maslin [8], illustrates the behavior of the various noise sources as a function of frequency based on data from CCIR. We can see that man-made noise could well predominate, especially in mobile platforms, where noise levels could be above the CCIR business level. Figure 2.39 also shows the rather insignificant contribution of galactic noise.

The important point about the nature of the noise observed on HF channels is that it cannot be modeled as Gaussian. In fact, HF noise is composed of impulses of significant amplitude with durations ranging from a few tens of milliseconds to more than 1 sec. To illustrate this behavior, Figure 2.40 shows a typical probability density function (pdf) of HF noise voltage measured in the Australian region on a frequency of about 5 MHz. Overlaid on Figure 2.40 is the pdf of Gaussian noise of the same noise power as the measured noise. (The standard deviation was 20 units on this scale.) It can be seen from this figure that the measured noise exhibited a broad range of amplitudes not characteristic of Gaussian noise. The cumulative density function (cdf) for this noise sequence is shown in Figure 2.41(a) and is again contrasted with Gaussian noise. Note the dominant accumulation of low-amplitude impulses as well as the significant probability associated with high-amplitude impulsive noise excursions from the otherwise quiet background. The distribution of noise burst duration, shown in Figure 2.41(b), is particularly valuable in assisting in the design of error-control coding schemes for HF radio (see Chapter 4).

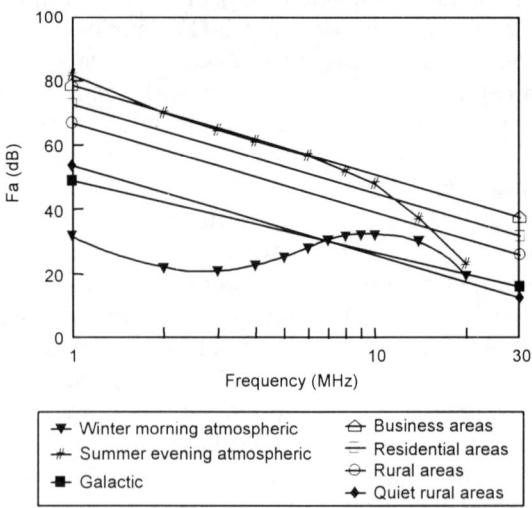

Figure 2.39 Antenna noise figure dependence on frequency. (*After:* [8].)

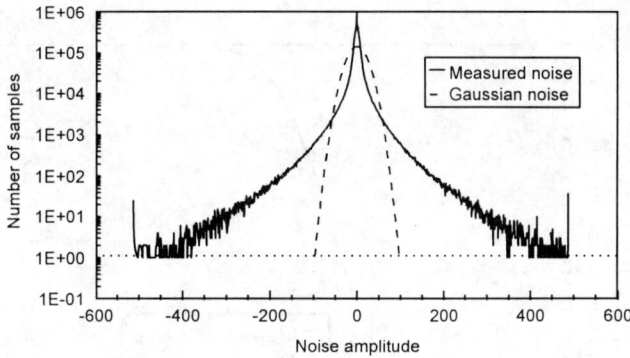

Figure 2.40 Example of pdf of noise amplitude at 5.4765 MHz compared with pdf of Gaussian noise of same noise power.

2.4.3 Galactic Noise

Galactic noise originates from celestial radio sources with a clustering of sources found along the galactic plane. These are located throughout the sky, but the spatial density is not evenly distributed over the celestial sphere. Taylor [87] has presented maps of galactic noise temperatures in celestial coordinates, right ascension and declination, for 136 and 400 MHz. The noise temperature is a function of frequency, increasing with a decrease in frequency. The noise temperature at frequencies other than 136 or 400 MHz can be obtained by scaling the mapped temperatures by $(f/f_{ref})^{-2.3}$, where f is the frequency of interest, f_{ref} is either 136 or 400 MHz, depending on which map is used. The 136-MHz map is useful for computation of antenna noise temperatures in the low VHF and HF frequencies for antennas with low or moderate gain. The antenna temperature T_a component from galactic noise sources is computed by integration over the antenna aperture according to

$$T_a = \frac{(f/136)^{-2.3} \int_0^{\pi/2} \int_0^{2\pi} L_i(\phi, \theta) G(\phi, \theta) \, T_G(\phi, \theta) d\phi d\theta}{\int_0^{\pi/2} \int_0^{2\pi} G(\phi, \theta) d\phi d\theta} \qquad (2.3)$$

where f is the frequency of interest in MHz, dq and df are increments of q and f, and $T_G(\phi, \theta)$, $L_i(\phi, \theta)$, and $G(\phi, \theta)$ are the antenna gain, the noise temperature from the galactic map, and the ionospheric loss in the direction determined by θ and ϕ. The earth's rotation will sweep the antenna aperture "across" the noise map, and antennas with moderate to high gain will see a corresponding

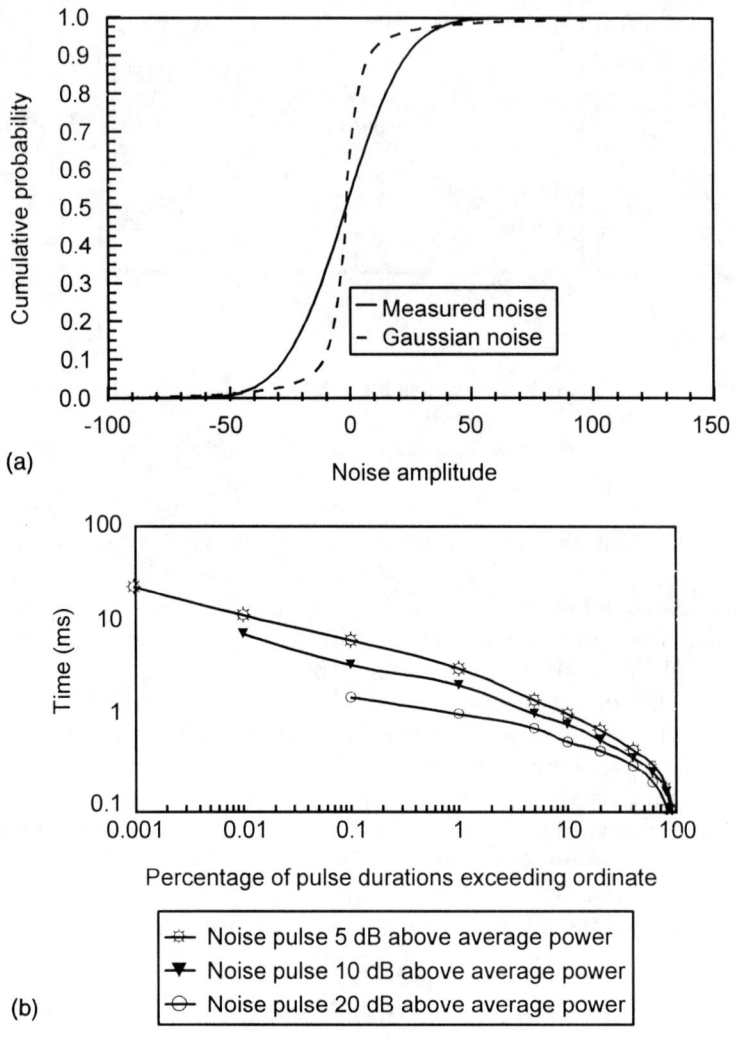

Figure 2.41 Noise sample analysis: (a) cumulative density function for noise shown in Figure 2.40 and (b) pulse duration distributions for noise used in part (a).

diurnal variation of the noise temperature, whereas low-gain antennas will see little variation. Also, the reflectivity of the ionosphere determines the noise temperature contribution through the $L_i(\phi, \theta)$ term. Typical F_a values of galactic noise were shown in Figure 2.39 for a vertical antenna [88] were derived from measurements by Cottony and Johler [89].

2.5 RADIO INTERFERENCE

An HF channel is said to be available, or open, whenever the minimum required SNR can be achieved anywhere in the HF band. In practice, however, even large national systems have a restricted number of frequencies available to meet their communication requirements. Unfortunately, there is a significant probability that interfering signals will prevent the use of the assigned frequencies because of HF spectral congestion in many parts of the world [90]. In this regard, the lack of interference modeling is an important shortcoming of available HF link performance prediction programs.

As with the desired signal, interfering signals are propagated by ground-wave or sky-wave modes from distant transmitters. In principle, it should therefore be possible to use characteristics of the transmitters available from the International Frequency Registration Board of the ITU, in conjunction with a ionospheric prediction program such as IONCAP, to compute the level of interference in any HF communication channel. However, the proliferation of unlicensed radio transmitters at HF, as well as the diurnal and seasonal variability of the ionosphere, has complicated such computations to a degree that this approach had to be abandoned. Rather, efforts have been focused on spectrum occupancy measurements and statistical modeling of the results. Fundamental work in this area was presented by Middleton [91], Spaulding and Hagn [92], and others.

Spectrum occupancy measurements are often performed with a spectrum analyzer or a programmable HF receiver. The data acquired are stored on tape or disk for later analysis. Instantaneous bandwidths of 1 and 4 kHz are common, the 4-kHz value being used for examination of voice channels and the 1-kHz value for FSK or other digital modes used at HF [93,94]. A range of measurement schedules and channel dwell times is needed to overcome problems with excessive amounts of data, while still retaining the capability to cover the variation in sunspot number and seasonal and diurnal effects. Also, spectrum "occupancy" data are often acquired to examine mitigation technologies such as new modems or modulation schemes [88,95] (see Chapter 4).

The algorithm and model to estimate spectral congestion described by Laycock and Gott has gained acceptance in several countries [88]. Their data collection scheme employed a 1-kHz resolution bandwidth and a frequency dwell time of 1 sec, which resulted in slow data acquisition at this value of frequency resolution. For this reason, measurement periods near local noon and midnight were chosen when ionospheric conditions were known to be most stable. To include the whole HF band, the measurement campaign lasted for three successive days, with each day devoted to one-third of the band. A 1-kHz channel was considered "occupied" at a specific signal level if the signal in the channel exceeded the designated level for the entire 1-sec dwell time.

The percentage of such 1-kHz channels occupied in a user's allocation defines the congestion for that allocation. Results of modeling congestion in 95 allocations, each representing broadcast, fixed service, maritime mobile, aeronautical mobile, or amateur segments of the HF spectrum, have been presented. The model seeks to express values of congestion as an analytical function $f(x_1, x_1, \ldots, x_1)$ with arguments such as frequency, time, bandwidth, geographical location, sunspot number, etc.

By linear regression, a set of coefficients was determined to model $Q = \sum_{i=1}^{n} a_i x_i$. Because Q is a probability, its values must lie in the interval from zero to one. This constraint is obtained with the so-called "Logit" transform given by $Q = \exp(y)/[1 + \exp(y)]$, where $y = \sum_{i=1}^{n} a_i x_i$. Extensive numerical analysis of data acquired in England through years with sunspot numbers from 2 to 139 yielded a function of the form $y = A_k + BP_r + (c_0 + c_1 f_k + c_0 f_k^2) n_{SSN}$. The coefficient A_k has 95 values corresponding to the 95 ITU allocations with the argument f_k chosen as the center frequency in each allocation. The coefficient B is a single value to be multiplied by the desired receiver threshold level, P_r. There is no facility for changing geographic location, but experience has shown that interference remains relatively unchanged in areas within a 100-km radius. The accuracy of the model is claimed to be 0.01% at the 63% level of congestion, increasing to 0.1% at the 99% level of congestion. In densely populated and highly developed parts of the world, such as Europe and North America, it is increasingly difficult to obtain frequencies in the lower HF spectrum for a 3- to 6-kHz voice or high-speed digital channel that is free of interference. This limitation is especially true during low sunspot years, when all users of the HF spectrum pursue lower frequencies due to the lower values of f_oF_2.

Two interference mitigation techniques have emerged in recent years, narrowband frequency-agile FSK and spread-spectrum modulation. These techniques are introduced in this chapter to explain techniques used to assess spectral occupancy. In particular, the work of Gott and Laycock [88] has been advanced in an effort to design HF modems for FSK modulation incorporating narrow shift and frequency diversity. Interest in such schemes was inspired by the discovery of narrow, interference-free frequency bands within an active voice channel. Such fine structure in the spectrum occupancy can be utilized for slow data transmission, such as 75-bps FSK with an 85-Hz shift, provided problems with channel duration and frequency-selective fading can be mitigated. The probability that such a narrow band will be free of interference for an extended time is negligible. Therefore, a sixfold frequency diversity utilizing six 100-Hz-wide bands within a 6-kHz bandwidth has been considered with an adaptive combiner capable of rejecting signals impaired by interference. The scheme showed a 90% availability under severe congestion condition, that is, $Q = 0.5$. For lower values of Q, less diversity is needed; hence, signaling speed can be raised by using dual or triple diversity while maintaining six FSK channels.

Perry and Abraham [95] have developed an algorithm and associated model to examine spectrum occupancy. A 1-MHz segment of the HF band is sampled at a rate of 2.5 M samples/s for a duration of 105 ms. Four such samples are taken every minute in six different segments of the HF spectrum ranging from 5 to 23 MHz. The data samples are stored and later analyzed with digital transforms. The resolution of the resulting spectra is 9.5 to 159 Hz, dependent on grouping of the samples in 1, 2, 4, 8, or 16 samples for each point in the discrete Fourier transform (DFT). Distributions of occupancy are then constructed from the spectral points after a model by Perry and Rifkin [96]. In this profile, p_{min} is the level of the external noise floor and p_{max} is the level of the highest interference level in a single cell of the transformed spectrum. The model depicts a occupancy/power log-log relationship between p_{min} and p_{max}.

The Perry and Abraham model was built from measured data samples and can be used to derive the percentage of cells occupied by interference, once the intersection of a best linear fit to the log-log distribution has been determined. Typical values of 10% to 40% are found in rural areas on the east coast of the United States. Their results indicates that, on average, 20% of the resolution cells were occupied by interference at all power levels. The dynamic range of the interference, $D(p_{max})$, can also be derived from the distributions of measured values of 10 $\log(p_{max}/p_{min})$, which has shown dynamic ranges of 40 to 70 dB. Since the noise power in a 1-MHz bandwidth is 50 dB above the noise in a resolution cell of 9.5 Hz, dynamic ranges of 50 dB or below represent cells free of interference. The maximum total power is estimated to exceed p_{max} by 10 dB to yield a total dynamic range requirement for p_{max} of 10 to 50 dB. For the measurements performed in the referenced survey, the resulting dynamic range of the wideband signal was 16 to 30 dB. These results provide a starting point for the evaluation of HF interference at any given location.

References

[1] Wayne, R. P., *Chemistry of Atmospheres*, Oxford: Clarendon Press, 1991.
[2] Silberstein, R., "The Origin of the Current Nomenclature of the Ionospheric Layers," *J. Atmos. Terrest. Phys.*, Vol. 13, 1959.
[3] Shen, J. S., et al., "Ionization Layers in the Nighttime E-region Valley Above Arecibo," *J. Geophys. Res.*, Vol. 81, 1976, pp. 5517–5526.
[4] Davies, K., *Ionospheric Radio Waves*, Waltham, MA: Blaisdell Publishing Company, 1969, pp. 39–40.
[5] Weeks, W. L., *Antenna Engineering*, New York: McGraw-Hill Book Company, 1968.
[6] Hayt, W. H., *Engineering Electromagnetics*, New York: McGraw-Hill Book Company, 1981.
[7] Ganguly, S., "Diagnostics for Ionospheric Modification," *1993 Ionospheric Effects Symp.*, J. M. Goodman (editor), Washington, DC: Naval Research Laboratory, May 4–6, 1993, pp. 635–643.

[8] Maslin, N., *HF Communications: A Systems Approach*, New York: Plenum Press, 1987.
[9] Goodman, J. M., *HF Communications: Science and Technology*, New York: Van Nostrand Reinhold, 1992.
[10] Tascione, T. F., *Introduction to the Space Environment*, Malabar, FL: Orbit Book Company, 1988.
[11] Krall, N. A., and A. W. Trivelpiece, *Principles of Plasma Physics*, San Francisco Press, 1986.
[12] Chen, F. F, *Introduction to Plasma Physics*, New York: Plenum Press, 1974.
[13] Rishbeth, H., and O. K. Garriott, *Ionospheric Physics*, New York: Academic Press, 1969.
[14] Piggot, W. R., and K. Rawer (editors), *Report UAG-23A: URSI Handbook of Ionogram Interpretation and Reduction*, 2nd ed., World Data Center A (WDC-A) for Solar-Terrestrial Physics, Nov. 1972, p. 15.
[15] Hagn, G. H., "MUFs and MOFs and LUFs and LOFs," *IEEE Ant Prop. Mag.*, Vol. 34, No. 6, Dec. 1992, pp. 68–74.
[16] Teters, L. R., et al., "Estimating the performance of telecommunication systems using the ionospheric transmission channel," *Ionospheric Communications Analysis and Prediction Program User's Manual*, Institute for Telecommunication Sciences NTIA Report 83-127, Boulder, CO: U.S. Department of Commerce, National Telecommunications and Information Administration, Institute for Telecommunication Sciences, July 1983.
[17] CCIR, "CCIR Atlas of Ionospheric Characteristics," Recommendation 434-4, *Recommendations of the CCIR, 1990*, Vol. VI, *Propagation in Ionized Media*, Geneva: International Radio Consultative Committee, 1990, p. 49.
[18] Davies, K., *Ionospheric Radio*, London: Peter Peregrinus, Ltd., 1990.
[19] Ichimaru, S., *Basic Principles of Plasma Physics: A Statistical Approach*, Reading, MA: Benjamin/Cummings Publishers, 1973.
[20] Nicholson, D. R., *Introduction to Plasma Theory*, New York: John Wiley & Sons, 1983.
[21] Schmidt, *Physics of High Temperature Plasmas*, New York: Academic Press, 1979.
[22] Stix, T. H., *The Theory of Plasma Waves*, New York: McGraw-Hill Book Company, 1962.
[23] Piggot, W. R., and K. Rawer (editors), *Report UAG-23A: U.R.S.I. Handbook of Ionogram Interpretation and Reduction*, 2nd ed., World Data Center A (WDC-A) for Solar-Terrestrial Physics, Nov. 1972, pp. 19, 21.
[24] Piggot, W. R., and K. Rawer (editors), *Report UAG-23A: U.R.S.I. Handbook of Ionogram Interpretation and Reduction*, 2nd ed., World Data Center A (WDC-A) for Solar-Terrestrial Physics, Nov. 1972, p. 22.
[25] Goodman, J. M. (editor), *Proc. 1993 Ionospheric Effects Symp.*, jointly sponsored by the Naval Research Laboratory, Office of Naval Research, USAF Phillips Laboratory, Los Alamos National Laboratory, US Army CECOM, Voice of America, Defense Nuclear Agency, and the Naval C^2 and Ocean Surveillance Center, Alexandria, VA, May 1993.
[26] Piggot, W. R., and K. Rawer (editors), *Report UAG-23A: U.R.S.I. Handbook of Ionogram Interpretation and Reduction*, 2nd ed., World Data Center A (WDC-A) for Solar-Terrestrial Physics, Nov. 1972, pp. 161–171.
[27] Submarine Electromagnetic Systems Department, *High Frequency (HF) and Meteor Burst Communications in a Polar Environment*, NUWC-NPT Technical Document 10,375, New London, CT: Naval Undersea Warfare Center Detachment, Sep. 30, 1993.
[28] Birdsall, C. K., and A. B. Langdon, *Plasma Physics via Computer Simulation*, New York: McGraw-Hill Book Company, 1985.
[29] Wheeler, J. L., "Transmission Loss for Ionospheric Propagation Above the Standard MUF," *Radio Sci.*, Vol. 1, 1303, 1966.
[30] CCIR, "Characteristics and Applications of Atmospheric Radio Noise Data," *Reports to the CCIR, 1983*, Report 322-2, Geneva: International Radio Consultative Committee, 1983.

[31] deBlasio, L., "New features in VOACAP," *HFMAP Newsletter*, Vol. 1, No. 2, Washington, DC: U.S. Information Agency, Summer 1994, p. 2.
[32] Parker, J., "Development of ICEPAC," *HFMAP Newsletter*, Vol. 1, No. 2, Washington, DC: U.S. Information Agency, Summer 1994, p. 1.
[33] Powers, R. A., M. J. Packer, and A. P. Tsitsopoulos, "User-Friendly Software System for Analyzing Antennas and Radio Wave Propagation Links," *IEEE MILCOM'92 Conf. Proc.*, Vol. 1, San Diego, CA, Oct. 1992, pp. 73–76.
[34] Ionospheric Prediction Service, *Advanced Stand Alone Prediction System (ASAPS) Users Guide*, West Chatswood, New South Wales, Australia, 1993.
[35] Kelley, M. C., *The Earth's Ionosphere*, New York: Academic Press, 1989.
[36] Proakis, J. G., *Digital Communications*, New York: McGraw-Hill Book Company, 1983, pp. 459–463.
[37] Wagner, L. S., *Morphology and Characteristics of Disturbed HF Channels*, Transmissions Technology Branch, Information Technology Division, Washington, DC: Naval Research Laboratory, Sep. 1993, p. 1.
[38] Wagner, L. S., et al., "Delay, Doppler, and Amplitude Characteristics of HF Signals Received Over a 1300-km Transauroral Sky-Wave Channel," *Radio Sci.*, Vol. 30, No. 3, May–June 1995.
[39] Wagner, L. S., and J. A. Goldstein, *Channel Spread Parameters for the High-Latitude, Near-Vertical-Incidence-Sky-Wave Channel: Correlation with Geomagnetic Activity*, NRL Report NRL/FR/5550-95-9772, Washington, DC: Naval Research Laboratory, 1995.
[40] Rishbeth, H., and O. K. Garriott, *Introduction to Ionospheric Physics*, New York: Academic Press, 1969.
[41] Jursa, A. S. (editor), *Handbook of Geophysics and the Space Environment*, ADA 167000, Washington, DC: National Technical Information Service, 1985.
[42] Wagner, L. S., et al., *Northern Exposure 92: An Investigation of Transauroral HF Radio Sky-Wave Propagation*, NRL Report NRL/FR/5554—93-9575, Washington, DC: Naval Research Laboratory, 1993.
[43] Wagner, L. S., et al., "Measurements of Delay and Doppler Spread on a 1300-km Transauroral Channel," *Proc. 1993 Ionospheric Effects Symp.*, John M. Goodman (editor), jointly sponsored by the Naval Research Laboratory, Office of Naval Research, USAF Phillips Laboratory, Los Alamos National Laboratory, US Army CECOM, Voice of America, Defense Nuclear Agency, and the Naval C^2 and Ocean Surveillance Center, Alexandria, VA, May 1993.
[44] Basler, R. P., et al., *Ionospheric Distortion of HF Signals*, DNA-TR-87-247, 1987.
[45] Basler, R. P., et al., "Ionospheric Distortion of HF Signals," *Radio Sci.*, Vol. 23, No. 4, July-Aug. 1988, pp. 569–579.
[46] Cannon, P. S., et al., "Initial Results from DAMSON_A System to Measure Multipath, Doppler Spread and Doppler Shift on Disturbed HF Channels," *IEE 9th Int. Conf. Antennas and Propagation*, Vol. 2, Eindhoven, Netherlands, April 1995, pp. 104–108.
[47] Fitzgerald, J., *Channel Probe Measurements for the American Sector Experiment, 1994*, Report LA-UR-94-1811, Los Alamos, NM: Los Alamos National Laboratory, 1995.
[48] Ostergaard, J. C., et al., *Effects of Absorption on High Latitude Meteor Scatter Communication Systems*, PL-TR-91-2197, Environmental Research Papers, No. 1092, Hanscom AFB, MA, July 31, 1991.
[49] Farley, D. T., et al., "Equatorial Spread-F: Implications of VHF Radar Observations," *J. Geophys. Res.*, Vol. 75, 1970, p. 7199.
[50] Kelley, M. C., and J. P. McClure, "Equatorial Spread-F: A Review of Recent Experimental Results," *J. Atmos. Terr. Phys.*, Vol. 43, 1981, p. 427.
[51] Robinson, R. M., et al., "Sources of F-Region Ionization Enhancements in the Nighttime Auroral Zone," *J. Geophys. Res.*, Vol. 90, 1985, pp. 7533.

[52] Szuszczewicz, E. P., "Theoretical and Experimental Aspects of Ionospheric Structure: A Global Perspective on Dynamics and Irregularities," *Radio Sci.*, Vol. 21, No. 3, 1986, pp. 351–362.
[53] Proakis, J. G., *Digital Communications*, New York: McGraw-Hill Book Company, 1983, pp. 462–463.
[54] Flaherty, J. T., et al., "Simultaneous VHF and Transequatorial HF Observations in the Presence of Bottomside Equatorial Spread-F," *Geophys. Res. Lett.* (in press).
[55] Osterman, G. B., R. A. Heelis, and G. J. Bailey, "Modeling the Formation of the Intermediate Layers at Arecibo Latitudes," *J. Geophys. Res.*, Vol. 99, 1994, p. II357.
[56] Brown, T. L., "The Chemistry of Metallic Elements in the Ionosphere and the Mesosphere," *Chem. Revs.*, Vol. 73, 1973, p. 645.
[57] Owen, M. R., "The Great Sporadic-E Opening of June 14, 1987," *QST*, May 1988, pp. 21–29.
[58] Davies, K., *Ionospheric Radio Waves*, Waltham, MA: Blaisdell Publishing Company, 1969, pp. 337–342.
[59] Wagner, L. S., Unpublished data, Washington, DC: Naval Research Laboratory, 1996.
[60] Feldstein, Y. I., "Peculiarities in Aurora and Magnetic Disturbances Distribution in High Latitudes," *Planet. Space. Sci.*, Vol. 14, 1966, pp. 121–130.
[61] Feldstein, Y. I., and G. V. Starkov, "Dynamics of Auroral Belt and Polar Geomagnetic Disturbances," *Planet. Space Sci.*, Vol. 15, 1967, pp. 209–229.
[62] Hartz, T. R., and N. M. Brice, "The General Pattern of Auroral Particle Precipitation," *Planet. Space Sci.*, Vol. 15, 1967 pp. 301–329.
[63] Whalen, J. A., "The Aurora." Chapter 12 in *Handbook of Geophysics and the Space Environment*, A. S. Jursa (editor), ADA 167000, Washington, DC: National Technical Information Service, 1985.
[64] Tsunoda, R. T., "High Latitude F Region Irregularities: A Review and Synthesis," *Rev. Geophys.*, Vol. 26, No. 4, 1988, pp. 719–760.
[65] Bowles, K. L., "Doppler Shifted Radio Echoes from Aurora," *J. Geophys. Res.*, Vol. 59, 1954, pp. 553–555.
[66] Dyce, R. B., "Auroral Echoes Observed North of the Auroral Zone on 51.9 Mc/sec," *J. Geophys. Res.*, Vol. 60, No. 3, 1955, pp. 317–323.
[67] Booker, H. G., "A Theory of Scattering by Nonisotropic Irregularities," *J. Atmos. Terr. Phys.*, Vol. 8, 1956, pp. 204–221.
[68] Bates, H. F., "Direct HF Backscatter from the F Region," *J. Geophys. Res.*, Vol. 65, No. 7, 1960, pp. 1993–2002.
[69] Muldrew, D. B., "F Layer Ionization Troughs Deduced from Alouette Data," *J. Geophys. Res.*, Vol. 70, No. 11, 1965, pp. 2635–2650.
[70] Sharp, G. W., "Mid-Latitude Trough in the Night Ionosphere," *J. Geophys. Res.*, Vol. 71, No. 5, 1966, pp. 1345–1356.
[71] Knudsen, W. C., "Evaluation and Demonstration of the Use of Retarding Potential Analyzers for Measuring Several Ionospheric Quantities," *J. Geophys. Res.*, Vol. 71, 1966, pp. 4669–4678.
[72] Spiro, R. W., R. A. Heelis, and W. B. Hanson, "Ion Convection and the Formation of the Mid-Latitude F Region Ionization Trough," *J. Geophys. Res.*, Vol. 83, No. A9, 1978, pp. 4255–4264.
[73] Malaga, A., *Theory of VHF Scattering by Field-Aligned Irregularities in the Ionosphere*, RADC-TR-86-118, Sep. 1986.
[74] Wagner, L. S., and J. A. Goldstein, "Response of the High Latitude HF Sky-Wave Channel to an Isolated Magnetic Disturbance," *Proc. 5th Int. Conf. HF Radio Systems and Techniques*, Edinburgh, UK: IEE, 1991.
[75] Bates, H. F., "The Height of F Layer Irregularities in the Arctic Ionosphere," *J. Geophys. Res.*, Vol. 64, No. 9, 1959, pp. 1257–1265.
[76] Proakis, J. G., *Digital Communications*, New York: McGraw-Hill Book Company, 1983, p. 465.
[77] Cormier, R. J., *Thule Riometer Observations of Polar Cap Absorption Events (1962-1972)*,

AFCRL-TR-73-0060, Bedford, MA: U.S. Air Force Systems Command, U.S. Air Force Cambridge Research Laboratories, 1973.

[78] McDonough, A. K., J. R. Katan, and R. I. Desourdis, "Blackout of Simultaneous HF Sky-Wave and Meteor Burst Communication Links," *MILCOM '93 Conf. Proc.*, Boston, MA: IEEE, October 1993, pp. 402–406.

[79] Ostergaard, J. C., "Short Range Communications System, Design and Operation at High Latitudes," *Proc. 4th Int. IEE Conf. HF Radio Systems and Techniques*, London: IEE, April 11–14, 1988, pp. 177–181.

[80] CCIR, "HF Ionospheric Channel Simulators," Report 549-2, *Reports to the CCIR, 1986*, Vol. III, Part A, Geneva: International Radio Consultative Committee, 1986, pp. 59–67.

[81] CCIR, "Use of High Frequency Ionospheric Channel Simulators," Recommendation 520-1, *Recommendations of the CCIR, 1986*, Vol. III, Part B, Geneva: International Radio Consultative Committee, 1986, pp. 57–58.

[82] Lauber, W. R., "Radio Noise Surveys at Canadian HF Communication Sites," *IEEE Trans. Electromagnetic Compatibility*, Vol. EMC-19, May 1977.

[83] CCIR, "Man-Made Radio Noise," *Reports to the CCIR, 1986*, Report 258-3, Geneva: International Radio Consultative Committee, 1986.

[84] Spaulding, A. D., and R. T. Disney, *Man-made Radio Noise I: Estimates for Business, Residential, and Rural Areas*, NTIA Report 74-38, Washington, DC: U.S. Department of Commerce, National Telecommunications and Information Administration, 1974.

[85] Spaulding, A. D., and J. S. Washburn, *Atmospheric Noise: Worldwide Levels and Other Characteristics*, NTIA Report 85-173, Washington, DC: U.S. Department of Commerce, National Telecommunications and Information Administration, 1985.

[86] Spaulding, A. D., and F. G. Stewart, *An Updated Noise Model for IONCAP*, NTIA Report 87-212, Washington, DC: U.S. Department of Commerce, National Telecommunications and Information Administration, 1987.

[87] Taylor, R. E., "136 MHz/400 MHz Radio-Sky Maps," *Proc. IEEE*, 1973.

[88] CCIR, "Characteristics and Applications of Atmospheric Radio Noise Data," *Reports to the CCIR, 1983*, Report 322-2, Geneva: International Radio Consultative Committee, 1983, p. 5.

[89] Cottony, H. V., and J. R. Johler, "Cosmic Radio Noise Intensities in the VHF Band," *Proc. IRE*, Vol. 40, p. 1053.

[90] Gott, G. F., and P. J. Laycock, "HF Spectral Occupancy and Its Impact on Modem Design," *Proc. HF'89 Nordic Shortwave Conf.*, 1989.

[91] Middleton, D., "Statistical-Physical Models of Electromagnetic Interference," *IEEE Trans. Electromagnetic Compatibility*, Vol. EMC-19, No. 3, 1977.

[92] Spaulding, A. D., and G. H. Hagn, "On the Definition and Estimation of Spectrum Occupancy," *IEEE Trans. Electromagnetic Compatibility*, Vol. EMC-19, No. 3, 1977.

[93] Gibson, A. J., and L. Arnett, "New Spectrum Occupancy Measurements in Southern England," *Proc. 4th Int. Conf. HF Radio Systems and Techniques*, IEE Conference Publication No. 284.

[94] Laycock, M. Morrell, G. F. Gott, and A. R. Ray, "A Model for HF Spectral Occupancy," *Proc. 4th Int. Conf HF radio Systems and Techniques*, IEE Conference Publication 284.

[95] Perry, B. D., and L. G. Abraham, "A Wideband HF Interference and Noise Model Based on Measured Data," *Proc. 4th Int. Conf HF radio Systems and Techniques*, IEE Conference Publication 284.

[96] Perry, B. D., and R. Rifkin, "Interference and Wideband HF Communication," *Proc. 5th Ionospheric Effects Symposium*, Springfield, VA, 1987.

High-Frequency Antennas[1]

3

3.1 INTRODUCTION

In general, radio system performance strongly depends on the interface between the transmitter and receiver and the radio-wave propagation as depicted in the hierarchical HF system structure shown in Figure 3.1. This interface is embodied in the transmitting antenna, which controls the direction and magnitude of the transmitted signal, and the receiving antenna, which determines the directional sensitivity applied in the direction of the arriving received signal. In this regard, antennas provide one of the greatest sources of man-made control over HF link performance.

The transmitting antenna contributes significantly to the EIRP radiated by an HF station at the azimuth and elevation angles needed to close a desired HF link to a receiving antenna at a second station (see Fig. 3.2). If the transmitting antenna is properly designed and installed, it can be made to provide some peak gain value G_t in order to meet link power budget (EIRP) requirements (see (1.2) in Chapter 1) on the necessary radio frequencies. This required gain must be "pointed" so as to launch the desired propagating rays intended to arrive at the receiving site of the second station. Similarly, the ideal receiving antenna would maximize its power gain G_r in the direction of the arriving radio rays after traversing the intended propagation path (e.g., ground wave, tropospheric wave, or sky wave). Ideally, the receiving antenna system would also minimize

---

[1]Many of the antenna structure and wire diagrams used as figures in this chapter were created using the Electromagnetic Antenna Modeling (EAM) Microsoft Windows-based graphical user interface for the Numerical Electromagnetics Code (NEC). The EAM-NEC program was developed by Science Applications International Corporation (SAIC) under contract to the Rome Laboratory and is commercially available from the Telecommunications and Information Systems Division of SAIC in Marlborough, Massachusetts.

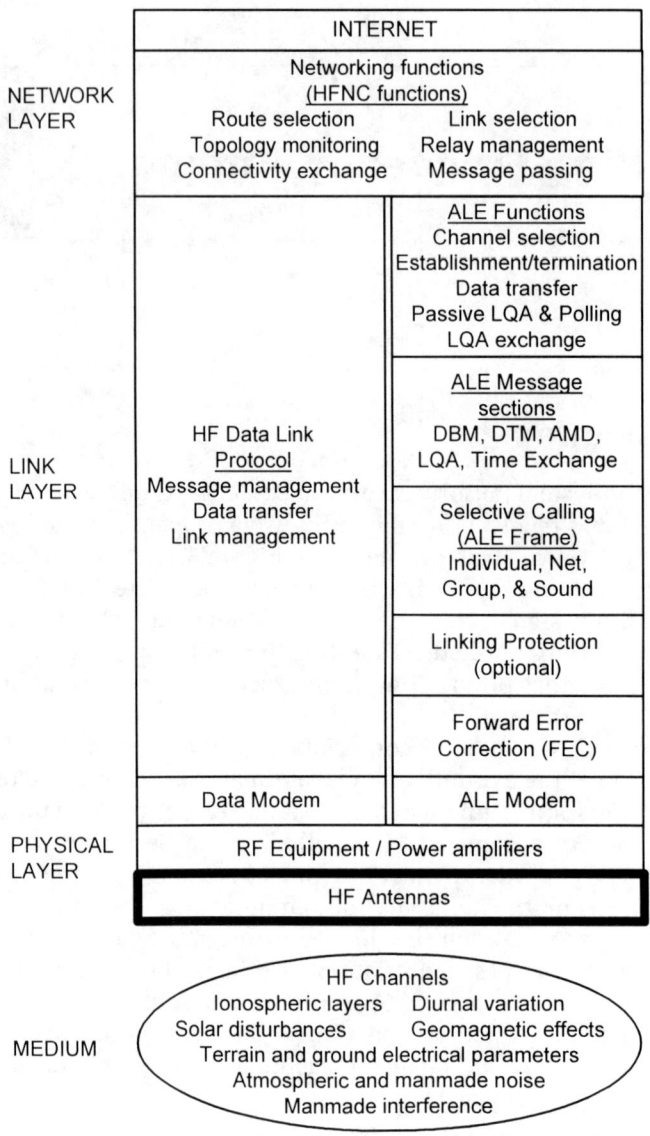

Figure 3.1 HF antennas in the hierarchical structure of HF communication systems.

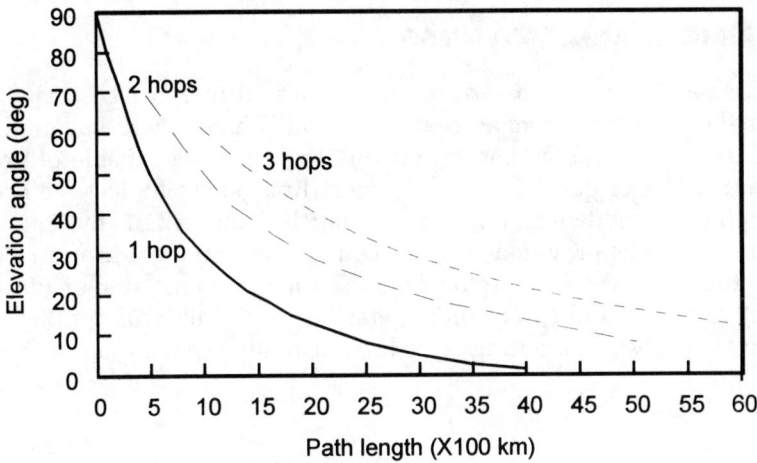

Figure 3.2 Elevation angle versus path length for 320-km F_2-layer reflection.

gain in the direction of significant noise signals arriving from distant thunderstorms, man-made interference, or local man-made noise sources [1].

In general, antenna performance is highly dependent on the near-field environment, such as antenna mounting location and the electrical characterization of nearby material. These environmental characteristics must be known as well as the electromagnetic performance requirements (e.g., desired pattern shape, peak gain, and bandwidth) to design an optimal antenna. In this regard, the antenna and its environment define an antenna *system*. The characteristics of the antenna system, not the antenna alone, determine HF link performance. In this context, each antenna system refers not only to the antenna per se, but to any other structures or objects within a few wavelengths of the antenna. These structures, particularly conducting structures, can have a significant effect on antenna performance and are therefore included in the concept of an antenna system.

Correct antenna selection for a particular communication link is a compromise among many factors, such as performance, size, cost, maintainability, and availability. Antenna selection is a difficult task because although many antennas meet certain requirements, no one antenna is ideal for all situations. An experienced antenna engineer with knowledge of antennas and ionospheric behavior, skilled in the use of the appropriate numerical modeling tools, can optimize antenna performance for specific performance requirements. This chapter describes typical HF antennas and discusses the influence of the antenna environment on performance. The discussion of broadband antennas is particularly important in the context of ALE-equipped HF radio.

3.2 NARROWBAND ANTENNAS

The advantage of ALE is based on the use of multiple spaced frequencies to increase the probability that at least one "good" channel will be found for any desired link. The antenna chosen for this link must be capable of providing the required power gain, G_t or G_r, to meet link power budget requirements, both for the ALE modem and the desired data link modem, if any. These power gain values must be provided at each frequency within the timing allocated for antenna tuning by the ALE protocol (see Chapter 6). Some antenna designs are inherently narrowband and cannot operate on ALE links without these tuners, whereas other designs are inherently broadband and may not need the benefit of tuners.

3.2.1 Vertical Polarization: The Whip Antenna

An HF *whip* is a thin monopole antenna typically between 2 and 10 m in length that is designed for operation in the HF band. For fixed installation, the whip is mounted above a ground screen formed from radial wires to provide a pattern approaching that of Figure 3.3. The whip antenna provides low-angled gain that supports both ground-wave and long-range sky-wave propagation modes. As shown in the figure, however, the pattern has a null at zenith. This null results in poor performance for high-angled propagation modes such as near-vertical-incidence sky-wave (NVIS) or even multihop sky-wave links.

The whip is a standard mobile antenna, but its pattern produces too little gain near zenith to support effective high-angled modes. The whip provides a purely vertically polarized pattern when mounted over the ground screen, but the diffraction and scattering produced by typical mobile platforms results in electric fields in both polarizations. Like the monopole, the whip is inherently capacitive if less than $\lambda/4$ in length and therefore requires inductive loading to tune the antenna once mounted.

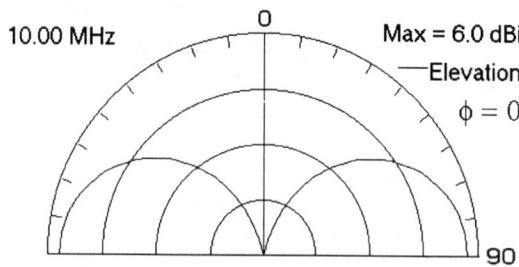

Figure 3.3 Monopole antenna pattern over perfect ground.

3.2.2 Horizontal Polarization

3.2.2.1 The V-Whip Antenna

To overcome the NVIS null, whip mounts on vehicles are often modified to tilt the whip either toward the front or the rear of the vehicle. This configuration produces vertical polarization at low elevation angles off the ends of the antenna and horizontal polarization overhead, thus "filling in" the zenith null to support NVIS propagation. The *V-whip* takes this bent whip concept one step further by providing two unequal length whips that diverge from a standard whip mount to the front and back of the vehicle. This configuration provides good performance for NVIS, intermediate, and long-range sky-wave and ground-wave propagation modes.

3.2.2.2 Half-Wavelength Dipole

The horizontal dipole antenna of total length $\lambda/2$, mounted at a 10-m height above ground, provides high-angled radiation at low frequencies (2 to 10 MHz) and mid-angled gain (e.g., $20 \leq \theta \leq 60$) as the frequency is increased. At 2 MHz, this dipole is 75 m in length. The high-angled radiation from the dipole supports NVIS and medium-range sky-wave communications up to about the 500-km range. As shown in Figures 3.4(a), (b), and (c), the dipole pattern is nearly omniazimuthal at 2 MHz. As the frequency is increased, the pattern of a fixed-length horizontal dipole becomes bidirectional and nulls become apparent both at zenith and off the ends of the dipole. The antenna now exhibits a directional pattern, with maximum gain now directed above 0° elevation (see Fig. 3.4[c]). To achieve omniazimuthal radiation patterns at these higher frequencies, a useful variant of the dipole antenna, called the *crossed dipole,* is employed. The crossed dipole is composed of two equal-length horizontal dipoles lying in a horizontal plane and crossed at their midpoints to form right angles. This configuration eliminates the nulls of the single dipole antenna but reduces the maximum gain by about 3 dB (factor of 0.5).

3.2.3 Antenna Tuners

All narrowband antennas, such as the monopole and dipole antennas described earlier, require tuning circuits to match the transmission line characteristic impedance to the impedance of the antenna. Typically, a simple "T" or "Π" network can allow maximum power to be transferred to the antenna at a single frequency. These circuits employ capacitors and inductors that can be hand-tuned with switches or rollers. More sophisticated (and expensive) antenna *couplers* are controlled by microprocessors and offer tuning times of typically

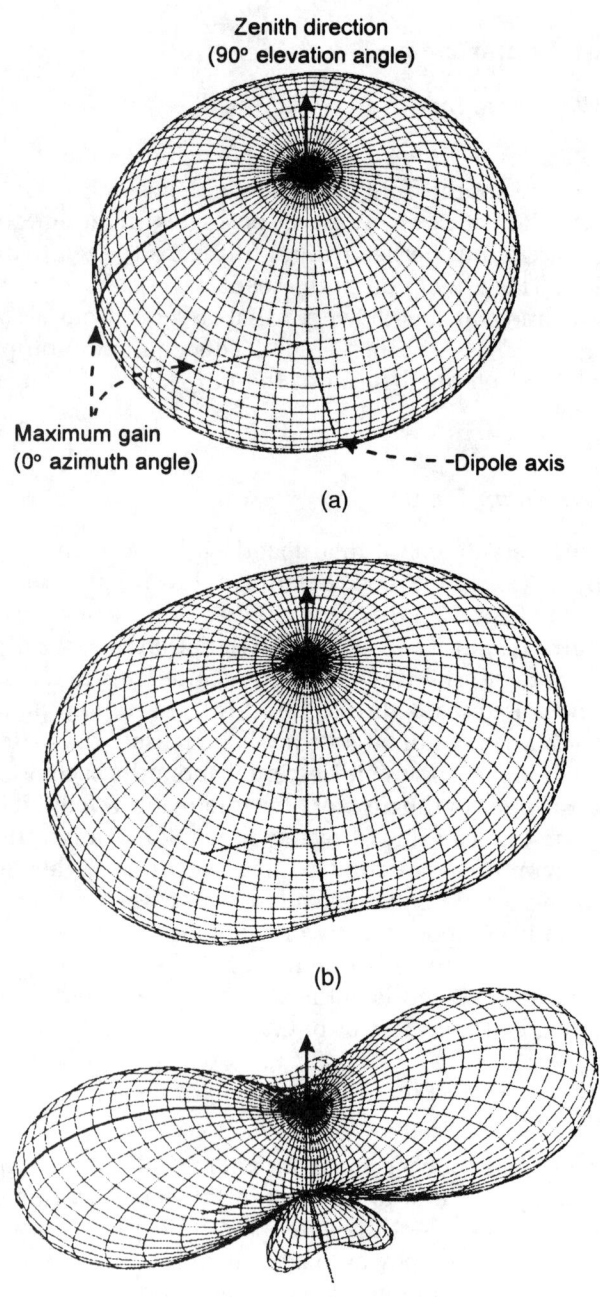

Figure 3.4 Horizontal dipole radiation pattern at (a) 2 MHz, (b) 12 MHz, and (c) 20 MHz.

less than 25 ms with high reliability. As discussed in Chapter 5, antenna tuning time on the order of 25 ms provides the reaction time needed by ALE-equipped HF stations employing narrowband antennas. Sophisticated couplers can match a wide variety of whip, dipole, and other wire antennas over the 1.6-to 30-MHz frequency range.

Table 3.1 describes the performance of an 18.5-m horizontal dipole at 10-m height above good ground showing the impact of an antenna coupler. The gain at zenith exceeds 0 dBi across the band, but drops at 12 MHz due to pattern shape. (Antenna efficiency is a function of conductor losses and ground losses, but not mismatch loss.) At 2 MHz, the antenna resistance is low with a significant negative reactance (capacitance), which results in almost total reflection of the transmitted power by the antenna. Near resonance at 8 MHz, the reflected loss is at a minimum then increases with frequency. The coupler's tuning efficiency is shown in the next row, which shows that "tuning out" the large capacitance required significant inductance and the associated resistance ("Henry") losses. The final row of the table shows the effective gain calculated by attenuating the power gain by the coupler's efficiency. Note that although this narrowband antenna has been matched to the transmission line, the 3-dB bandwidth of the power gain has been reduced.

3.3 BROADBAND ANTENNAS

To be classified as broadband, an antenna must have adequate impedance, gain, and pattern bandwidth. As shown in Table 3.1, the use of an antenna tuning circuit, or *impedance transformer,* raises the quality (Q_a)-factor of the antenna and thus decreases its bandwidth. One well-known compromise to this dilemma is the arraying of many high-Q elements to form a single antenna structure. In other words, rather than employing couplers to enable narrowband antennas to span the necessary frequencies, it is often possible to employ antennas whose inherent electrical characteristics are retained over a wide frequency band. Rumsey [2] was the first to postulate this theory, which is based on the *principle*

Table 3.1
Effective Gain at Zenith for a Horizontal Dipole Antenna With Coupler

Frequency (MHz)	2	4	6	8	10	12
Gain at zenith (dBi)	3.2	6.3	6.5	6.2	6.4	1.1
Antenna efficiency (η_a)	31%	67%	78%	82%	80%	81%
Input impedance (Ω)	$1.8 - j2500$	$10.5 - j1020$	$36.7 - j395$	$92 + j37$	$190 + j430$	$372 + j892$
Coupler efficiency (η_m)	4%	49%	83%	99%	98%	85%
Effective zenith gain (dBi)	−10.8	3.2	5.7	6.2	6.3	0.4

of similitude [3]. This principle states that if the dimensions of a given antenna are increased by a constant factor, and the operating frequency is decreased by this factor, then antenna performance is unchanged.

One way to meet the necessary broadband characteristics is to combine multiple narrowband elements, each designed for a different frequency, into a single antenna structure. This approach constitutes the basic design principle of the *log-periodic antenna* (LPA). The RF signal travels along the antenna transmission line until it finds antenna elements that are approximately resonant at the driving frequency. These elements, which are close to a half wavelength in length at each frequency, are excited by the RF energy to produce the desired radiation pattern. Thus, log-periodic structures, such as the LPA, employ multiple high-Q dipole elements to cover the frequency band of interest. If the number of high-Q elements approaches infinity and the relative sizes of adjacent elements are *differentially* close, then the antenna structure is appropriately described in terms of angles rather than linear dimensions. The LPA elements and interelement spacings are then specified in terms of angles, and will therefore exhibit *frequency-independent* characteristics [4–8]. In other words, the characteristics of an antenna are frequency independent if its dimensions in wavelengths remain constant. In addition to the combination of multiple-element antennas to form frequency-independent structures, *traveling-wave antennas* such as *long-wire*, *V,* and *rhombic* antennas exhibit frequency-independent characteristics.

3.3.1 Vertical Polarization

3.3.1.1 Conical Monopole

A *conical monopole* is an arrangement of multiple monopole wire elements that forms a cone shape as shown in Figure 3.5(a). As with a monopole, ground radials are often used to minimize ground losses. All elements are fed at one end and driven against the center point of the ground radials. The basic conical shape is supported mechanically by either (1) multiple nonconductive towers around the cone lip or (2) a single tower at the cone center. The multiple-tower configuration is usually referred to as an *inverted-cone* antenna. The single-tower configuration allows a second cone to be placed above the inverted cone, to form a conical monopole. This single tower version has the advantage that it can be configured to meet transportable station requirements.

Figure 3.5(a) shows a 2- to 30-MHz inverted-cone antenna employing six towers with catenary supports to hold the radiating elements. This 36-element cone is 22 m in height and 34 m in breadth with 75-m-diameter ground radials. The single-tower version shown in Figure 3.5(b) has a usable frequency range from 1.6 to 32 MHz. The apex of this antenna is 26 m above the ground and

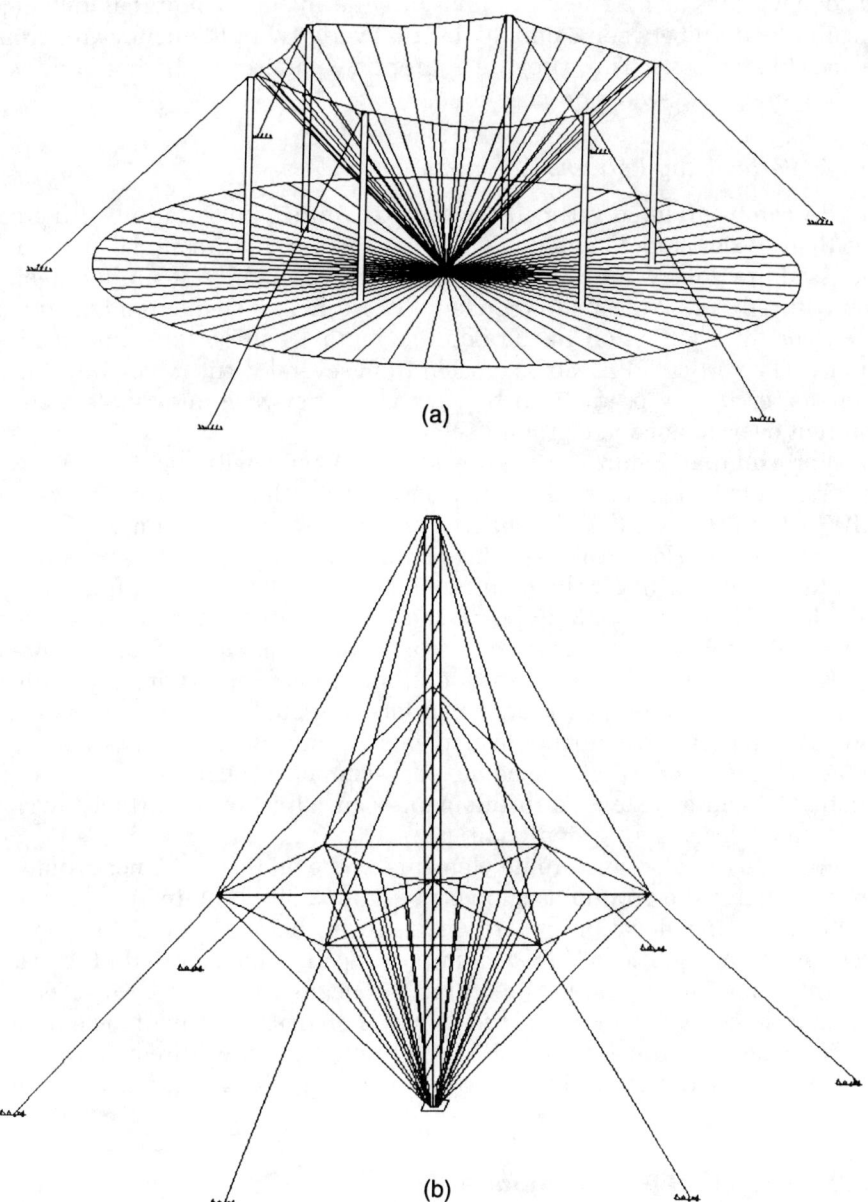

Figure 3.5 Conical monopoles: (a) six-tower inverted conical and (b) single-tower conical.

the cone width is 38 m. These antennas provide an omni-azimuthal low-angled elevation pattern between 5 and 30 deg that varies with frequency and ground electrical parameters. They provide appropriate patterns for both ground-wave and long-range sky-wave propagation.

3.3.1.2 Vertical Log-Periodic Antennas

An LPA can be oriented vertically to provide vertical polarization with greater gain than the monopole antennas distributed in a unidirectional pattern. As discussed previously, an LPA is constituted from a family of similar elements each with a different resonant frequency. These similar elements add extra gain at the cost of a reduction in the omni-azimuthal pattern to a unidirectional pattern. The vertical LPA can be created from several vertical half-wave dipole elements, or it may be realized by several quarter-wave monopole elements mounted over an adequate ground screen.

For example, Figure 3.6(a) shows an LPA built with eight quarter-wave monopole elements. This LPA provides a 2:1 voltage standing wave ratio (VSWR) from 2.5 to 32 MHz with more than 6-dBi peak gain over average ground. Monopole length and spacing requirements to achieve this performance include a 43-m height for the tallest element and a 64-m length from the first to the last monopole. The 3-dB beam width in elevation angle for this antenna is contained below an elevation angle of 30 deg, making it ideal for one-hop sky-wave communication ranges beyond about 1,000 km. As with any vertically polarized antenna mounted near the surface, a ground screen must be placed below the quarter-wave monopole elements for improved gain at low elevation angles. This ground screen need not be symmetric relative to the antenna structure. It can be extended in the main beam direction to further lower the elevation angles showing significant gain.

The use of half-wave dipole elements in the vertical LPA necessitates its elevation above the ground as shown in Figure 3.6(b). The increased antenna height lowers the elevation pattern and reduces the need to employ a ground screen to attain significant low-angled power gain. This elevated LPA reaches a height of just under 100 m if required to operate as low as 2 MHz. The use of dipole elements has improved the gain by 3 dB over the gain of the monopole LPA. In general, vertical dipole LPAs are limited to a lower frequency limit of 4 or 5 MHz to maintain the same vertical height as the quarter-wave monopole LPA.

3.3.2 Horizontal Polarization

3.3.2.1 Traveling-Wave Antennas

Traveling-wave antennas are simple low-cost broadband antennas that are many wavelengths in length. Although their radiation pattern changes with frequency,

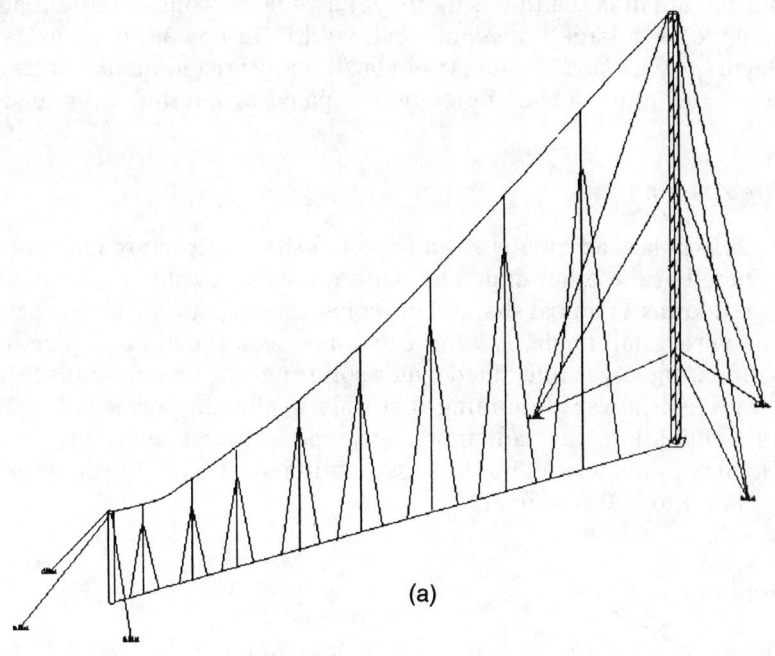

(a)

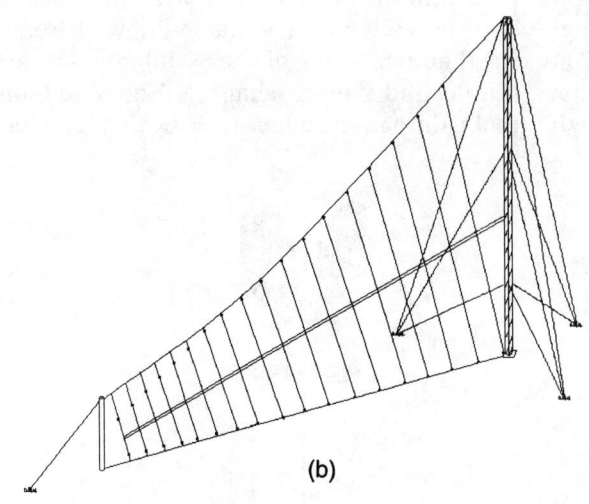

(b)

Figure 3.6 Vertical log-periodic antennas: (a) monopole elements and (b) dipole elements.

thus limiting antenna bandwidth, they have been considered broadband because they offer a large impedance bandwidth. In operation, signal energy travels down the wire surface and is radiated. Energy not radiated after traveling the length of the antenna is completely dissipated by a terminating load.

Long-Wire Antenna

A horizontally polarized long-wire antenna is a simple structure fed at one end and terminated at the other end. The pattern created by this antenna, shown in Figure 3.7, forms a conical shape. This cone-shaped pattern can be characterized by its vertex half-angle, θ_w, and calculated as a function of wire length. The radiation intensity is integrated over a solid angle of 4π steradians to obtain the total radiated power. Assuming that only negligible power is lost in the long-wire termination, the radiation resistance is found using the resulting value of total radiated power. For this assumption to be valid, the antenna must operate at close to 100% efficiency.

The V Antenna

The V antenna is an arrangement of two horizontal equal-length long-wire antennas oriented to form a unidirectional radiation pattern. Each long-wire antenna is terminated with a resistive load to prevent reflections that would result in a standing-wave pattern. The two long-wire antennas are arranged to form the half-angle θ_V at the vertex of a V shape (viewed from above) where the two wires converge. The two sides of the V antenna are fed 180 deg out of phase to achieve the desired pattern shaping. The V antenna produces a unidirectional azimuthal gain pattern when $\theta_V = \theta_w$, that is, when the V vertex

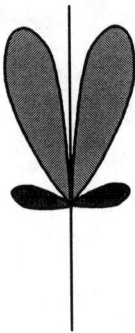

Figure 3.7 Radiation pattern for a two-wavelength long-wire antenna.

half-angle θ_V equals the half angle θ_W of the conical pattern of each long-wire antenna in isolation. This arrangement results in the addition of the fields along the line that bisects the V and cancellation of fields in other directions as depicted in the diagram shown in Figure 3.8.

Rhombic Antenna

A rhombic antenna is a traveling-wave antenna in the shape of a rhombus (viewed from above), and appears to be two V antennas placed end to end. The rhombic consists of two equal-length long-wire antennas that diverge from the feed point and bend at the center to converge again at the end point, where the two wires are each terminated with a 600- to 800-Ω resistor at the converging vertex. When the vertex angles are chosen correctly, this configuration of long-wire antennas forms a unidirectional gain pattern similar to the V antenna. The rhombic antenna offers slightly greater gain than the V antenna and with the same total wire length. In addition, the rhombic antenna pattern is less frequency sensitive than the pattern of the V antenna. For these reasons, radio amateurs often prefer the rhombic antenna to the V antenna.

3.3.2.2 Horizontal LPA

A horizontally polarized LPA consists of horizontal half-wave dipole elements mounted along a transmission line and elevated above the ground. The hori-

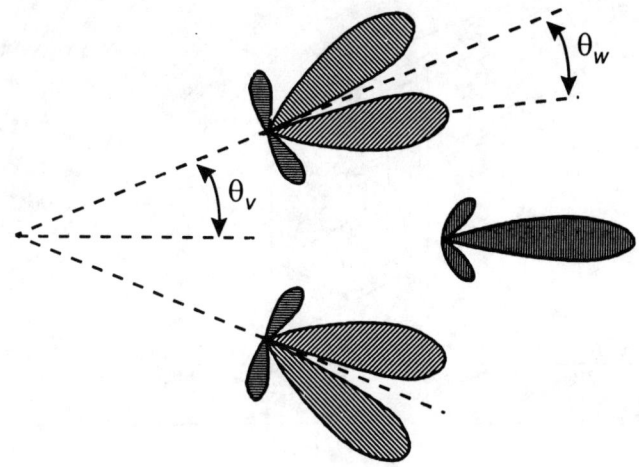

Figure 3.8 Composite nature of a V antenna radiation pattern.

zontal LPA is one of the most popular and versatile broadband HF antennas. This general-purpose broadband antenna is typically employed in transportable, rotatable, or fixed configurations. The horizontal LPA is typically deployed as a rooftop antenna (see Fig. 3.9[a]), providing effective broadband (remember ALE) point-to-point communications capability. Figure 3.9(b) displays the radiation patterns for 5 to 20 MHz. At 10 MHz, the main beam is at a 16-deg elevation with a peak gain of 7.4 dBi. As the frequency is increased to 20 MHz, several lobes appear with equal magnitude, the lowest lobe is at a 10-deg elevation with a gain of 9.8 dBi. In this configuration, the rotatable horizontal LPA can be installed on roofs with significant pitch and many roof-mounted structures, such as chimneys or ventilation ducts. Many rotatable LPAs are designed with fully efficient half-wave dipoles as radiating elements, while others are designed with foreshortened elements to minimize the risk of physical interference with nearby obstructions or trees. A rotatable LPA also requires a motor-driven rotor unit that mounts at the top of the mast tower. A remote

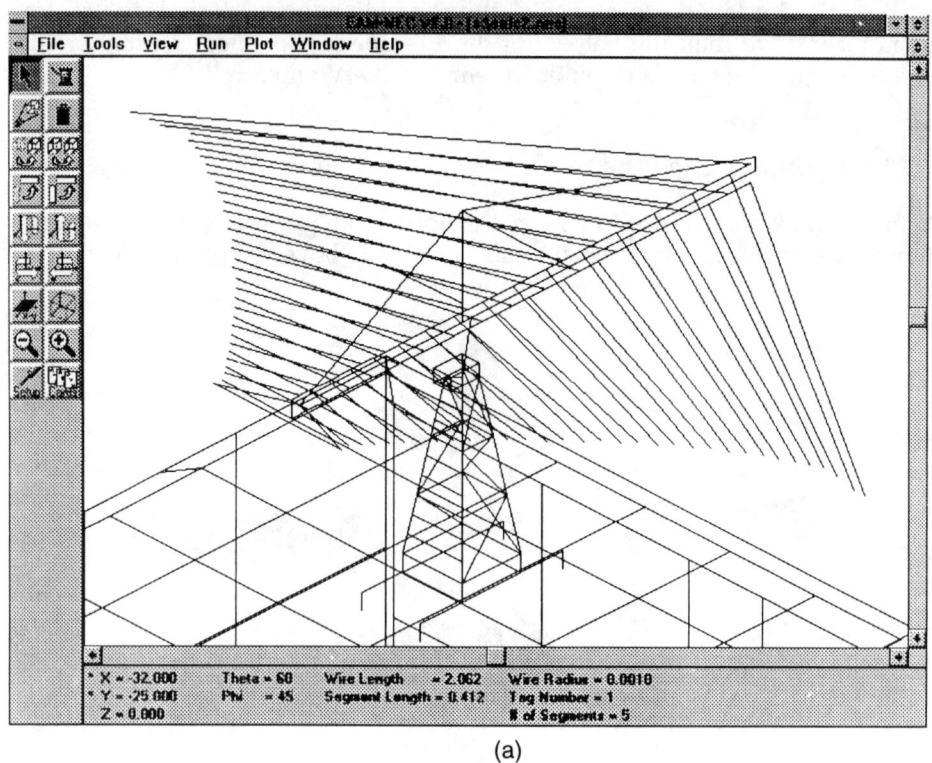

(a)

Figure 3.9 Roof-mounted rotatable horizontal log-periodic dipole antenna: (a) wire grid antenna model and (b) radiation patterns at 5, 10, 15, and 20 MHz.

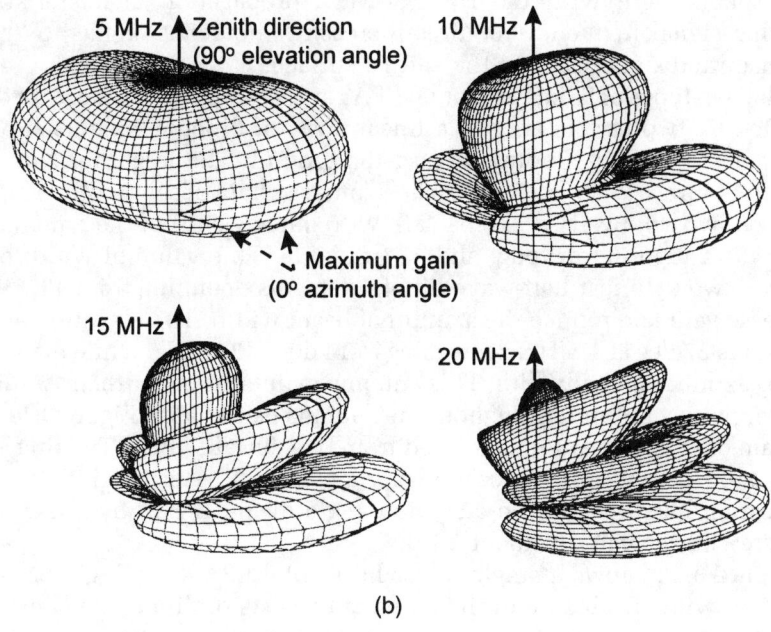

Figure 3.9 (continued)

control panel can be connected by multiple-core power and control cables to the rotator unit.

Ground reflections determine the elevation angles of peak gain for any antenna elevated above the earth. This effect is more pronounced for horizontal than vertical polarization, because the horizontally polarized electric field has a 180-deg phase reversal at the ground. For this reason, the long-wavelength dipole elements are furthest from the ground and the short wavelength dipole elements are closest to the ground. In this way, the forward tilt of the horizontal LPA will somewhat mitigate the changes in pattern shape expected with frequency variation for a horizontal LPA above ground. The LPA tilt angle may not be sufficient to maintain each LPA element at exactly the same electrical height (in wavelengths) above ground. As a result, some pattern-shape change occurs with changing frequency. For example, the shorter wavelength elements could be further from the ground in wavelengths than the longer wavelength elements. As a result, the low-frequency pattern would provide the high-angled radiation needed for NVIS operation. As the frequency was increased, the smaller LPA elements would be excited at a higher electrical height above ground, so the LPA pattern would move toward lower elevation angles. This result is consistent with the HF sky-wave propagation characteristic that increasing signal frequency requires increasing incidence angles to "reflect" the transmitted signal back to the earth's surface.

The two types of fixed horizontal LPAs are (1) a planar "curtain" of dipoles supported by two towers and (2) a linear array of dipole elements supported by a single tower and sloping toward the ground. The large horizontal LPA shown in Figure 3.10 consists of one horizontal curtain of saw-toothed elements supported by two towers at the back (low-frequency end) of the antenna. This LPA is 43 m high, 84 m long, and 122 m wide. This value of width enables the use of two collinear half-wave dipole elements operating from 4 to 30 MHz to increase gain and reduce the azimuthal beam width. The elevation angle for peak gain is 27 deg at 4 MHz, decreasing to 14 deg at 30 MHz, while maintaining a 38-deg azimuthal bandwidth. This antenna maintains better than 15-dBi peak gain over perfect ground throughout the antenna bandwidth. Figure 3.11 shows a diagram of the same antenna formed from two curtain LPAs. The first curtain is at a 26-m height and the second curtain is at a 51-m height. This major modification has lowered the elevation angle for peak gain by a few degrees and increased the gain by about 2 dB.

Figure 3.12 shows a single tower horizontal LPA 43 m high, 95 m long, and 123 m wide. Each side of the antenna consists of dipole elements fed by a diverging transmission line. This antenna provides an ideal pattern shape from 2 to 30 MHz, with 10-dBi peak gain at an elevation angle of 50 deg and an 80-deg azimuthal 3-dB beam width.

3.3.3 Multimode Antennas

Several HF antennas can produce two radiation patterns, depending on the feed method. For example, a *conical spiral antenna* is an elliptically polarized

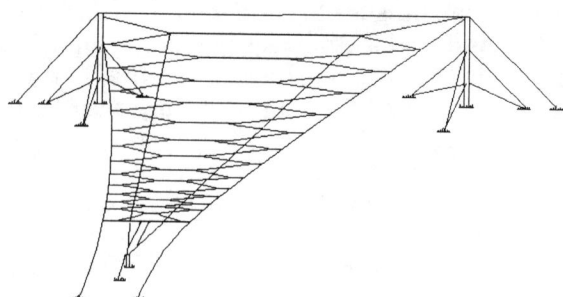

Figure 3.10 Single-curtain horizontal log-periodic antenna.

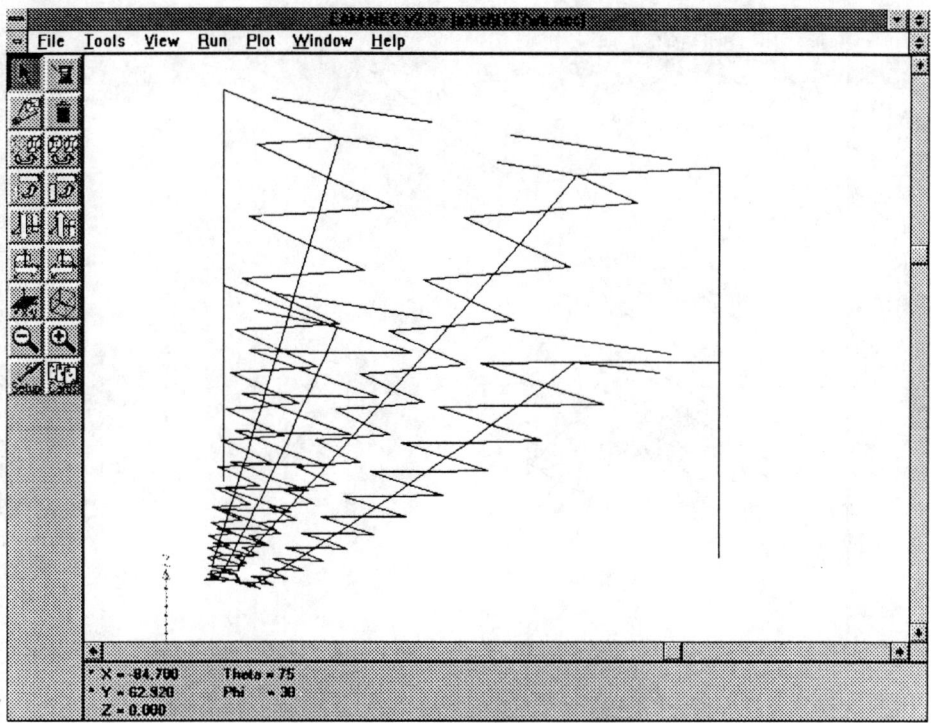

Figure 3.11 Wire model of a double curtain horizontal LPA.

omnidirectional antenna. As shown in Figure 3.13(a), the conical spiral antenna consists of four adjacent radiating elements wound into an inverted cone. Two different feed methods are employed, resulting in the two pattern shapes shown in Figure 3.13(b). The high-angled mode provides pattern coverage from 45 deg to the zenith, while the low-angled mode provides a 30-deg elevation beam width at 4 MHz diminishing to 10 deg at 30 MHz. Clearly, the high-angled mode supports NVIS and multihop HF propagation paths, while the low-angled mode supports frequency-dependent propagation path geometries.

3.4 "SMALL" HF ANTENNAS

3.4.1 Overview

The need for HF band coverage for ALE radio combined with the need to minimize the impact on platforms or real estate suggests the use of small broadband antenna designs. In general, small antennas fit into three principal catego-

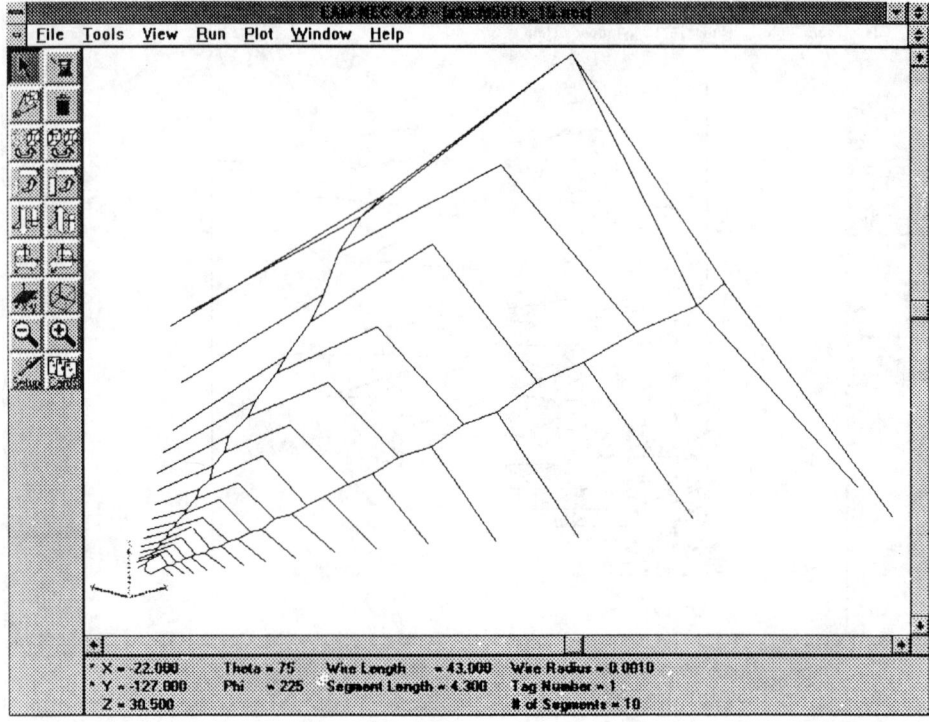

Figure 3.12 Single-tower LPA with two sets of half-wave dipole elements.

ries: (1) *electrically small,* which refers to antennas that are physically bounded by a sphere of radius $\lambda/2\pi$, where λ is the free-space wavelength; (2) *physically constrained* antennas, which are electrically small in at least one plane but which may be relatively large in other planes, e.g., a *conformal* antenna; and (3) *physically small* antennas, which do not fall into the other categories, but whose physical dimensions cause them to be considered small, such as a 40-GHz horn antenna. From an HF perspective, in which signal wavelengths vary from 10 to 150 m, a small antenna must be either electrically small or physically constrained to provide operational flexibility.

3.4.2 Receive-Only Antennas

As discussed previously, operational constraints may place significant limits on the HF antenna size for a particular application. Moreover, operating requirements may limit the HF station to receive-only operation. In this case, an

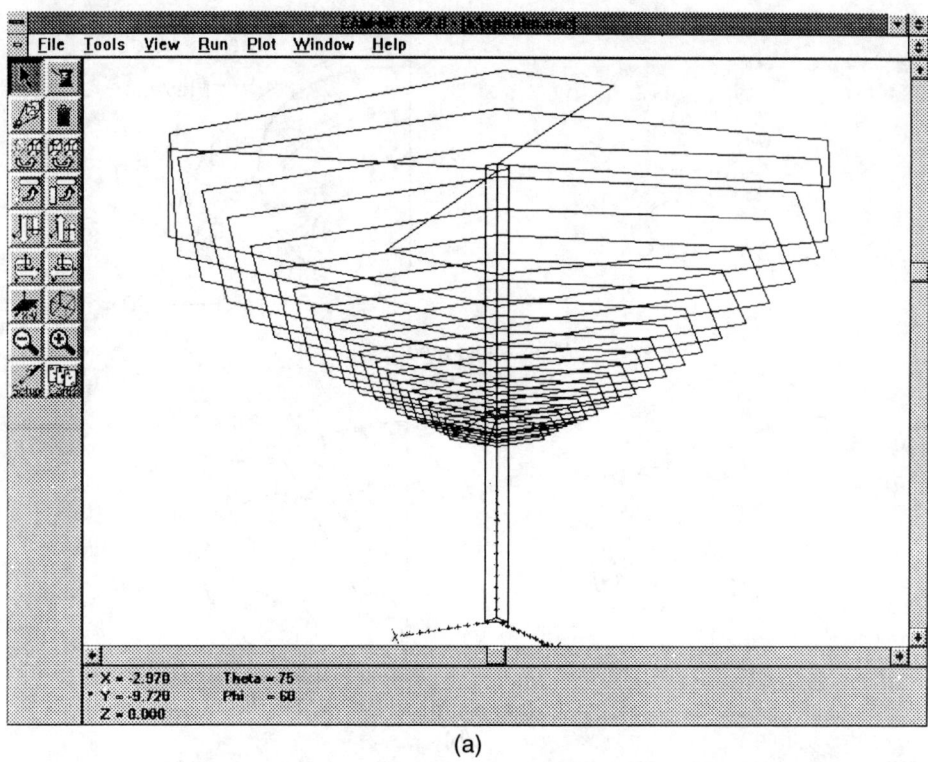

(a)

Figure 3.13 The conical spiral antenna: (a) wire grid model of the multimode conical spiral antenna and (b) feed methods and resulting radiation patterns.

electrically small antenna can be designed to meet both operational and operating objectives. Although this small antenna will not be usable as a transmitting antenna, it will nonetheless provide adequate SNR for effective operation as a receive-only antenna. As stated earlier, this antenna may have a VSWR of greater than ten to one and still provide effective operation as a receiving antenna. This outcome is possible because atmosphere noise levels in the HF band measured on an efficient receiving antenna far exceed thermal noise generated by the receiver. For this reason, a tuned or resonant antenna will deliver atmospheric noise levels and interfering signals typically as much as several tens of decibels above the receiver noise floor. In this case, the receiving system is said to be *externally noise limited*.

The signal and noise power delivered to the receiver by an antenna is determined by its effective area or aperture as well as the intended operating frequency. This aperture determines the amount of energy intercepted by the antenna from incident signal and noise energy. The size of this aperture is a

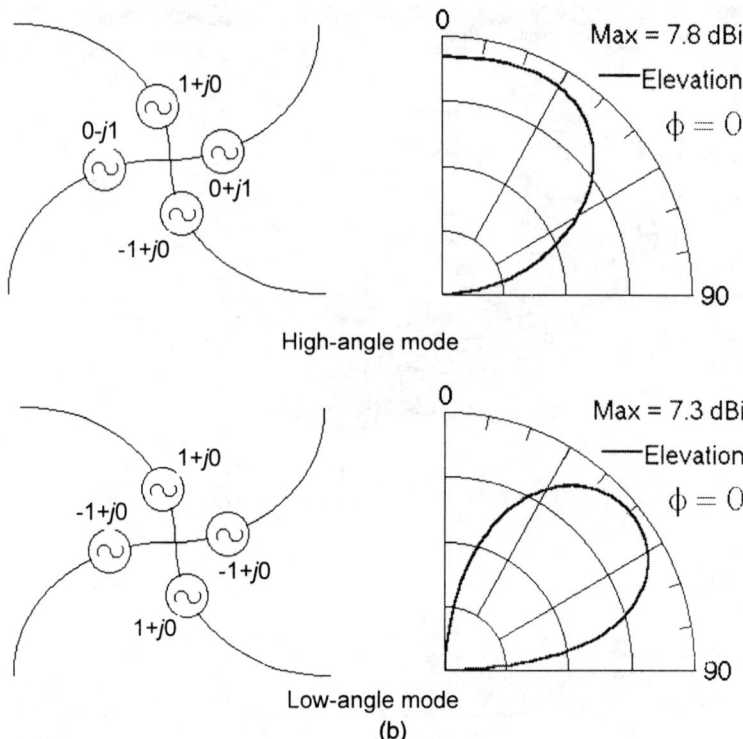

Figure 3.13 (continued)

function of antenna orientation in the arriving field. For example, a small loop antenna is planar, so if the arriving field is propagating normal to the loop, then the loop will extract maximum energy from the incident wave. If the loop is parallel to the direction of wave propagation, then minimum (near-zero) wave energy is intercepted by the loop. It is independent of orientation for an ideal omnidirectional antenna and highly sensitive to orientation for a unidirectional antenna such as an LPA. If the antenna is electrically small on a given frequency, it will have a small aperture for that frequency and therefore deliver less power to the receiving system than a larger resonant antenna. Of course, reduction in antenna size also impacts the antenna pattern and its impedance. To keep the impedance approximately constant across the frequency band of interest, additional mismatch loss must be added. By properly configuring the shape and orientation of the small-aperture antenna, and making the correct selection of mismatch loss versus frequency, the additional loss can be engineered to reduce atmospheric noise without impacting the available SNR.

If both the signal and noise are attenuated equally, the SNR of the received signal will remain constant until the external noise level approaches the receiver noise level. Typically, reduction in SNR becomes significant once the external noise is attenuated to within 6 to 10 dB of the receiver noise. This result is true if the external noise remains above the internal receiver noise floor, which will be a function of antenna pattern as well as electrical size. The correct amount of attenuation should be the difference between the atmospheric noise power and the internal receiver noise power on the same frequency. Two HF small aperture antennas are shown in Figure 3.14. The small conical monopole antenna has a typical monopole pattern and is therefore primarily vertically polarized. The horizontal *bow-tie* antenna is horizontally polarized and is well suited for receiving NVIS and medium-range sky-wave signals.

3.4.3 Small Transmit Antenna Design

When the electrical size (i.e., wavelength-scaled) of an antenna is reduced below the natural resonance point, the radiation resistance, R_{rad}, is decreased and the antenna impedance is increased relative to their respective resonant-length

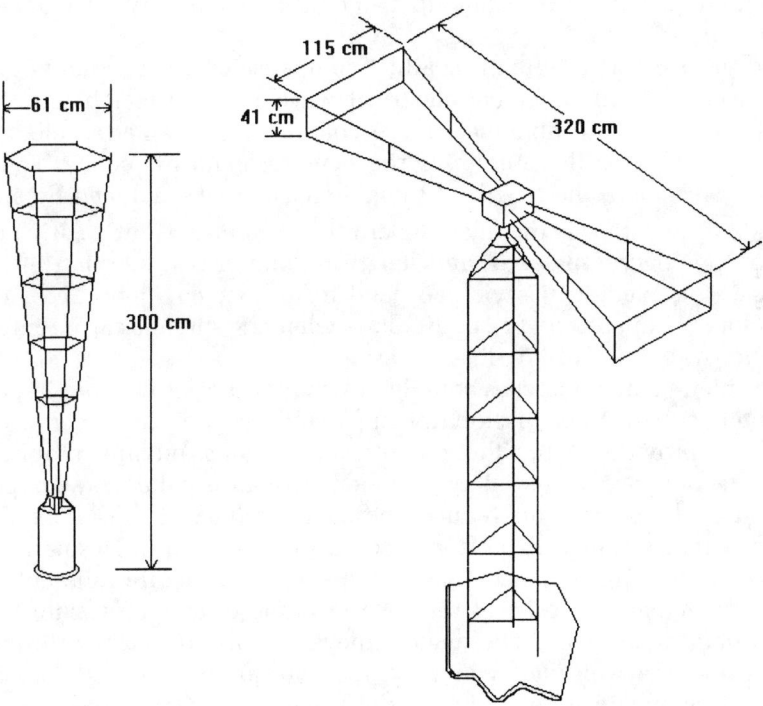

Figure 3.14 Vertical and horizontal small-aperture antennas.

values. The decrease in radiation resistance decreases antenna resistance R_a, which in turn increases the quality factor Q_a; thus, the impedance bandwidth of the antenna is reduced. The concomitant change in antenna impedance could then result in antenna/transmission line mismatch, so that less power is transferred to or from the antenna. In practice, four techniques have been used to compensate for this mismatch effect: (1) addition of "lumped" or "distributed" impedance loads to the antenna, (2) encapsulation of the antenna in dielectric or ferrite materials, (3) insertion of the appropriate antenna tuner, or (4) modification of the antenna structure to reestablish a self-resonant design. Each of these matching techniques can further reduce antenna bandwidth.

Impedance loading [9–11] is the placement of one or more reactive loads to reduce the size of both straight- and circular-element antennas. It is the simplest technique used to reduce antenna size while maintaining a suitable impedance match with the transmission line. In this case, analytic solutions exist that determine the necessary load for any wire antenna. It is well known, however, that antenna bandwidth is reduced as the electrical size of an antenna is reduced. A small antenna, even if it were self-resonant (no reactance), often requires an impedance transformer to match the antenna resistance to the transmission line resistance.

Dielectric or ferrite antenna coatings compensate for reduction in physical size by decreasing the effective electrical wavelength to match the physical antenna size. For wire antennas, these coatings allow antenna elements to be shortened while still maintaining the same resonant frequency. To remain electrically resonant, the length of the antenna must be reduced by approximately $\sqrt{\epsilon_r \mu_r}$, where ϵ_r is the relative permittivity and μ_r is the relative permeability of the coating material, and length reduction is a function of coating thickness. This method has yielded good results for *cavity-backed* antennas such as slots and apertures [12], particularly when the antenna can be embedded in a sufficient volume of low-loss material.

An antenna tuner matches both the resistive and reactive input impedance of the antenna to the transmission line impedance. High-speed tuners are available that can provide rapid, although not instantaneous, antenna retuning. This matching technique reduces antenna efficiency because of the lossy components employed in the tuner. This reduced efficiency reduces total antenna power gain and degrades radio link performance. Alternatively, optimal small antenna designs employ self-resonance, that is, the antenna structure inherently compensates for its small electrical size. An example of a self-resonant antenna is the *normal mode helix*. The helix achieves natural resonance through the inductance of the *windings,* which compensate for the increased capacitive reactance inherent in its electrically short design. Natural resonance is also achieved in a *top-loaded* monopole, in which a large circular plate is attached at the top of a short monopole.

The use of antenna loads, coatings, or tuners to design small antennas sacrifices radiation efficiency to maintain acceptable antenna-input impedance. By comparison, dielectric or ferrite material coating achieves greater efficiency than impedance loading or tuners for the same antenna size reduction. If the designer is free to suggest any small antenna structure, then self-resonant designs are optimum because no lossy components or materials are used. For example, consider the interaction between the antenna and matching efficiencies (η_a and η_m, respectively) for an electrically short dipole antenna (length l much less than half-wave) and a normal mode helix antenna of the same physical length. The results are shown in Table 3.2. Both antennas are 1/20-wavelength in length but require significantly different matching networks. Because a short helix is designed to be self-resonant, its impedance has no reactive component. The dipole impedance, on the other hand, is a large capacitive reactance. As shown in the table, the dipole antenna has greater efficiency than the helix. With the addition of matching efficiency, the overall helix efficiency ($\eta_s = \eta_a + \eta_m$) drops from 80% to 75%. The dipole, however, requires lossy components to negate the capacitance and thus reduce the overall efficiency to less than 20%.

The equiangular spiral and log-periodic structures shown in Figure 3.15 exhibit the characteristics of a small broadband antenna design defined by the radial r angle ϕ. If the spiral antenna is mounted on dielectric material with a conductive backplane, it will produce a unidirectional pattern. In addition, the physical size can be reduced by embedding the entire structure within a low-loss material. For example, a typical planar spiral antenna covering the 30- to 100-MHz band is contained within a 50-cm-radius disk. This area can reduced to a 10-cm-radius disk when embedded in the correct low-loss material. Analogous results can be achieved in the HF band.

3.5 ENVIRONMENTAL EFFECTS

3.5.1 Overview

Antenna performance is often analyzed in free space to avoid the complications of the earth, mounting platform (e.g., building, ship, aircraft, etc.), or nearby

Table 3.2
Comparison of Short-Helix and Short Dipole Antennas

Antenna characteristic	Short helix ($l = 0.05\lambda$) [13]	Short dipole ($l = 0.05\lambda$)
Antenna impedance	$11.7 + j0.0$	$0.5 - j900$
Antenna efficiency (η_a)	80%	91%
System efficiency (η_s)	75%	5% @ $Q_m = 100$; 18% @ $Q_m = 400$

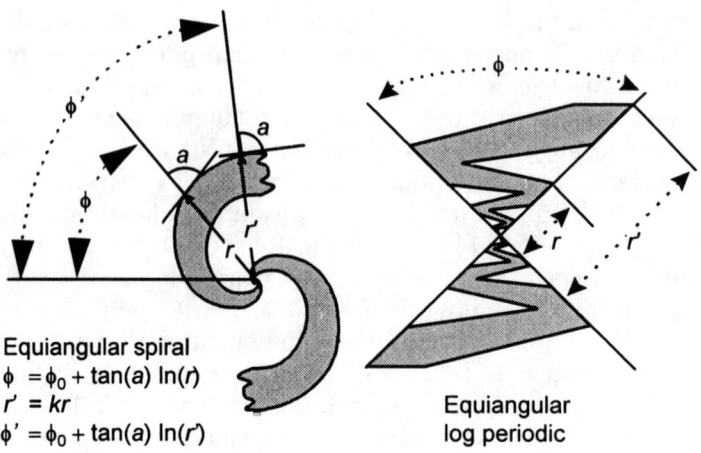

Figure 3.15 "Angular" antenna designs.

structures. These objects can significantly affect antenna impedance, particularly when they are located within one wavelength of the antenna and are on the order of a wavelength or more in size. At 3 MHz, the near field of the antenna extends to the wavelength of 100 m and beyond. Except for airborne platforms at high altitude, the HF antenna system is typically within one wavelength of the earth's surface, so the effects of the earth's surface are generally important in evaluating HF antenna-system performance.

3.5.2 Ground Effects

As stated earlier, the proximity of the earth's surface to ground-mounted antennas will modify their electrical performance. The effects of the earth's surface are highly dependent on the ground's electrical parameters, specifically the relative permittivity (ϵ_r) and conductivity (σ). The relative permeability (μ_r) is usually assumed to correspond to the free-space value. Ground electrical parameters have been measured worldwide [14–17]. Figures 3.16(a) and (b) present plots of the values for typical relative permittivity and conductivity, respectively, across the HF band from "dry" earth to seawater [18].

To demonstrate effects on a vertically polarized antenna, consider different ground electrical parameters on the gain pattern of a simple quarter-wave monopole antenna. Figures 3.17(a) through (d) plot the elevation radiation patterns of identical resonant monopoles mounted over perfect ground ($\sigma = \infty$); seawater [$\epsilon_r = 81$, $\sigma = 4.0$ siemens/meter (S/m)]; "good" earth ($\epsilon_r = 12$, $\sigma = 0.01$ S/m); and dry earth ($\epsilon_r = 4$, $\sigma = 0.001$ S/m), respectively. Figure 3.17(a) shows that a

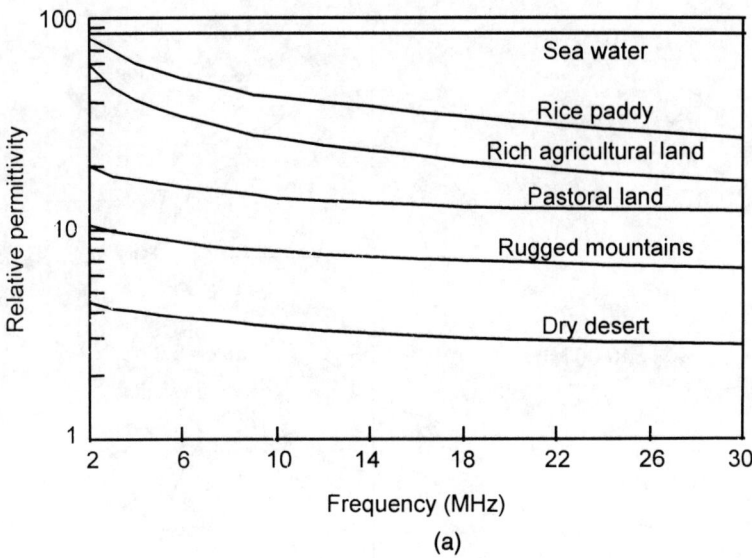

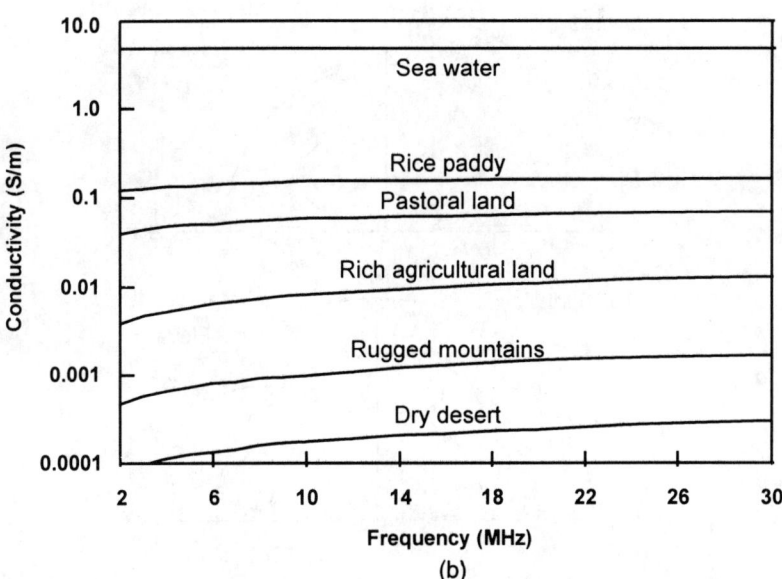

Figure 3.16 Typical values of surface electrical constants across the HF band: (a) relative permittivity and (b) conductivity in siemens per meter.

154 Advanced High-Frequency Radio Communications

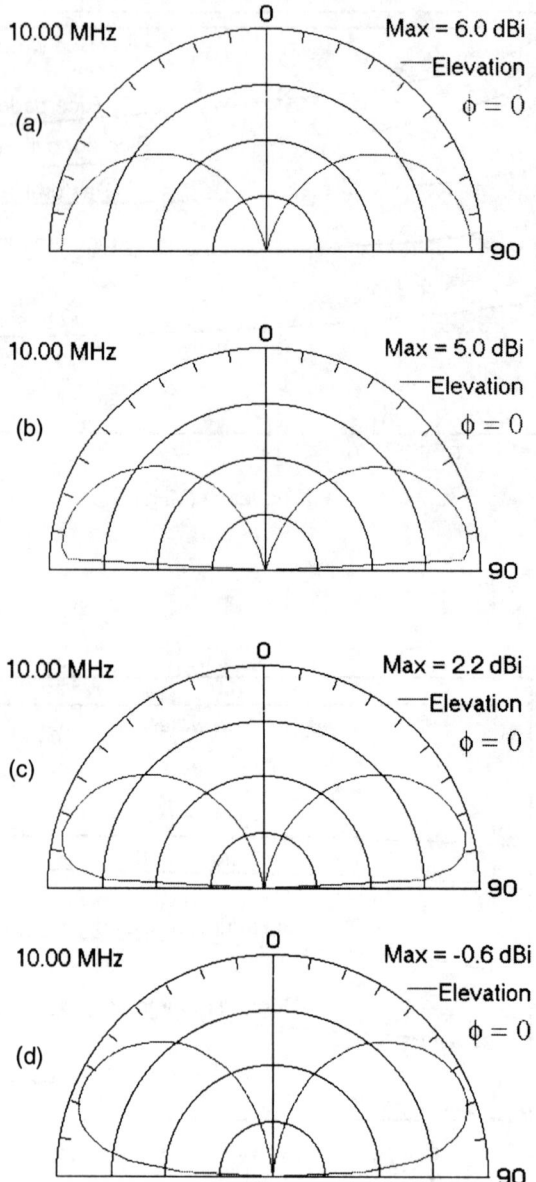

Figure 3.17 Monopole antenna pattern over various surfaces: (a) "Perfect" ground ($\sigma = \infty$), (b) Seawater ($\epsilon_r = 81$, $\sigma = 4.0$), (c) "Good" earth ($\epsilon_r = 12$, $\sigma = 0.01$), and (d) "Dry" earth ($\epsilon_r = 4$, $\sigma = 0.001$).

peak gain of 6.0 dBi is achieved at the horizon for the perfect-ground case. Over seawater (Fig. 3.17[b]), the peak gain has dropped to 5.0 dBi with significant pattern "tuck" noticeable at the lowest elevation angles. Over good earth (Fig. 3.17[c]), the gain has further decreased to a peak value of 2.2 dBi with severe pattern tuck extending to higher elevation angles than the seawater case. Over dry earth (e.g., desert), the directivity has dropped below 0.0 dBi as shown in Figure 3.17(d). Thus, if the monopole was to be erected over a dry earth, the antenna pattern would vary significantly from the perfect-ground case. These ground effects demonstrate the importance of adequate ground conductivity for vertically polarized antennas.

Clearly, the gain of vertically polarized antennas is improved as the conductivity of the earth's surface is increased. This effect can be produced artificially with the addition of a ground screen (ground plane) or ground radials below the antenna and in the desired direction of propagation. The ground screen reduces ground reflection losses; therefore, the minimum improvement is seen near the horizon (θ = 90 deg), while maximum improvements happen at the take-off angle where the reflected and direct waves add constructively. Figure 3.18 shows the theoretical increased gain due to use of a ground screen [19] over good earth (ϵ_r =10, σ = 0.01S/m) at a take-off angle of 15 deg versus

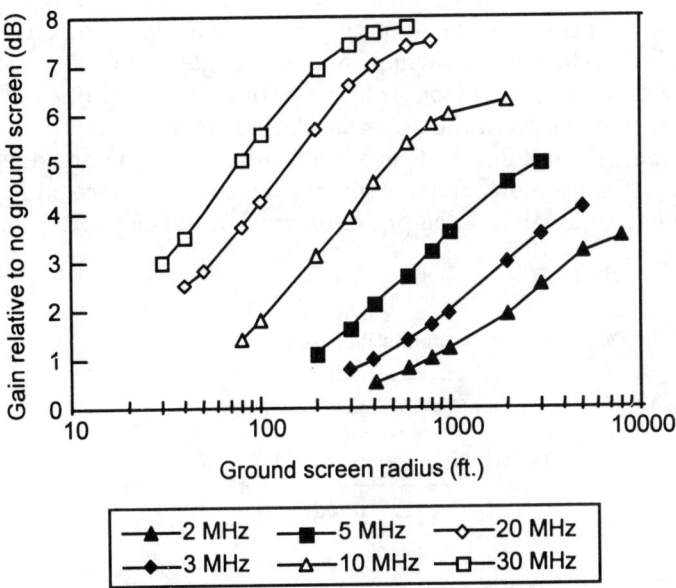

Figure 3.18 Gain improvement for a monopole antenna provided by ground screens of different radius at a 15-deg take-off angle.

radial length and frequency. These curves apply to short monopoles, but can also be applied to quarter-wave monopole antennas. From these results, it is apparent that even a small ground screen can provide significantly enhanced performance for vertically polarized antennas.

Typically, ground screens are made from radial wires diverging from the base of the monopole and resembling bicycle spokes. Several optimized ground radial systems and the concomitant ground-loss values are shown in Table 3.3. The high gain at low elevation angles ($\theta \to 90$ deg) achieved by vertical antennas over highly conductive surface materials, including ground screens, suggests that these antennas should be used for HF links requiring low elevation angles.

The effects of ground electrical parameters on horizontally polarized antennas differ from those of vertically polarized antennas. For example, consider a half-wave resonant dipole over the earth's surface. Figure 3.19 displays four peak-gain curves for a 2-MHz half-wave dipole versus electrical height above seawater, good earth, dry earth, and pure water ice ($\epsilon_r = 1$, $\sigma = 0.0001$ S/m). In general, the maximum gain at zenith ($\theta = 0$ deg) occurs when the zenith direct and reflected waves add in phase. Because horizontal polarization results in a 180-deg phase reversal of the reflected wave relative to the incident wave at the earth's surface, the carrier wave must undergo an additional 180-deg phase offset to add constructively with the incident wave at zenith. This phase offset can be obtained by increasing the path length traversed by the reflected wave, that is, by increasing the antenna height above ground. The maximum gain at zenith therefore occurs when the path length from the dipole's phase center to ground is 0.25 wavelengths (λ) or 90 deg, for a total phase shift of 180 deg for two-way travel from the antenna phase center to the ground and then back to the phase center. For finite ground conductivity, the downward-propagated signal will be somewhat absorbed by the earth's surface and will not in general provide an ideal 180-deg phase reversal at the air/earth interface.

Table 3.3
Ground Radial Design

Ground Radial System		
Number of Radials	Radial Length (wavelengths)	Ground Loss (dB)
16	0.100	3.0
24	0.125	2.0
36	0.150	1.5
60	0.200	1.0
90	0.250	0.5
120	0.400	0.0

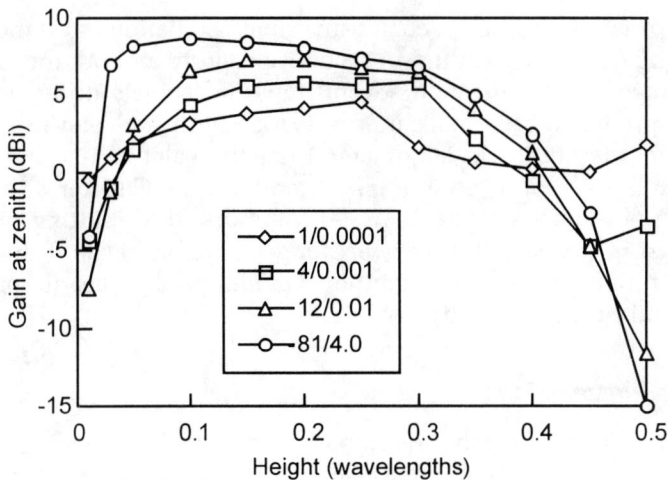

Figure 3.19 Ground effects on a horizontal dipole at 2 MHz (e_r/s).

For example, the presence of good earth below a dipole at just less than a quarter-wavelength height above ground results in maximum constructive gain with minimum ground losses (see Fig. 3.19). In this case, the good earth ground type resulted in imperfect wave reflection (some energy is absorbed by the ground), such that the wave penetrates the surface. This ground penetration further increases the effective path length traveled by the reflected wave. As surface conductivity is increased, the losses for horizontally polarized antennas below a 0.1-wavelength height are also increased. As the dipole is raised above quarter-wavelength height, the decreased gain at zenith is evident in the shape of the peak gain curves plotted in the figure. As the permittivity and conductivity are decreased, however, the antenna directivity (peak gain) decreases and ground losses at low heights are less pronounced. The figure demonstrates that a highly conductive ground or ground plane is detrimental to achieving maximum gain at high elevation angles ($\theta \rightarrow 0$ deg) for horizontally polarized antennas near the earth's surface. For this reason, ground screens are not recommended for near-earth horizontal antennas designed for high-angled HF ray paths, such as those needed by NVIS communications. If the antenna is more than a quarter-wavelength above a low-conductivity ground, then a ground screen can improve the gain at these angles.

3.5.3 Antenna Mounting Effects

In general, HF antenna performance is strongly coupled with the local environment, including ground electrical parameters, mounting height above ground, foreground terrain irregularity, co-site antenna installations, and the proximity of conducting structures. Failure to account properly for both the antenna and its environment could result in several tens of decibels difference between actual and predicted antenna patterns for azimuth and elevation angles of interest. This variability is far greater than the relatively marginal benefit achieved by the use of a more efficient modem (see Chapter 4) or changing transmitting power by several dBW. Furthermore, the resulting distortion in the relationship between link power budget and ray take-off angles would result in significant error in the computer model predictions of optimum link frequency and time availability.

3.5.3.1 Fixed Installations

Large metal objects like rebar-reinforced concrete buildings, towers, power lines, fences, or even other antennas (see Fig. 3.20) can act like *parasitic radiators* and perturb antenna behavior and, ultimately, its performance. At 3 MHz, the 100-m wavelength can encircle many nearby objects, such as fence posts or small buildings or enclosures, expected to affect the antenna. On the other hand, HF radio waves are not significantly affected by small metal objects, such as nails in wooden buildings. All significant antenna effects must be considered in practical HF system design and performance prediction. Otherwise, pattern nulls might be formed in the desired direction of propagation and unwittingly degrade or prevent HF communications.

3.5.3.2 Mobile Platforms

HF antennas are often mounted on mobile platforms, such as automobiles, trucks, aircraft, or ships. If a platform's size is significant relative to the radio wavelength, the antenna system includes the platform as well as the antenna. In fact, the radiation pattern, VSWR, and other performance parameters of an antenna can differ significantly after mounting on the platform. Fundamentally, an antenna mounted on or near a conductive surface can induce currents that flow in the surface material. These flowing currents produce reradiation of signal energy, thus determining the performance of the entire antenna system. Examples of a mobile platform modeled as wire grids used for numerical modeling with NEC [20] are shown in Figures 3.21(a) and (b). Figure 3.22(a) shows a symmetric three-dimensional radiation pattern from a 16-ft (5.8-m) monopole antenna over 16 ground radials; Figure 3.22(b) shows the radiation pattern of the same monopole antenna mounted on the side of the vehicle shown in 3.21(a). The vehicle has clearly altered the pattern relative to the ground-mounted case.

The use of ALE HF radio on mobile platforms requires the use of broadband antennas combined with special mounting considerations. If the platform is

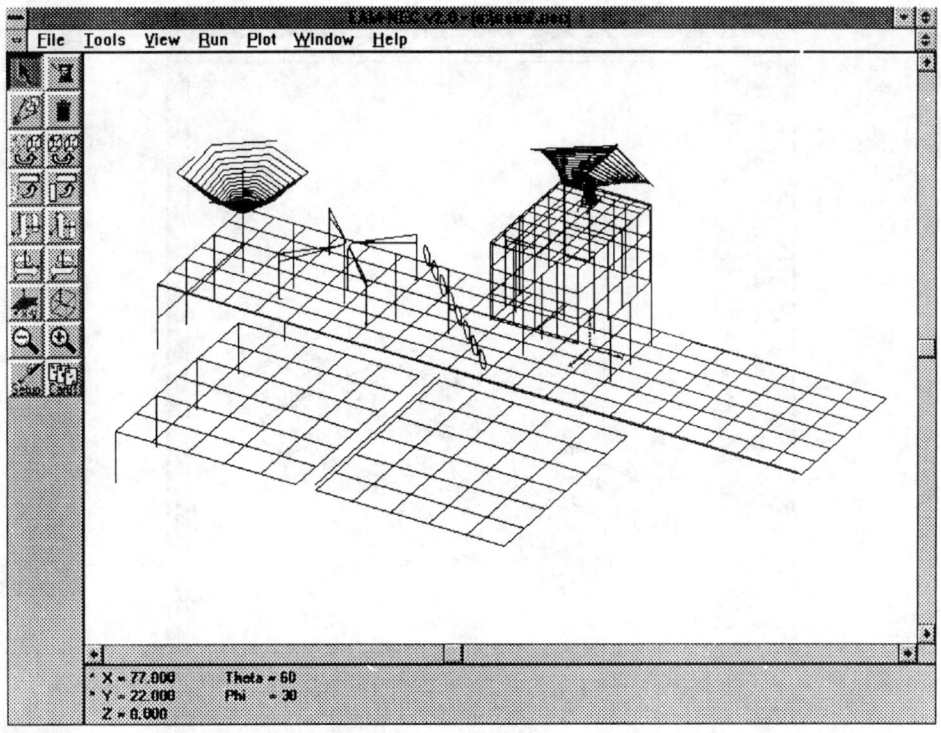

Figure 3.20 Wire model of rooftop with multiple antennas.

not much larger than the signal wavelength in the frequency range of interest, then small antenna designs should be considered. From the previous discussion, it is apparent that optimum small antennas are (1) designed for self-resonance and (2) integrated into the surface structure of the mobile platform. At best, the antenna would be conformal to the platform surface; that is, it would not create a significant disruption to the physical appearance or mechanical performance of the platform. In this context, mobile platforms may include human or animal subjects as well as vehicles, ships, or aircraft. The physical and electrical characteristics of these platforms must be included in the small antenna design process to meet platform operational constraints.

On mobile platforms, conformal mounting of the antenna system requires a compromise between antenna visibility and electromagnetic performance. With the addition of dielectric encapsulation, a portion of the mobile platform is replaced with a low-observable dielectric material containing an embedded antenna. In this configuration, antennas may be positioned to form parallel, collinear, or crossed-element arrays. Dielectric mounting can serve a dual pur-

160 Advanced High-Frequency Radio Communications

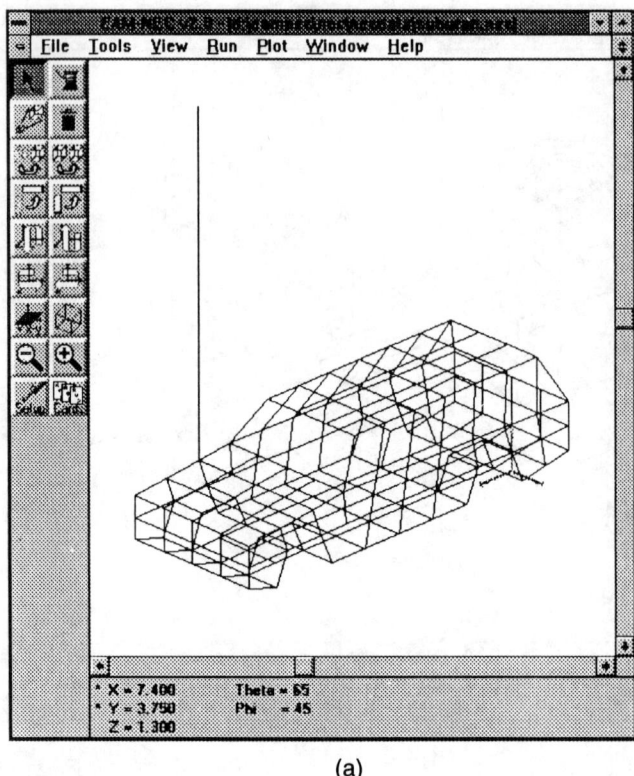

(a)

Figure 3.21 Wire "grid" models of antennas and mobile platforms: (a) vehicle with whip antenna and (b) helicopter with towel bar antenna.

pose, allowing a horizontal wire antenna to be mounted near a horizontal surface without major disruption to the radiated field while still providing the necessary size-reduction factor.

3.6 CONCLUSIONS

HF transmit and receive antennas provide the critical interface between the HF communication system and the propagation path; therefore, selection of the optimal antennas plays a major role in determining overall system performance. Several common HF antenna types have been discussed in this chapter, but the best antenna for a particular communications requirement is strongly application-dependent. Three antenna electrical parameters are critical to HF link

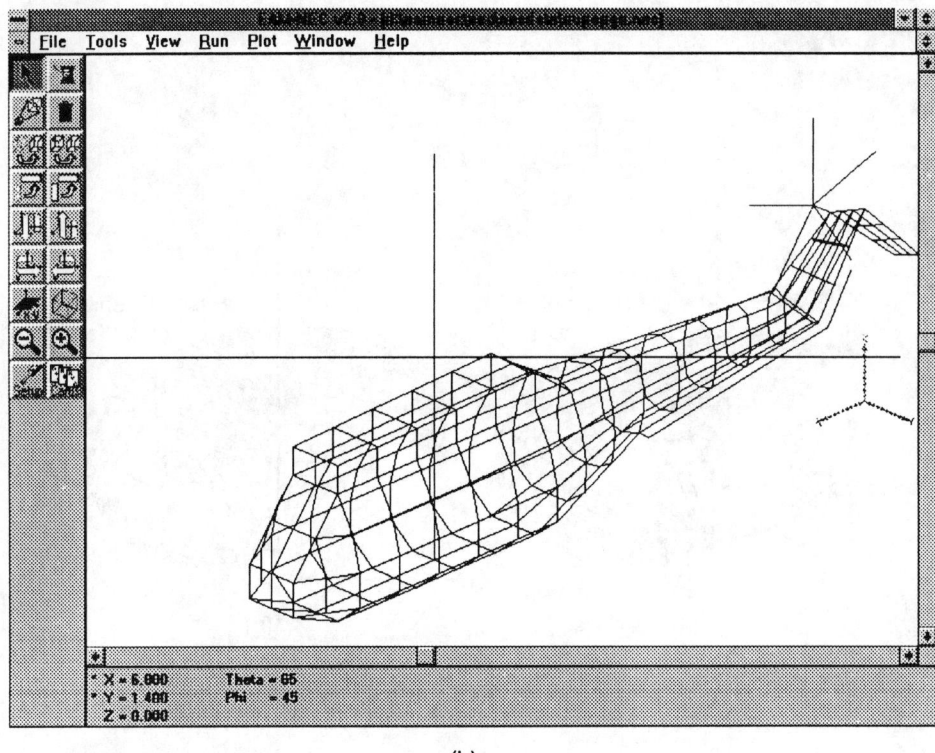

(b)

Figure 3.21 (continued)

and network performance: polarization, gain, and pattern. Antenna selection must also include trade-off analysis for physical size, cost, erection time, availability, and maintainability. Theory and practice demonstrate that vertically polarized antennas are most appropriate for ground-wave communications, long-range sky-wave links requiring low take-off (elevation) angles, or for simple omnidirectional antennas. Horizontally polarized antennas are ideal for NVIS and medium-range sky-wave links or to meet directional high-gain requirements.

The physical size of HF antennas makes them difficult and expensive to test for a given HF link or network application. Advances in numerical modeling techniques and processor power in the past two decades have made it possible to predict the performance of antenna structures in a simulated environment. Numerical modeling provides a versatile cost-effective approach for determining link reliability resulting from use of alternative transmit and receive antennas for particular propagation paths.

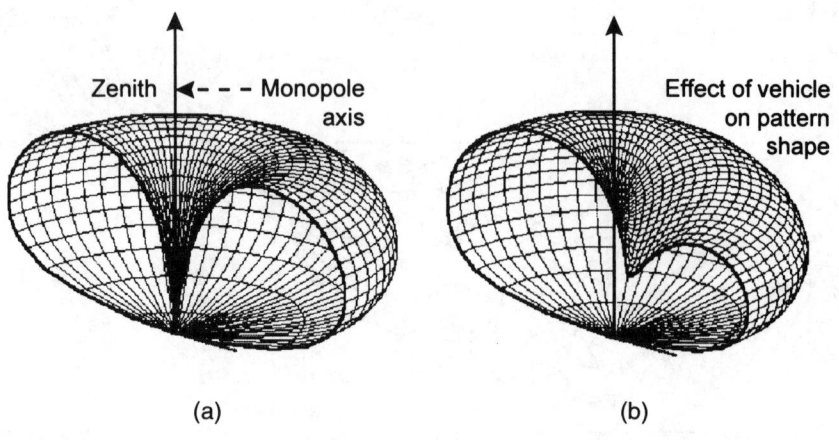

Figure 3.22 Distortion of radiation pattern from an asymmetric platform: (a) monopole on ground plane and (b) monopole on vehicle in Figure 3.21(a).

References

[1] Webb, J. K., "Electronic Steering of Antenna Nulls for HF Interference Reduction," *2nd IEE Conf. HF Communication Systems and Techniques,* Feb. 15–17, 1982.

[2] Rumsey, V. H., "Frequency Independent Antennas," *IRE Int. Conf. Rec.,* Part I, 1957, pp. 114–118.

[3] King, W. P., and S. Prasad, *Fundamental Electromagnetic Theory and Applications,* Englewood Cliffs, NJ: Prentice Hall, 1986.

[4] Carrel, R. T., *Analysis and Design of the Log-Periodic Dipole Antenna,* Ph.D. dissertation, Urbana, IL: University of Illinois, 1961.

[5] Stutzman, W. L., and G. A. Theile, *Antenna Theory and Design,* New York: John Wiley & Sons, 1981.

[6] Balanis, C. A., *Antenna Theory, Analysis and Design,* New York: Harper and Row, 1982.

[7] Mayes, P. E., "Frequency Independent Antennas and Broad-Band Derivatives Thereof," *Proc. IEEE,* Vol. 80, No. 1, Jan. 1992, pp. 103–112.

[8] Rumsey, V. H., *Frequency Independent Antennas,* New York: Academic Press, 1966.

[9] Harrison, C., "Monopole with Inductive Loading," *IRE Trans. Ant. Propagat.,* Vol. AP-11, July 1963, pp. 394–400.

[10] Harrington, P., and J. Maute, "Straight Wires with Arbitrary Excitation and Loading," *IEEE Trans. Ant. Propagat.*, Vol. AP-15, July 1967, pp. 502–515.
[11] Hirasawa, K., "Design of Arbitrarily Shaped Thin Wire Antennas by Passive Impedance Loading," *Elect. Commun.*, 61-B, 1978, pp. 55–63.
[12] Packer, M. J., *Investigation of Two Silo-Integrated UHF Antennas for Propagation Through Various Debris Layers*, GTE Internal Report, July 1987.
[13] Fujimoto, K., et al., *Small Antennas*, Letchworth, Hertfordshire, England, Research Studies Press, 1987.
[14] Longmire, L. L., and K. S. Smith, *A Universal Impedance for Soils*, Report DNA-3788T, Defense Nuclear Agency, Oct. 1975.
[15] CCIR, "Electrical Characteristics of the Surface of the Earth," Recommendation 527-3, *Recommendations of the CCIR*, Geneva: International Radio Consultative Committee, 1992.
[16] Messier, M., "Another Soil Conductivity Model," *JAYCOR*, June 1985.
[17] CCIR, "World Atlas of Ground Conductivities," Recommendation 832, *Recommendations of the CCIR*, Geneva: International Radio Consultative Committee, 1992.
[18] Hagn, G. H., "Ground Constants at High Frequencies (HF)," *1987 Applied Computational Electromagnetic Society Conf. Proc.*, Mar. 1987.
[19] Wait, J. R., and L. C. Walters, "Influence of a Sector Ground Screen on the Field of a Vertical Antenna," NBS Monograph 60, Apr. 1963.
[20] Burke, G. J., *Numerical Electromagnetics Code—NEC-4, Method of Moments, Part I: User's Manual*, Lawrence Livermore National Laboratory, Jan. 1992.

Digital Modems for High-Frequency Radio[1]

4.1 INTRODUCTION

4.1.1 Background

Although the global information age is exemplified by worldwide Internet access, many important communication services still rely on HF sky-wave radio. In fact, an increasing number of remote "surfers" access the Internet via HF. For many years, the maximum throughput over voiceband (3-kHz) channels has been limited to 2400 bps. More recently, high-speed HF modem developments have focused on minimizing the average error rate, albeit often at the expense of increased transmission delay. A good review of the state of HF modem development in the late 1980s is given in Pennington [1]. This important work described the variety of modem waveforms developed over the years to counter HF channel conditions and enumerated the issues that need to be considered when selecting an HF modem for a particular application. Much as antennas serve as the interface between the transmission equipment and the ether, modems provide the interface (see Fig. 4.1) between the digital global information infrastructure and signaling waveforms designed to mitigate the distortion and noise characteristics of HF channels described in Chapter 2.

This chapter seeks to update Pennington's work by including results from subsequent modem designs and considering additional performance issues. The chapter begins with a description of the design approaches that resulted in the three 2400-bps MIL-STD-188-110A HF modems [2]. The review of these

---

[1]This chapter was adapted from "Advances in High-Speed HF Radio Modem Design" by Stephen C. Cook of the Australian Defence Science and Technology Organisation (DSTO), P.O. Box 1500 Salisbury, South Australia, published in *Proc. Nordic Shortwave Conf. HF'95*, Faro, Sweden: Nordic Radio Society, Aug. 1985, and reproduced by permission. It represents the results of extensive modem evaluations performed at DSTO.

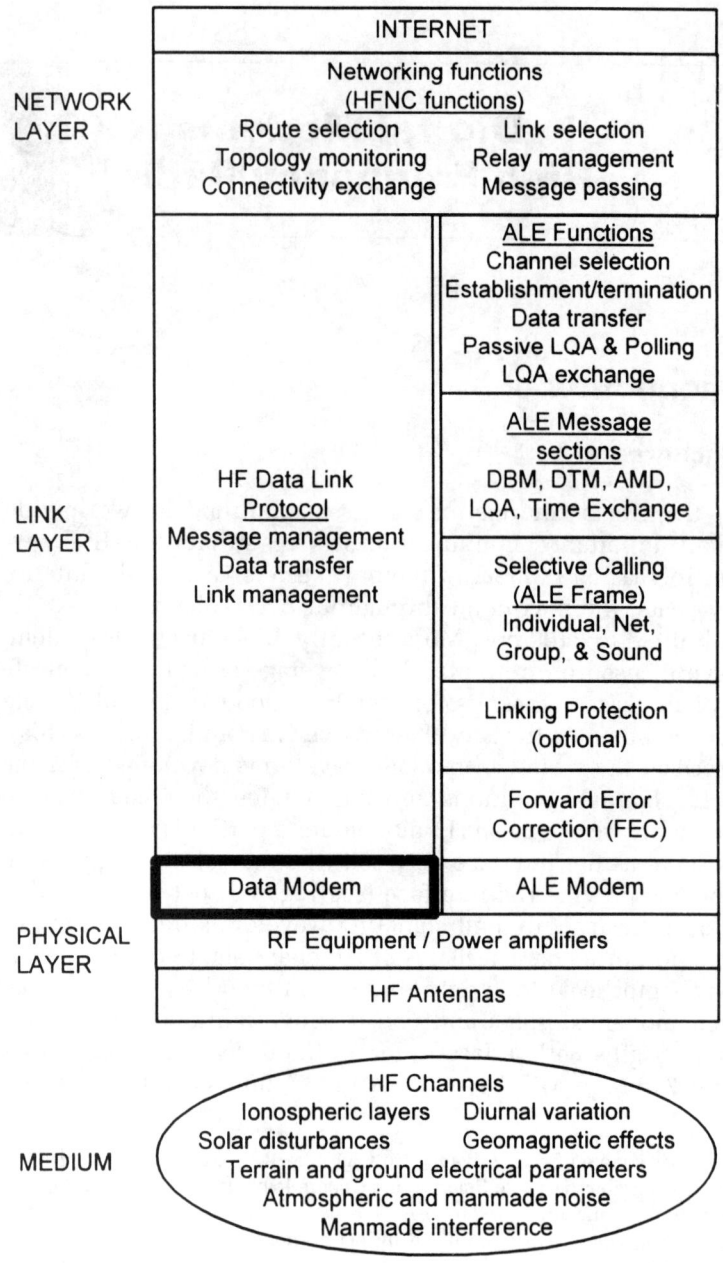

Figure 4.1 Digital HF modems in the hierarchical structure of HF communication systems.

HF modems is followed by a comprehensive set of comparative results between the 39-tone, 16-tone, and single-tone waveforms in four environments: CCIR channels, CCIR channels with impulsive noise, carrier-wave (CW) interference, and pulsed sine-wave interference. We show that the parallel-tone waveforms give better subjective performance for digital voice transmission, whereas the single-tone modem is definitely superior for data transmission when long interleaving is employed. The chapter also presents several issues that affect the design of improved-performance modems, including coded modulation design and specially designed detection metrics to counter non-Gaussian noise and interference.

The chapter concludes with a brief description of two new modems: (1) a 52-tone modem that provides a useful throughput of 4800 bps with an error performance at least as good as the best contemporary 2400-bps modems and (2) a 3200-bps single-tone modem that outperforms its 2400-bps counterpart. The development of these advanced modems was initiated by the Australian Defence Science and Technology Organisation (DSTO) to increase the usable throughput over voiceband HF channels. This increased throughput was to be achieved while minimizing both the errors and circuit delay, so that multimedia communications such as digitized voice, image, and interactive data transfer could be effectively supported over HF sky-wave links. Many of these services have real-time delivery requirements, thus placing limits on the degree of time diversity that can be employed to overcome fading and impulsive noise.

4.1.2 Channel Distortion

The ionosphere is a dispersive medium that spreads the signal in both time and frequency as described in Chapter 2. In addition, received radio signals have usually been reflected from more than one ionospheric layer, a phenomenon known as *multipath propagation*. The enormous variability of the ionosphere makes it difficult to adequately predict short-term variability in HF channels. To provide a quantitative baseline for equipment evaluation, the CCIR [3,4] has proposed three standard channel models termed "Good," "Moderate," and "Poor." These characteristic channel models have been widely adopted as a basis of channel simulation for the comparison of HF communications equipment in general and HF modems in particular. These simulation channels are each composed of two equal-amplitude, independent Rayleigh-fading sky-wave paths representing reflections from different layers in the ionosphere. Testing is performed with the time delays and average fade rates given in Table 2.2 in the presence of different amounts of additive white Gaussian noise (AWGN). These tests investigate the capability of the modem under test to mitigate the effects of multipath propagation, namely, frequency-selective fading and a time-varying impulse response, and have served as the mainstay of modem comparisons for many years.

4.1.3 Additive Noise

As shown in Chapter 2, impulsive noise can be a dominant source of disruption in HF radio channels. In this regard, the distribution of noise burst duration shown in Figure 2.41(b) is particularly useful in the design of error-control coding schemes for digital HF radio. The figure can be used to estimate the minimum number of sequential transmitted symbols received in error over the channel. If necessary to achieve a certain quality of service, this symbol error distribution aids selection of a suitable error-control code to correct those errors caused by the impulsive noise-impaired channel. In conjunction with the pdf, shown in Figure 2.40, the burst error rate and duration can be approximated to predict an average symbol error rate.

For example, Figure 2.41(b) indicates that the necessary error-correction capability of the code must span approximately 1.4 ms to correct symbol errors produced by more than 99.9% of the expected noise bursts. This capability is necessary even if the modem is operating at signal levels of 30 dB above the average noise power, in which case only noise impulses exceeding 20 dB above the average noise power need be considered. At 2,400 symbols per second, this error-correction capability must handle three erroneous symbols and if three bits per symbol were transmitted, a code capable of correcting nine consecutive erroneous bits is required to guarantee error-free transmission. Codes of this strength are rarely seen in practice, hence, postdecoder errors can be expected for this modem even for the otherwise adequate SNR value considered.

4.1.4 Cochannel Interference

The discussion of noise sources in the HF band is often of purely academic interest to HF users, because the limiting factor is the interference from other users in the allocated channels. In general, it is possible to cope with very high levels of interference when analog voice is being used, especially if it is narrowband. Moreover, intermittent interferers such as radars and chirp sounders pose no great problem to this mode of communications. In contrast, quite low levels of cochannel interference can significantly degrade the performance of digital communication systems. As discussed in Section 2.5.4, interference tolerance is a significant consideration in HF system design and, more specifically, in the selection of an appropriate HF modem.

4.2 TRADITIONAL DESIGN APPROACHES

4.2.1 Background

The comparison of different modulation schemes is usually based on two well-known measures of merit. One measure is the baseband value of average signal energy per bit, E_b, to noise power spectral density, N_0, measured at the input of the receiver; that is, E_b/N_0. For each modulation, a given value of this ratio will yield a corresponding value of bit error rate (BER). In this chapter, modem performance is measured by the relationship between average SNR and BER. A modem exhibiting a lower BER value than a second modem for the same average SNR value as well as all other channel conditions would therefore be considered "better" at the specified SNR. Note that better performance at one SNR value does not imply that the modem is better at other values of SNR. Another measure of merit is a modulation's speed, defined as R/W, where R is the bit rate and W is the IF bandwidth. A number of digital modulations such as binary phase shift keying (BPSK) or frequency shift keying (FSK) have theoretically infinite sidebands, so postmodulation filters are usually included to limit the interference to neighboring channels. In general, band-limiting the transmitted spectrum causes a degradation in the performance of these modulations [5].

An effective method to improve the performance of modems operating in a fading environment is to employ time diversity in the transmitted signal. In this way, periods of high symbol (or bit) error rate are distributed evenly throughout a block of correctly received symbols, thus lowering the peak error rate to a value correctable by the associated error-control code. This diversity is achieved, in practice, by blocking data into a two-dimensional matrix. The data bits are then fed into the rows of the matrix and read out to the modulator one column at a time. After transmission, reception, and demodulation, the data bits are read into a matrix of identical proportions by columns and passed to the error-correction unit by rows. As a result, the received bits are recovered in the original order presented to the modulator. Error dispersion techniques of this type are known as *interleaving*.

If a received signal fade corrupts the data bits in the received sequence for less time than it takes to fill a column, the error-control decoder only sees one error per row. As a result, all of the demodulated bit errors can be corrected. The only drawback with this approach is the consequent time delay of at least twice the time needed to fill the matrix. The maximum interleaver delay is 9.6 sec for both the single-tone and 39-tone modems. If the effects of the noise impulses characterized in Figure 2.41(b) were to be dispersed in time by an interleaver, the postdecoder error rate would approach zero, providing the code can correct at least one erroneous symbol. This conclusion is based on the negligible probability of the noise exceeding 20 dB above the average amplitude (see Fig. 2.41[b]), so the few noise-affected symbols would be adequately dispersed.

4.2.2 FSK Modems

Before the advent of inexpensive digital signal processing hardware, low-cost modems were limited to using simple waveforms such as BPSK and FSK. For example, the modem used in the U.S. military and federal standards for automatic link establishment is an 8-ary FSK design that can be implemented with a bank of 8 analog filters. To overcome channel multipath effects, this modem uses an 8 ms baud, resulting in a relatively low 125 symbols per second, or 375 bps. The waveform achieves fairly robust performance in low-SNR skywave channels, and the low data rate is adequate for the small data bursts used to establish links (see Chapters 5 and 6). However, higher-speed modems are preferred for conveying user voice and data traffic.

4.2.3 Parallel-Tone Modems

The nature of the HF channel necessitates specially designed signaling waveforms to achieve a 2400-bps transmission rate in a standard 2700-Hz HF voice-band channel. In this regard, the first modem design objective is to mitigate channel multipath effects observed as time spread in the received signal around some value of nominal propagation delay. The time spread is measured as the duration of significant signal energy receptions from the transmission of a wideband pulse over the channel. The root mean square (rms) delay value computed over many time-spread samples is called the *delay spread*. Delay spread arises when a signal arrives at the receiver via more than one propagation path (i.e., multipath), or if the symbol transmission rate is near the maximum (Nyquist) rate of 2700 baud. In these cases, the duration of any arriving symbol could be expected to span many transmitted symbols, and the resulting superposition of multiple symbols would then disrupt correct demodulation of any one symbol.

An early approach devised to overcome delay spread was to lower the transmission rate so that the time spread comprised only a small fraction of the symbol duration. Low-rate signals occupy correspondingly small bandwidths, so the overall bit rate can be recovered by utilizing multiple low-rate signals placed in the available channel bandwidth. Overlength symbols are then employed to overcome multipath-induced time spread. The demodulator discards the first part of each received symbol, known as the *guard time*, and integrates the remainder of the symbol. In this way, the demodulator can combine the received signal energy from all of the possible signal paths.

Demodulation of these parallel-tone waveforms was originally achieved by effectively employing one matched filter and one demodulator per tone. Recent designs employ orthogonal tone spacing, that is, frequency spacing set to the reciprocal of the symbol period, which enables a simple demodulation using a fast Fourier transform (FFT). Figure 4.2 is an illustration of a parallel-tone waveform received via two ray paths with a path delay difference of about 10% of the symbol duration. Each symbol period contains a single phaser for each tone, so the addition of time-shifted signals simply results in a phase shift that varies little from one symbol to the next. These phase shifts can then be accommodated using differential phase modulation.

The most widely known modems of this type are the 16-tone and 39-tone modems described in MIL-STD-188-110A. The 16-tone modem, which has its origins in the 1950s, employs 16 tones each with DQPSK modulation, a 75-baud symbol rate, and a 4.2-ms guard time. No error-control coding is employed and the resulting BER is often quite poor at 10^{-2} to 10^{-3}. This BER performance can be improved by using space diversity reception [6]. Space diversity combines signals from two or more receiving antennas spaced by several wavelengths to achieve uncorrelated fading. Space diversity reception can be implemented with minimal additional complexity in parallel-tone modems, but it requires the use of multiple receiver/antenna pairs. The newer 39-tone modem uses more signaling tones and a frame rate of 44.44 baud with a guard time of 3.5 ms, resulting in an uncoded throughput of 3466.66 bps. Forward error correction (FEC) in the form of a (14,10) shortened Reed-Solomon block code (code rate 10/14), with four bits per code symbol, is employed to

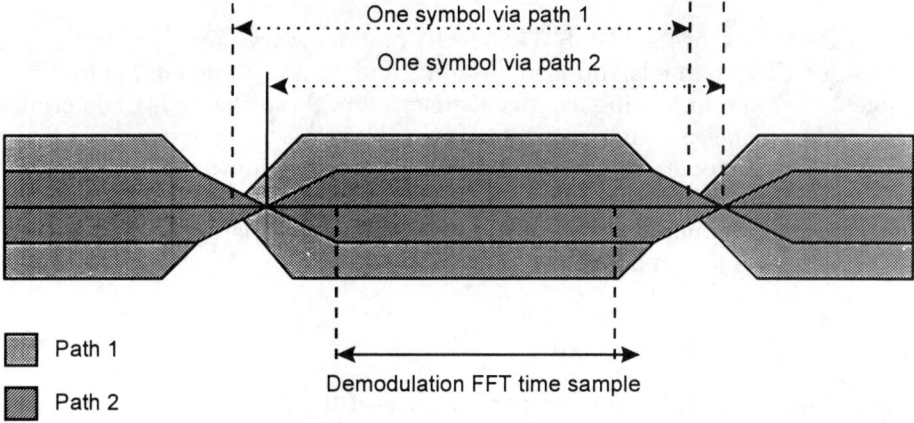

Figure 4.2 Parallel-tone signaling waveform arising from two propagation paths showing demodulation time-sample position to achieve detection without intersymbol interference.

yield a useful throughput of 2400 bps. The performance of this 39-tone modem is clearly superior to the 16-tone device.

4.2.4 Single-Tone Modems

The early 1980s saw the development of the first single-tone HF modems. These modems transmit near the Nyquist rate and employ a variety of detector structures to overcome any intersymbol interference introduced by the channel. The single-tone modem described in MIL-STD-188-110A represents a mature development of this form of modem. It operates at 2400 baud with regular training sequences used by the receiver to *equalize* the channel [7]. These training sequences occupy one-third of the total transmission time. The modulation employed is coherent 8-PSK, giving a raw bit rate of 4800 bps. When the constraint-length-seven, rate 1/2, convolutional error control code is employed, the useful data rate drops to 2400 bps. An important feature of this modem is its ability to cope with rapid fading. In common with all detectors with training memory, however, it has the potential to produce prolonged error bursts whenever the detector fails to adapt to the channel. Modems conforming to this standard have been in continuous development in recent years and the results presented later reflect recent results.

4.3 RELATIVE MERITS OF CONTEMPORARY DESIGNS

4.3.1 CCIR Channel Performance

The performance of the MIL-STD-188-110A modems was analyzed in the two extreme CCIR channels, dubbed "Good" and "Poor," and summarized in Table 2.2. Two interleaving lengths were tested in these channels to determine the performance of different interleaving lengths. In general, minimum interleaving is necessary for delay-sensitive information such as voice and interactive data, whereas maximum-length interleaving is necessary to exhibit the best performance (e.g., throughput) for delay-insensitive traffic such as broadcasts and nonemergency messages.

4.3.1.1 Minimum Interleaving

Figures 4.3(a) and (b) show the performance of the three MIL-STD-188-110A waveforms in CCIR Poor and Good channels, respectively, with minimal time delay and no interleaving. Note that in these and subsequent modem performance curves, the 40-dB SNR value actually represents an effectively infinite SNR. The simulation results of the parallel-tone modems exhibited a seemingly continuous flow of errors that increased somewhat during received signal level

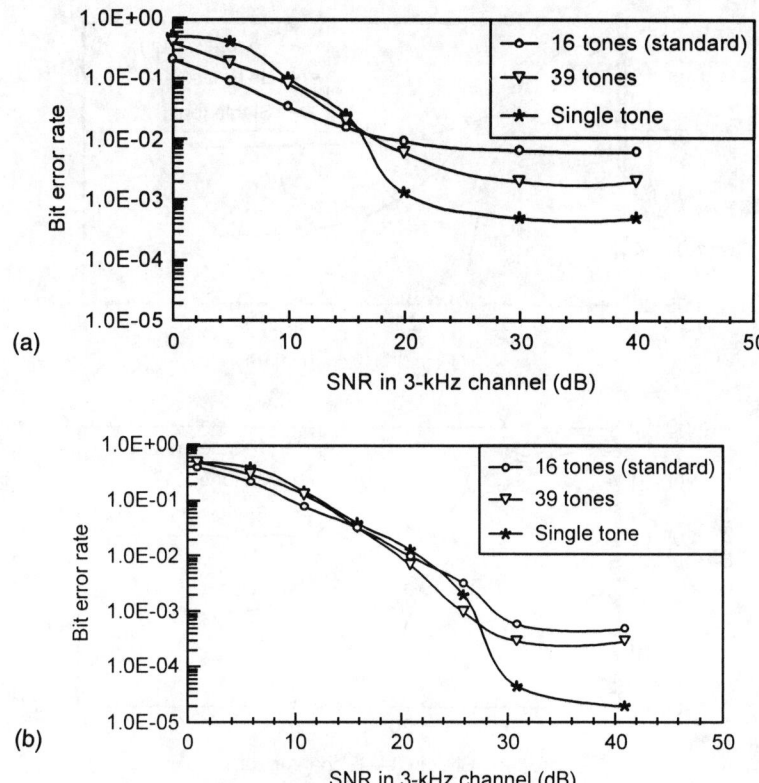

Figure 4.3 Performance of 2400-bps modems with no interleaving: (a) CCIR Poor channel and (b) CCIR Good channel.

fades. In contrast, the single-tone modem exhibited clumps of errors, or error bursts, interspersed with prolonged periods of low-error-rate reception. The 16-tone modem exhibited the poorest performance at high SNR values, probably because it lacks any form of error-control coding. Interestingly, as SNR decreases, the FEC of the other tested modems will eventually fail, producing higher error rates than that of the 16-tone design. Thus the 16-tone design can exhibit better performance in low-SNR channels when minimal interleaving is employed.

4.3.1.2 Maximum Interleaving

Figures 4.4(a) and (b) show that maximum interleaving improves the performance of the two modems that employ error-control coding. The performance

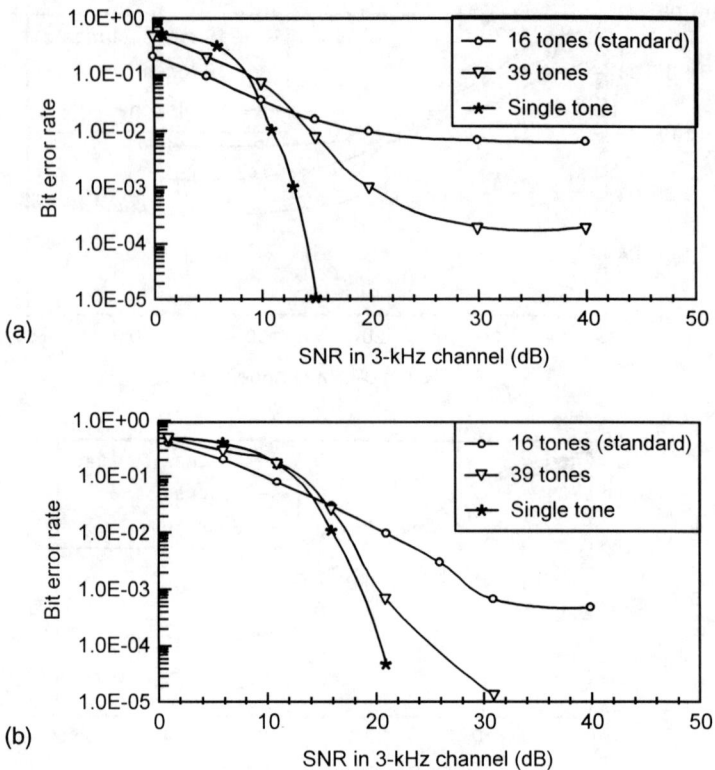

Figure 4.4 Performance of 2400-bps modems with 9.6-sec interleaving: (a) CCIR Poor channel and (b) CCIR Good channel.

curve for the 16-tone modem is also shown to highlight the superior performance possible with the newer modems when the 9.6-sec interleaving delay can be tolerated. At workable SNR values and error rates, the single-tone modem is the best performing device in both channels.

4.3.2 Performance in Impulsive Noise

The nature of the cumulative distribution function (cdf) of the measured HF noise shown in Figure 2.41(a) has significant implications. For example, the noise level in the AWGN channel exceeds 30 units in amplitude about 10% of the time, whereas the 10th percentile for the measured noise is only 15 units. In low SNR values, therefore, better performance would be expected in the presence of measured noise versus Gaussian noise of the same average power. To investigate the impact of realistic HF radio noise on modem performance, a new test channel was employed that combined the CCIR Good channel fading model with measured noise rather than AWGN. The average power of the measured noise was then adjusted in order to produce correctly calibrated curves of BER versus SNR similar to those plotted in Figures 4.3 and 4.4.

Of the three fading channel models, the CCIR Good fading channel model was selected for these modem trials because (1) it was found to be representative of the multipath propagation commonly seen in the Austral-Asian region with single-hop circuits [8], and (2) it was arguably the most challenging of the CCIR channels for modems with error-control coding. Modem tests with this "new" channel model employed the same recording that produced the results plotted in Figures 2.41(a) and (b). In fact, limited comparisons showed that similar results were obtained for other man-made noise recordings. Error rates for the two most promising MIL-STD-188-110A waveforms were then measured over the new channel using 3-min transmission periods. The same additive noise was used to obtain each measurement over the range of SNR values. The performance curves for the MIL-STD-188-110A single-tone and 39-tone modems operating at 2400 bps without interleaving are given in Figure 4.5.

Figure 4.5 shows that the two modems exhibited similar performance. Although this observation is valid relative to the average BER versus SNR values, the grouping of errors is quite different between the two modem trials. This performance difference has significant implications for the transmission of different types of traffic. For example, digital voice traffic from an LPC-10 telephone can be passed over the test channel using the 39-tone waveform at a minimum SNR values of 10 dB. When using the single-tone waveform, how-

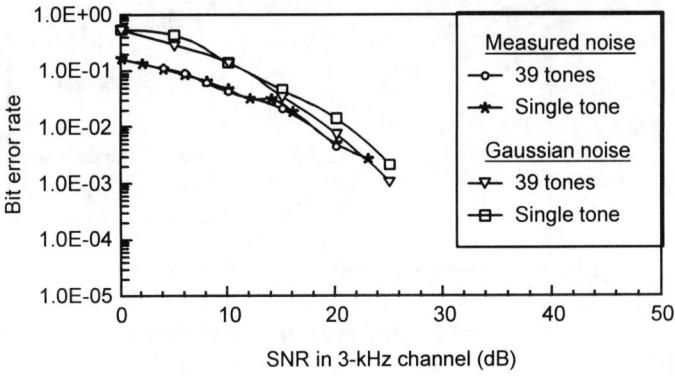

Figure 4.5 Performance of MIL-STD-188-110A 39-tone and single-tone modems at 2400 bps without interleaving in a CCIR Good channel with measured HF noise and AWGN.

ever, the transmissions become unusable at an SNR value of about 15 dB. This result occurs because the single-tone modem produces error bursts that cause the vocoder greater difficulty than the more continuous errors produced by the parallel-tone modem.

In contrast, consider longer duration data transmissions sent using an ARQ scheme to retransmit erroneous data blocks. In this case, the prolonged periods of error-free transmission characteristic of the single-tone modem would lead to fewer retransmitted packets and therefore greater throughput. Thus, the single-tone modem would be the preferred choice for the transmission of data with no maximum delay requirement.

The results displayed in Figure 4.6 demonstrate the performance improvement gained from interleaving as compared to the results presented in Figure 4.5. As postulated, the figures show that the performance of both modems was improved in the actual noise environment as compared to the results achieved in the AWGN channel. This comparison was consistent regardless of the interleaving depth. Clearly, the single-tone modem performed effectively in this environment because it required only 12-dB SNR to achieve a BER of 10^{-3}, a reduction of 6 dB from the results achieved with AWGN.

4.3.3 Performance With Cochannel Interference

4.3.3.1 Carrier-Wave Interference

Pennington [9] reported that the single-tone and 16-tone 2400-bps modems performed quite differently in the presence of cochannel interference. To verify

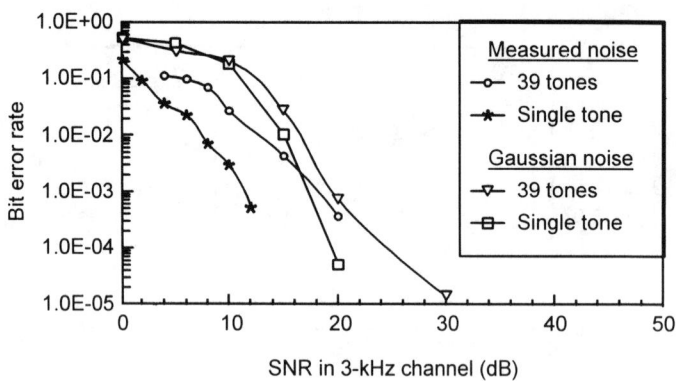

Figure 4.6 Performance of MIL-STD-188-110A 39-tone and single-tone modems at 2400 bps with 9.6-sec interleaving in a CCIR Good channel with measured HF noise and AWGN.

these results, Pennington's test for the tolerance of modems to CW interference was repeated for the three MIL-STD-188-110A modems. In this test, the interfering signal is simply a fixed-frequency sine wave with no channel distortion or noise included in the HF channel simulation. The results of this test are presented in Figure 4.7 and are in general agreement with those of Pennington. Note that only one single-tone curve is shown in the figure (for clarity's sake), because the single-tone modem proved insensitive to the frequency of the interferer. The 39-tone modem, in contrast, proved extremely sensitive to the position of the interferer. If the interferer was placed on the center of a signaling tone, the 39-tone modem performance was only just improved relative to the 16-tone modem results. However, if the interferer was conveniently placed between the signaling tones, then the 39-tone modem exhibited superior performance.

The performance of the 16-tone modem was arguably the simplest to interpret because no error-control coding or equalizer was employed. Whenever the strength of an interferer reaches about −24 dB relative to the modem ensemble power, the power of the interferer is just 11 dB below the power in the affected signal tone. In practice, this amount of interference is sufficient to cause errors in the DQPSK demodulator, compared with the 6-dB value predicted from theory. The reason for the difference between theory and practice results from the use of an FFT-based demodulator to recover each of the 16 tones. It was

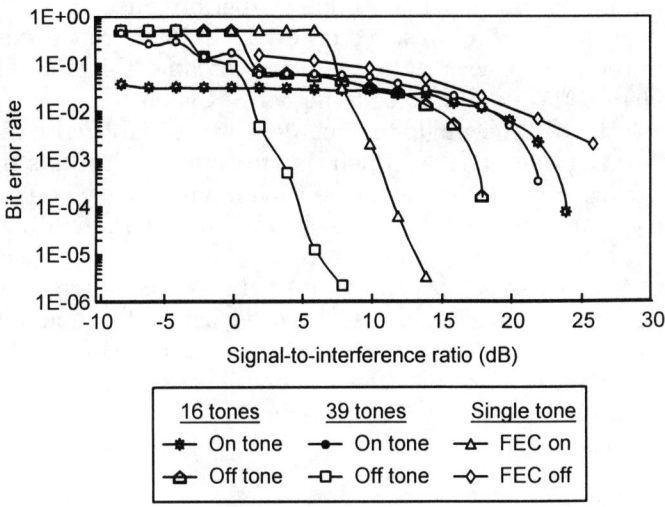

Figure 4.7 Performance of MIL-STD-188-110A modems in the presence of sine-wave interference.

conjectured that the modem FFT employed a rectangular time window because the signal is static for little more than the sampling interval.

In this case, the frequency transform of the sine-wave interferer is not concentrated into a single frequency bin, but instead becomes a sinc function spreading across all the frequencies. As the magnitude of the interferer increases, the interference-induced sidelobes will obliterate adjacent tones until the modem performance is completely degraded. When the interferer is centered on a tone, the maximum of the sidelobes falls between the signaling tones. As a result, the error rate remains at roughly at 1/32 (one tone obliterated) even at high interferer levels. Should the interferer be placed between tones, the onset of errors still occurs but requires increased interference levels. When the signal-to-interference ratio (SIR) reaches about 0 dB, the sidelobes are now coincident with the signaling tones, thus obliterating the received signal. Thus, if it were not for its error-control code, the 39-tone modem would be even more vulnerable than the 16-tone modem because each of its tones is transmitted with less power than those of the 16-tone modem. These results clearly show the power of error-control coding in improving modem performance.

The single-tone modem begins to fail at greater SNR values than the other modems for on-tone interference, but it fails completely at SIR values below about 7 dB, presumably due to equalizer failure. To test this hypothesis, the curve for the predecoder error rate was plotted from the raw 4800-bps stream. The validity of this hypothesis was also demonstrated in Figure 4.7, which revealed that the 16-tone modem outperformed the single-tone modem. It was concluded from these results that equalizer performance in the single-tone modem was degraded in the presence of low-level interference. Nevertheless, the error-control coding was sufficient to overcome this interference and improve modem better performance at higher SIR values.

Newer versions of the single-tone modem are available with interference-excision processing that behaves much like tunable notch filters in removing unwanted narrowband interference. One such device has been examined and it was found that it required about 5 sec to remove the offending interferer. Once adapted, however, this modem was capable of removing exceptionally large interferers and can remove up to four simultaneous narrowband interferers. Excision processing could also be used before the demodulator in a parallel-tone modem, as long as the error-control coding is sufficiently powerful to cope with the "erasure" of the affected tones. These results are particularly important given the significant HF interference observed in various parts of the world (see Sec. 2.5).

4.3.3.2 Pulse Interference

In addition to sine-wave interference, Pennington [9] employed pulsed sine-wave interference. In this case, similar tests were performed with a pulse length of 15 ms, the baseband interferer frequency set to 1800 Hz, and the pulse power equal to the modem signal power. The performance of the single-tone and 39-tone modems is given in Figure 4.8 for both interleaving lengths of 0.0 and 1.2 sec. Pennington's results were reproduced for the single-tone modem without interleaving. When this result was compared with the corresponding result for a 39-tone modem (rather than the 16-tone modem), the parallel-tone device was found to be the better performer. It is noteworthy that the 1800-Hz interferer coincided with one of the signaling tones of the 39-tone modem, thus making this interferer an arguably "severe" test for this modem. When a 1.2-sec interleaver was employed, so the resulting error bursts were dispersed into the symbol stream, error correction became more effective and the stronger code used on the single-tone modem prevailed.

4.3.4 On-the-Air Trials

Tilbrook and Cook [10] compared the on-the-air performance of two state-of-the-art MIL-STD-188-110A modems as part of an ALE performance evaluation for four days between Adelaide and Melbourne, Australia, in August 1993. Overall, the single-tone modem gave the better average performance regardless of the interleaving setting. It is interesting to note, however, that the 39-tone modem demonstrated better performance about 12% of the time. Clearly, there

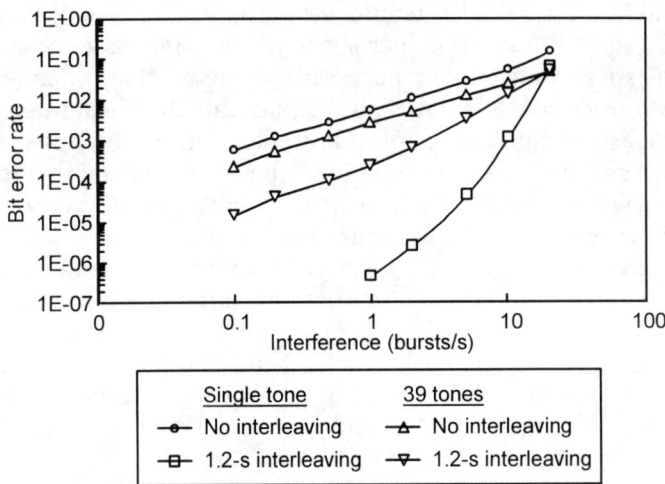

Figure 4.8 Performance of single-tone and 39-tone modems in the presence of tone burst interference with and without interleaving.

are still some circumstances when the parallel-tone waveform is superior even though it employs weaker error-control coding.

4.3.5 Other Attributes

4.3.5.1 Peak-to-Average Ratio

A parallel-tone modem waveform comprises the summation of a number of frequency-multiplexed signals, so its amplitude is continually changing. The central limit theorem [11] indicates that the sum of random variables should result in another random variable with a Gaussian probability density function (pdf). The signals are, however, not truly independent because all of the tones take on specific phases and amplitudes according to the coded modulation scheme in use. For this reason, the peak-to-average ratio is smaller than might be predicted by this simple statistical analysis. In practice, the 16-tone modem displays a peak-to-average ratio of 9 dB. This modem is normally operated at output levels such that the power amplifier is operated 6 dB below saturation to give the best compromise between signal distortion and received SNR values. The 39-tone modem has a soft limiter option designed so the peaks of the waveform are compressed to yield a peak-to-average ratio of 5 dB. The error-control code then compensates for the resulting signal distortion introduced by this peak clipping. Soft clipping was enabled when measuring all the performance curves presented in the previous figures.

A single-tone modem uses a single signaling tone, so it might be expected that the peak-to-average ratio would be unity. To achieve the near-Nyquist signaling rate of 2,400 symbols per second in a 2700-Hz voiceband channel, however, the single-tone modem necessarily employs steep channel filters that inescapably introduce intersymbol interference into the transmitted signal. This phenomenon is exhibited as amplitude fluctuation on the transmitted waveform, producing a peak-to-average ratio of about 3 dB. Given this result, the 39 parallel-tone waveform has about a 2 dB greater peak-to-average ratio than the single-tone waveform. If the transmitter is peak-voltage limited, then the increased peak-to-average ratio would yield an effective reduction of 2 dB in the available transmitted power. It is not uncommon, however, to find power transmitters designed for voice transmission that are average power limited from 3 to 6 dB below their peak power. Thus, the performance effect of the higher peak-to-average ratio exhibited by parallel-tone waveforms is dependent on power amplifier characteristics and may, in fact, be immaterial.

4.3.5.2 Equipment Specifications

Of perhaps historical significance, parallel-tone modems have been found more sensitive to exciter and receiver filter artifacts than the single-tone variants.

First, the overall spectrum of a parallel-tone waveform is nearly flat to the extremity of the signaling tones. As a result, parallel-tone modems are more adversely affected by poor frequency response. Some amateur HF equipment exhibits a frequency response that varies over the nominal 300- to 3000-Hz voice band, attenuating the peripheral tones by as much as 10 dB. Clearly, this inherent distortion would be reflected as an increase in the minimal SNR required for a given modem performance. Second, the group delay distortion of modest equipment can extend to some milliseconds. As a result, the center tones arrive at the demodulator before (or after) those at the band edges by a time difference that is a noticeable portion of a symbol duration. This delay reduces the available guard time of parallel-tone modems and degrades their multipath tolerance. The power spectral density of the single-tone waveform has minimal energy at the band edges, so it is inherently more robust to poor frequency response in the radio equipment. Also, the channel equalizer of modern single-tone designs can compensate for equipment filter frequency response and group delay distortion. While modern professional equipment has superb performance and eliminates these considerations in system design, it should be considered when older HF systems are being upgraded for use with modern HF modems.

4.3.5.3 Synchronization

Both coded modems use preambles for synchronization of both the demodulator and the block interleaver. There is no slow synchronization requirement stated for the single-tone modem in MIL-STD-188-110A. Early examples of this modem would not synchronize if the preamble was not received because of fading, noise, or simply because the receiver was not listening at the preamble time. Newer models have a slow synchronization capability, often referred to as "sync on data." As of this writing, this feature has been found to be unreliable in fading channels when long interleaving is selected. It was discovered that approximately one-third of the time, the modem failed to detect the correct bit rate of 2400-bps and preferred instead to select 75 bps; an error from which the modem never recovered. There is arguably no theoretical reason why these functions cannot operate reliably. In fact, maximum likelihood estimators are currently being implemented for symbol and interleaver synchronization and automatic bit rate determination for a DSTO single-tone modem design. In contrast, parallel-tone modems have always had reliable slow synchronization such that some applications have avoided use of the preamble entirely. In fact, it has been shown possible to detect automatically the bit rate of a parallel-tone modem and then switch to the appropriate demodulator and error-control decoder without use of the preamble.

4.3.5.4 Space Diversity Reception

Space diversity reception [6] is a proven technique to improve the performance of modems in fading channels. It is straightforward to implement space diversity when using parallel-tone modems, and it can result in a performance improvement approaching the theoretical optimum. For single-tone modems, the need for two equalizers, plus some method to produce a near-optimal combining of the signals after the equalizer, complicates implementation of space diversity reception. To date (July 1996), DSTO has not found a single-tone modem with this feature on the market. Because propagation effects best corrected by space diversity reception may endure for several hours, the lack of a modem option supporting space diversity reception could be a major disadvantage in a fixed-station scenario. Mobile stations will eventually change propagation path geometry, which reduces the need for space diversity reception.

4.4 ENHANCED HIGH-FREQUENCY MODEM DESIGN

4.4.1 Design Issues

Cook et al. [12] compared the predecoder error rates of the single-tone and 16-tone modems in CCIR channels and found that the 16 parallel-tone waveform was the better performer. Thus, it was concluded that the generally superior performance of the single-tone waveform was primarily a function of its error-control code, rather than the relative advantages of the modem waveform alone. This hypothesis is supported by comparing the predecoder error rates of the different modem types in different channels, such as the tone burst and cochannel interference tests described in earlier sections. Thus, it was concluded that the design of the coded modulation is crucial to the success of any high-performance HF radio modem.

In designing a coded modulation scheme, the fundamental parameter to maximize is the *free Euclidean distance* of the transmitted code sequence. This distance relates directly to the amount of signal distortion or noise that must perturb the waveform before the demodulator incorrectly demodulates the coded sequence. The distance can only be optimized if the design of the modulation and coding are combined and soft-decision decoding techniques are employed. Arguably, the best way to apply redundancy to a waveform is to increase the bandwidth. Bandwidth is a premium at HF, however, so it is therefore necessary to increase the constellation packing above one bit per signaling dimension. In other words, each transmitted symbol must represent multiple data bits. Standard binary error-control codes are designed to maximize the Hamming distance of the code, that is, the number of bits that separate valid code sequences. These codes are not designed to maximize the Hamming distance for a sequence of multibit symbols. Thus, the valid code sequences do not map onto the signal in a way that maximizes the Euclidean distance of the resulting coded waveform.

Ungerboeck [13] in his landmark paper showed how to construct trellis-coded modulation (TCM) schemes that maximize the desired Euclidean distance for multibit symbols received in AWGN. This development led to the work of Schlegal and Costello [14], which addressed multibit symbol coding for Rayleigh fading channels. These fading channel codes maximize the effective length of the code, that is, the number of symbols that need to be perturbed before errors result, while still providing near-optimal Euclidean distance. In practice, codes selected on this basis provide a few decibels of improvement over Ungerboeck codes [15] when used in the advanced HF modem described below.

The first application of TCM to a 16-tone modem employed a simple 8-state 8PSK Ungerboeck code [12] and produced very promising simulation results. Given that the HF channel suffers fading and impulsive noise, it was decided that the next step would be to employ a code with sufficient free distance to correct symbols that had been completely obliterated. This decision led to a digital signal processor (DSP) implementation of the so-called "TCM-16" modem. The TCM-16 used a 256-state Schlegal-Costello code in conjunction with a novel convolutional interleaver and careful mapping of the data onto the transmitted waveform [15]. The resulting modem has consistently outperformed other modems over CCIR channels.

We discussed earlier the fact that HF noise is non-Gaussian and it is, therefore, important to consider the design of the detector metric from the outset. In this regard, most communication textbook detectors are far from optimal for real-world situations. Giles [16] developed the theory for five purpose-designed detection metrics and tested their performance in measured noise using the TCM-16 coded modulation modem. Using his *adaptive* Gaussian metric, he was able to demonstrate a significant 6-dB reduction in required SNR for the same BER performance. This metric divides the conventional distance-squared metric by an estimate of the variance for all the symbols in the modem frame. Thus, as the variance increases as a result of a noise burst, the affected symbols are effectively erased (in a maximum likelihood sense), and they therefore contribute less than half the signal path error of the standard distance-squared metric.

Similarly, it is essential to address the vulnerability of the modem to narrowband interference. While the application of error-control coding improves the problem, it simply masks the underlying inadequacies of the detector design. Giles [17] showed that it was possible to use the adaptive Gaussian metric to correctly weight the quality of the received signal tones, thus incorporating inherent interference suppression into the demodulation/

decoding process. This metric, together with the use of a carefully designed FFT window, provided about 20-dB interference rejection performance improvement over the uncoded 16-tone modem. Since this FFT window is not an excision filter, this rejection is achieved without the time lag associated with the filter being "steered" onto the interferer (a significant result). Should it be deemed necessary, additional "resistance" to very large interferers could be achieved through the application of predemodulation excision filters.

In practice, a range of data rates is provided using the same basic waveform. Parallel-tone modems can be configured to work very well at low bit rates because the symbol timing recovery of the slow symbol rate is inherently robust at low SNR values. Gill et al. [18] showed that by changing the TCM coding scheme, it was possible to produce a modem based on the 16-tone format that outperformed all modems evaluated by DSTO at rates from 300 to 3600 bps. Despite its success, the TCM-16 modem is suboptimal because its underlying format does not fully utilize the available channel bandwidth. A modem that redresses this deficiency is outlined in the following section.

4.4.2 Enhanced Parallel-Tone Modems

An advanced 52-tone modem developed by DSTO occupies the entire HF voice-band channel of 2700 Hz and uses the same error-control code as the TCM-16 modem, culminating in a postdecoder throughput of 4800 bps. The performance of this practical 4800-bps HF modem is shown in Figure 4.9 compared with the single-tone modem at 2400 bps. The performance of the 52-tone device without interleaving shown in the figure is advantageous because it delivered

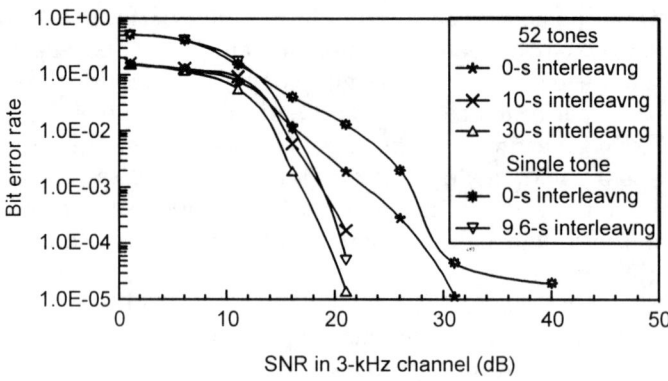

Figure 4.9 Performance of 4800-bps 52-tone modem in the CCIR Good channel with no interleaving and 10-sec interleaving.

twice the number of bits at the same error rate, but required about 2 dB less received signal power. When interleaving is used, again the 52-tone modem looks impressive, particularly when 30-sec interleaving is used. This 52-tone 4800-bps modem can be converted into a very robust 2400-bps modem by changing the coding rate from rate 2/3 to rate 1/3. Codes for D8PSK have not been published at this rate, so Gill [19] undertook to search for a suitable family of fading channel codes. The result of his efforts is shown in Figure 4.10. The ability of this strong error-control code to cope with multipath fading is quite apparent in this figure.

The 52-tone modem has been implemented in software using the TMS320C40 chipset. The modem incorporates the impulsive noise and in-band interference rejection techniques perfected in the TCM-16 modem. Space diversity reception and double-sideband operation are also supported. No space diversity BER/SNR plots for the 52-tone modem were available at the time of this writing. Nevertheless, a performance curve for the TCM-16 modem with space diversity is presented in Figure 4.10 to indicate the value of this feature when used with all the other enhancements already mentioned. A rate 3/4 16PSK coded modulation has also been applied to the 52-tone modem to give a usable throughput of 7200 bps on single sideband, or 14,400 bps on double sideband, albeit at higher error rates than the 4800-bps mode.

For data transmission, versions of the 52-tone modem have also been developed that have embedded ARQ capability. ARQ works by dividing the

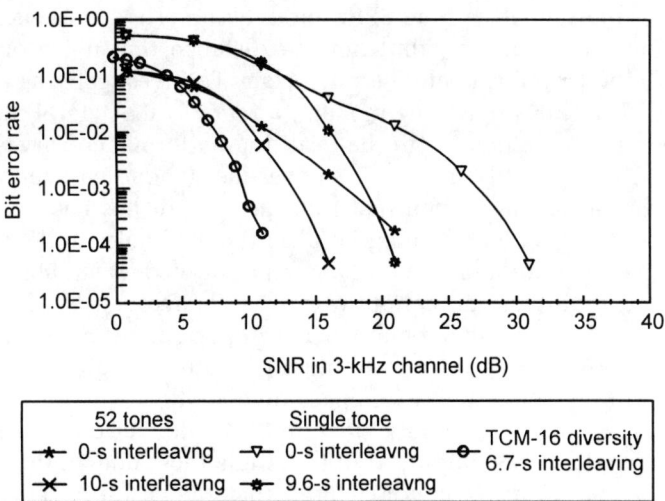

Figure 4.10 Performance of the 2400-bps 52-tone modem in the CCIR Good channel versus single-tone modem performance.

data stream into blocks and automatically retransmitting erred blocks. The parallel-tone structure of the modem is exploited such that blocks of data comprise just two modem frames (208 bits), thus making synchronization straightforward without the use of framing overhead bits. ARQ requires a reverse channel for block acknowledgments which, in the simplest case, can be provided by full-duplex operation giving bidirectional error-free throughput of up to 4400 bps.

A more ambitious objective would be to design a modem that formed the core of a half-duplex ALE system providing about 3000 bps in each HF link direction. This integrated ARQ modem has a much higher theoretical throughput than that produced by the simple application of a data link protocol (see Chapter 8) over standard HF modems. This higher throughput would be achieved because the system would minimize signaling overhead and retain frame synchronization between transmissions. An adaptive code-rate selection between 600 and 7200 bps would form an integral part of the signaling protocol to suit prevailing channel conditions.

4.4.3 Enhanced Single-Tone Modems

Single-tone modems have also benefited from recent research findings. First, the error-control code on the MIL-STD-188-110A single-tone modem is not well designed to suit fading channels. It uses a rate 1/2 convolutional code on a raw 4800-bps stream originating from 8PSK modulation. This code maximizes the number of bits it can correct, but not the distance that noise or distortion must perturb the waveform before errors occur. To investigate the potential of applying TCM to this single-tone modem, a rate 2/3 64-state 8PSK Schlegal-Costello code was employed with the resulting performance curves shown in Figure 4.11. The 4- to 5-dB improvement over the standard single-tone modem is substantial, particularly given that link throughput has been increased by 800 bps. This result is encouraging because the available implementation of the single-tone demodulator used a large amount of the available processing capacity. If the additional throughput were used for an *outer block code*, useful additional robustness would be obtained. It is important to note that the configuration just described uses the same amount of processing as the 14,400-bps dual-sideband 52-tone modem with space diversity.

The application of the rate 3/4 16PSK code, used in the 7200-bps 52-tone modem, has been employed in the single-tone modem to give a useful throughput of 4800 bps. It is conjectured that the resulting modem would perform just worse than the standard modem, but would provide twice the throughput in a fading channel. This code would, however, give substantially poorer immunity to impulsive noise and interference.

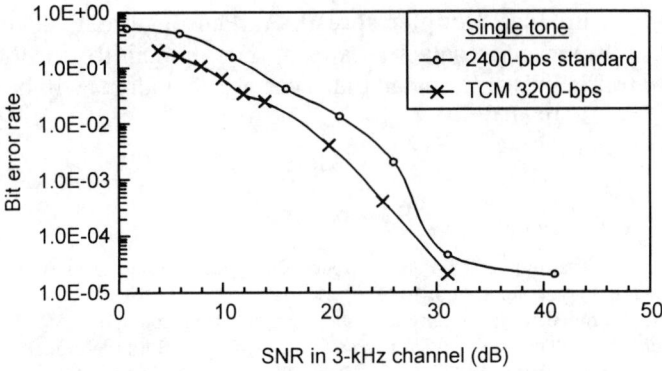

Figure 4.11 Performance of 3200-bps experimental single-tone modem versus performance of the standard 2400-bps single-tone modem in a CCIR Good channel.

Another avenue pursued to achieve improved HF modem performance was the use of a completely different detector structure. The most efficient detector for single-tone modems in fading channels is a maximum likelihood sequence estimator. Jayasinghe [20,21] has described BPSK, QPSK, 8PSK, and 16QAM versions of a modem using this type of detector that gives raw bit rates of 2400 to 9600 bps, including training sequences. Simulation results for the basic modem are quite promising, but two problems are still to be resolved before a practical modem can be built. The first problem results from implementation complexity, particularly when dense signal constellations are used. The second problem arises from the need for sharp channel filters to contain the signal within the 2,700-Hz band and provide partial-response signaling. These filters extend the channel impulse response to about 20 symbols, which mandates the use of a reduced-state detector. While this extension only marginally degrades the raw modem performance, it greatly reduces the coding gain of any TCM scheme operating on the waveform [21], thus reducing the potential for the complete detector.

4.5 CONCLUSIONS

Recent research indicates that parallel-tone modems offer significant future performance improvements over current MIL-STD-110A modems. These improved parallel-tone modems can provide the features needed for many time-critical applications, namely, high data rates and good SNR/BER performance, with minimal interleaving and the associated transmission delays. Single-tone waveforms will continue to be of interest for those concerned with communicating over paths with high Doppler spread. As Pennington stated in his 1989 paper [1], it is important to select an appropriate modem to suit the expected channels and traffic. It is anticipated that a variety of modems will be developed to fill these diverse needs.

References

[1] Pennington, J., "Techniques for Medium-Speed Data Transmission Over HF Channels," *IEE Proc.*, Vol. 136, Part I, No. 1, Feb. 1989, pp. 11–19.
[2] Department of the Army, Information Systems Engineering Command, *MIL-STD-188-110A: Interoperability and Performance Standards for Data Modems*, Philadelphia, PA: Naval Publications and Forms Center, Attn. NPODS, Sep. 1991. Available on the Internet at URL ftp://tracebase.nmsu.edu/pub/hf/pubs/mil—std—188—110a.
[3] CCIR, "HF Ionospheric Channel Simulators," *Report 549-2, Reports to the CCIR*, 1986, Vol. III, Part A, Geneva: International Radio Consultative Committee, 1986, pp. 59–67.
[4] CCIR, "Use of High Frequency Ionospheric Channel Simulators," *Recommendation 520-1, Recommendations of the CCIR*, 1986, Vol. III, Part B, Geneva: International Radio Consultative Committee, 1986, pp. 57–58.
[5] Proakis, J. G., *Digital Communications*, New York: McGraw-Hill Book Company, 1983, p. 140.
[6] Schwartz, M., W. A. Bennet, and S. Stein, *Communication Systems and Techniques*, New York: McGraw-Hill Book Company, pp. 418–421.
[7] Proakis, J. G., *Digital Communications*, New York: McGraw-Hill Book Company, 1983, pp. 357–385.
[8] Priess, M., "The Measurement and Analysis of Several HF Links Using the DSTO Replay Simulator," *Proc. 4th Int. Symp. DSP for Communication Systems*, Perth, Australia: Cooperative Research Centre for Broadband Telecommunications Networking, Sep. 1996.
[9] Pennington, J., "Interference Tests on HF Modems," *HF Communications Systems and Techniques*, IEE Conference Publication, 1987, pp. 318–320.
[10] Tilbrook, J. R., and S. C. Cook, "An On-Air Evaluation of ALE for High Speed Data Traffic," *IEEE MILCOM 94 Conf. Proc.*, Ft. Monmouth, NJ: IEEE, 1994, pp. 864–869.
[11] Papoulis, A. J., *Probability and Random Processes*, New York: McGraw-Hill Book Company, 1965, p. 266.
[12] Cook, S. C., et al., "Error-Control Options for Parallel-Tone HF Modems," *MILCOM 93 Conf. Proc.*, Boston: IEEE, 1993, pp. 933–938.
[13] Ungerboeck, G., "Channel Coding with Multilevel/Phase Signals," *IEEE Trans. Info. Theory*, Vol. IT-28, No. 1, 1982, pp. 55–67.
[14] Schlegal, C., and D. J. Costello, "Bandwidth Efficient Coding for Fading Channels: Code Construction and Performance Analysis," *IEEE J. Selected Areas Comm.*, Vol. 7, pp. 1356–1368.
[15] Cook, S. C., M. C. Gill, and T. C. Giles, "A High-Speed HF Parallel-Tone Modem," *Proc. 6th Int. Conf. HF Radio Systems and Techniques*, York, UK, July 1994, pp. 175–181.
[16] Giles, T. C., "Data Detection for a High Frequency Radio Modem in Impulsive Noise," *Proc. 3rd UK/Australian Int. Symp. DSP for Communication Systems*, University of Warwick, Coventry, UK, Dec. 1994.
[17] Giles, T. C., "A High Speed Modem with Built-In Noise Tolerance," *Proc. 6th Int. Conf. HF Radio Systems and Techniques*, York, UK, July 1994, pp. 164–169.
[18] Gill, M. C., S. C. Cook, T. C. Giles, and J. T. Ball, "A 300 to 3600 bps Multi-Rate HF Parallel-Tone Modem," *MILCOM 95 Conf. Proc.*, San Diego, CA: IEEE, 1995, pp. 1066–1070.
[19] Gill, M. C., *Trellis-Coded-Modulation for the DORIC Program: HF Channels Modem Technol-*

ogy, Ph.D. thesis, DSTO-RR-0041, Electronics and Surveillance Research Laboratory, Australian Department of Defense., University of South Australia, 1997.
[20] Jayasinghe, S. G., and A. P. Clark, "The Performance of a Modem Using 8-PSK Coded Modulation over HF Radio Channels," *Proc. IEE Conf. HF Radio*, Edinburgh, UK, July 1991, pp. 227–232.
[21] Jayasinghe, S. G., "A High-Speed Serial HF Modem for Use in Adaptive Rate Control," *MILCOM '92 Conf. Proc.*, San Diego, CA: IEEE, October 1992.

Automatic Link Establishment Signal Structure   5

5.1 INTRODUCTION

The purpose of automation in HF radio is to improve communication system performance while reducing the need for skilled manual operators, who are increasingly difficult to find. This ambitious goal requires both the establishment of a link, or linking, and the passing of traffic over challenging HF channels without manual effort. As noted previously, modern HF modems can pass data traffic over channels whose signal fading, noise, and interference characteristics render voice communications impossible, although skilled operators can often pass CW (Morse code) traffic through such channels.

In this chapter, we begin working our way up through the layers of protocols that build upon HF modems to provide nearly transparent communications over HF channels. Although a wide variety of HF protocol suites has been developed over the past decade, we focus our discussion on the U.S. standard technique because of its longevity and worldwide acceptance. Our journey "up the protocol stack" commences with techniques for automatically establishing HF links that are then used for voice or data communications.

Automatic link establishment (ALE) must equal or improve on the classic voice technique for establishing links, that is: "This is SAM, calling JOE, over." Note that this simple voice protocol quickly conveys the identity of the calling party, the party being called, and the end of the transmission. ALE works in much the same way, in that the identities of the calling and called parties are sent along with a few extra protocol bits to distinguish the *call sign* of the calling station from the call sign of the called station. To keep link establishment transmissions as short as possible, *station addresses* (the ALE term for call signs) are sent in variable-length fields in a standardized *ALE frame*. The ALE frame is an *allowable* sequence of *ALE words* (forming an *ALE Sentence*).

The basic quantum of address size could be as short as a single character. However, this choice would result in a more complex error-correction scheme, so the basic ALE word size in MIL-STD-188-141A [1] and FED-STD-1045 [2] was chosen to be three characters, along with a three-bit preamble that specifies the significance of those three characters. The three-character ALE word transmissions used by ALE protocols to automate, "This is SAM, calling JOE, over," that is, automatic HF link establishment, must be accurately and reliably communicated. ALE technology meets this challenging requirement with a robust combination of signaling techniques, including forward error correction (FEC) coding, interleaving, redundancy, and in-band diversity. This chapter (see Figure 5.1) presents the ALE signal structure that was developed to provide this robust communications capability, including (1) the ALE word structure, (2) coding, including FEC, (3) synchronization, (4) addressing, and (5) the ALE frame structure. Each of these subjects is addressed in the following sections.

A common requirement in HF networks is to address and interoperate simultaneously (or nearly simultaneously) with multiple stations. A prearranged group of stations with a common address is termed a *net*, and the common address in a *net call* is a *net address*. A group of stations whose interaction has not been prearranged, that is, a collection of stations lacking a predetermined common address, is termed a *group*. As with the net, a group is addressed in a *group call* by a *group address*. Net and group addressing structures are defined later. Note that any HF station may have more than one address identifying itself as an individual station, within a group, or within a net. Protocols for linking and networking with nets and groups are described in Chapter 6.

5.2 LINKING PERFORMANCE

The principal performance measure of an ALE system is the probability of successfully establishing links in degraded channels of known characteristics. This metric provides a useful benchmark for the evaluation of the ALE system in meeting its goal of automating the work of skilled human operators.

5.2.1 Minimum Required Performance

Performance criteria for ALE implementations are established in the standards as shown in Table 5.1. These criteria are stated in terms of minimum linking probability under specified channel conditions as well as a time limit for link establishment. This time limit is 14 s for a simple station-to-station call. The channel conditions are based upon the well-known Watterson model [3] for sky-wave links, with selected parameters for the CCIR Good and Poor channels as well as a purely additive white Gaussian noise (AWGN) channel.

Automatic Link Establishment Signal Structure 193

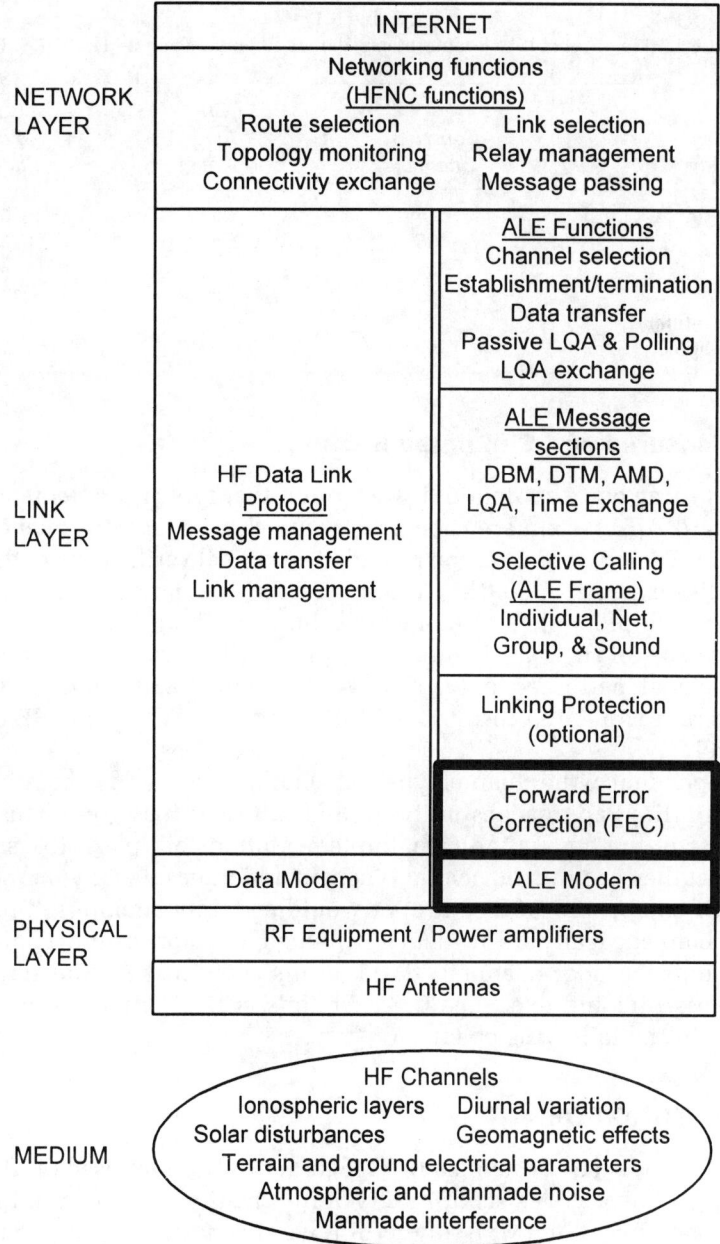

Figure 5.1 ALE Modem and FEC in the hierarchical-layered view of HF communications.

Table 5.1
ALE Linking Probability Requirements Versus Channel Type 3-kHz SNR (dB)

Linking Probability	Gaussian Channel	CCIR Good Channel	CCIR Poor Channel
0.25	−2.5	0.5	1.0
0.50	−1.5	2.5	3.0
0.85	−0.5	5.5	6.0
0.95	0.0	8.5	11.0
Multipath (ms)	0.0	0.52	2.2
Doppler spread (Hz)	0.0	0.10	1.0

5.2.2 Measured and Simulated Results

Through testing and simulation, it was found that typical implementations of the standard ALE system are capable of somewhat better performance than is required by Table 5.1. Simulation provided empirical verification of the linking probabilities expected from the military and federal standards. Figures 5.2(a), (b), and (c) are plots of the linking probability, P(link), for a simulated ALE modem versus 3-kHz SNR. Figure 5.2(a) plots P(link) versus 3-kHz SNR for a received signal perturbed only by AWGN. Figures 5.2(b) and (c) show the measured and simulated results for CCIR Good and CCIR Poor HF channels, respectively.

Independent of the channel type, both measurement and simulation results show that P(link) approaches unity for 3-kHz SNR values greater than 12 dB. As the data in the figures indicate, implementations of this ALE system work so well that they can link on channels considered unusable for voice communications. Of course, the ALE controller would certainly attempt to first link on the best channels, if any are available! Although operators cannot communicate by voice on such poor channels, these results show that ALE technology can nevertheless link and exchange the operator's critical orderwire messages as described in the following chapter.

5.3 ALE WORD FORMAT

The ALE frames used to establish links (Sec. 1.1.3) are composed of ALE words. The basic ALE word consists of 24 bits of information as shown in Figure 5.3. The most significant bit (MSB) of the ALE word is designated W_1 and the least significant bit (LSB) is designated W_{24}. This 24-bit ALE word is divided into a 3-bit *preamble* followed by a 21-bit data field, which *normally* contains three individual 7-bit character fields as shown in the figure.

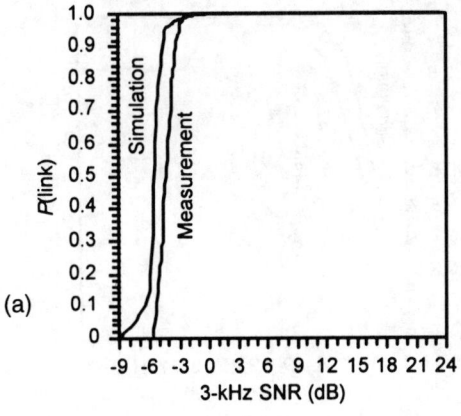

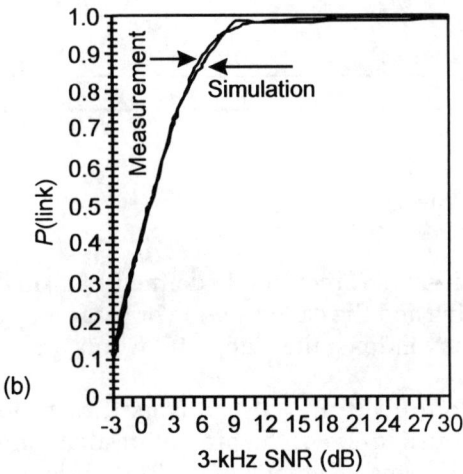

Figure 5.2 Channel performance models: (a) Gaussian channel, (b) CCIR Good channel, and (c) CCIR Poor channel.

5.3.1 Word Types

The leading three bits of an ALE word, W_1 through W_3, are, as mentioned, the preamble bits, P_3 through P_1, respectively. (Note the unfortunate reversal in bit numbering.) These preamble bits identify each ALE word as one of eight possible word types, each with a unique function in the ALE protocol. In the following discussion, each ALE word has been underlined to distinguish it from sur-

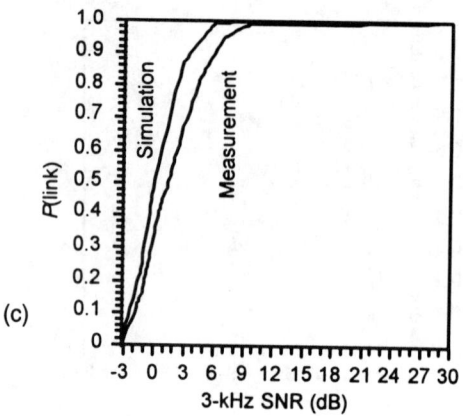

(c)

Figure 5.2 (continued)

MSB sent first	Preamble			Character 1		Character 2		Character 3		LSB sent last
	P_3	P_1	$C1_7$	$C1_1$	$C2_7$	$C2_1$	$C3_7$		$C3_1$	
	W_1	W_3	W_4	W_{10}	W_{11}	W_{17}	W_{18}		W_{24}	

Figure 5.3 ALE word structure.

rounding text. In addition, a *direct* link is defined as a single HF propagation path between the calling and the called station or prearranged group of stations or net. Given these conventions, the eight ALE words are

TO   A TO Word (P_3 P_2 P_1 = 0 1 0) is used as a *routing* designator to identify the address of the present destination station (or net) for whom the frame is intended. The TO word itself contains the first three characters of a station address. For extended addresses (more than three characters in length), the additional address characters are contained in alternating DATA and REP words (see below), which immediately follow the TO word. The sequence is TO, DATA, REP, DATA, and REP, using only as many words from this sequence as are required to contain the address. A maximum of five address words (15 characters) is permitted in the ALE protocol.

TIS   A TIS ("this is") word (P_3 P_2 P_1 = 1 0 1) is used as a routing designator that identifies the address of the station that is transmitting the frame. Either TIS (or TWAS, see below) is used to conclude all ALE frames. TIS indicates the continuation of a protocol or handshake, and directs, requests, or invites (depending on the specific protocol) responses or acknowledgments from other stations. The TIS word contains the first three characters of a calling station's address in its 21-bit (three-character) field. For extended addresses, the additional address characters are contained in alternating DATA and REP words, which immediately follow the TIS word, exactly as described for the TO word above.

TWAS   A TWAS ("this was") word ($P_3 \, P_2 \, P_1 = 0\ 1\ 1$) is used as a routing designator exactly like TIS, with the following variations. The TWAS word indicates the conclusion of an ALE protocol or handshake, and rejects, discourages, or does not invite (depending on the specific protocol) responses or acknowledgments from other stations. TIS and TWAS are never used in the same ALE frame; they are defined as mutually exclusive ALE protocol words.

DATA   The DATA preamble ($P_3 \, P_2 \, P_1 = 0\ 0\ 0$) is a special designator used to extend the data field of any previous ALE word type (except DATA itself) or to convey information in a message. When used with the routing designators TO, FROM, TIS, or TWAS, the DATA word performs address extension from three characters to six, nine, or more (in multiples of three) when alternated with REP words (see below). The limit for address extension is a total of 15 characters.

REP   The REP ("repeat") preamble ($P_3 \, P_2 \, P_1 = 1\ 1\ 1$) is a special designator used to duplicate any previous preamble function or ALE word meaning while changing the 21-bit data field contents (bits W_4 through W_{24}). The REP word is used to preclude uncertainty and errors in interpreting the ALE word stream, by forcing each new ALE word in an ALE frame to have a different preamble from its predecessor. That is, if the meaning of an ALE word is different from the preceding word, *even if the data field is identical to that in the previous word*, the preamble must change to clearly designate an ALE word change. Therefore, if adjacent words carry the same preamble, the later is known to be a repeat of the former.

CMD   The CMD word ($P_3 \, P_2 \, P_1 = 1\ 1\ 0$) is a special orderwire designator that is used for HF system-wide coordination, command, control, status, information, interoperation, and other special purposes.

CMD is used in any combination between ALE stations and operators. CMD is an optional designator that is used only within the message section of the ALE frame.

THRU   The THRU word ($P_3\ P_2\ P_1$ = 0 0 1) is used in the *scanning call* section of the *calling cycle* (see below) only with group call protocols as described in Chapter 6. In this case, the THRU word is used alternately with REP, if necessary, to indicate the first ALE word in the address of each station in a group of stations to be collectively addressed. (The THRU word was originally intended for future implementation of indirect and relay protocols, but these functions have been implemented at the network layer in MIL-STD-187-721C as discussed in Chapter 9.)

FROM   The FROM word ($P_3\ P_2\ P_1$ = 1 0 0) is an optional designator that can be used to identify the address of the transmitting station early in the ALE frame. If FROM is used, the whole address of the transmitting station must be sent. If the transmitting station has an extended address, DATA and REP words are used to send the complete address using the convention established for the TO address structure defined above. The FROM word should be used only once in each ALE frame, and it is used only immediately preceding a CMD in the message section. Under direction of the operator or controller, it is used to provide *quick identification* of the transmitting station when the normal ALE frame conclusion will not be sent for a considerable time, such as when a long message section is carried by an ALE frame. (The FROM word was originally intended for implementation of indirect and relay protocols, but these functions have been implemented at the network layer in MIL-STD-187-721C as discussed in Chapter 9.)

5.3.2 Valid Sequences

The valid sequences of these eight ALE words types (preambles) are shown in Figure 5.4. Invalid sequences are usually assumed to indicate uncorrected errors, and therefore cause an ALE-equipped receiver to abort reception of an ALE frame or discontinue a linking attempt.

5.4 CODING

Communication over adverse HF channels must overcome short error bursts due to impulsive noise as well as longer bursts due to fading and persistent

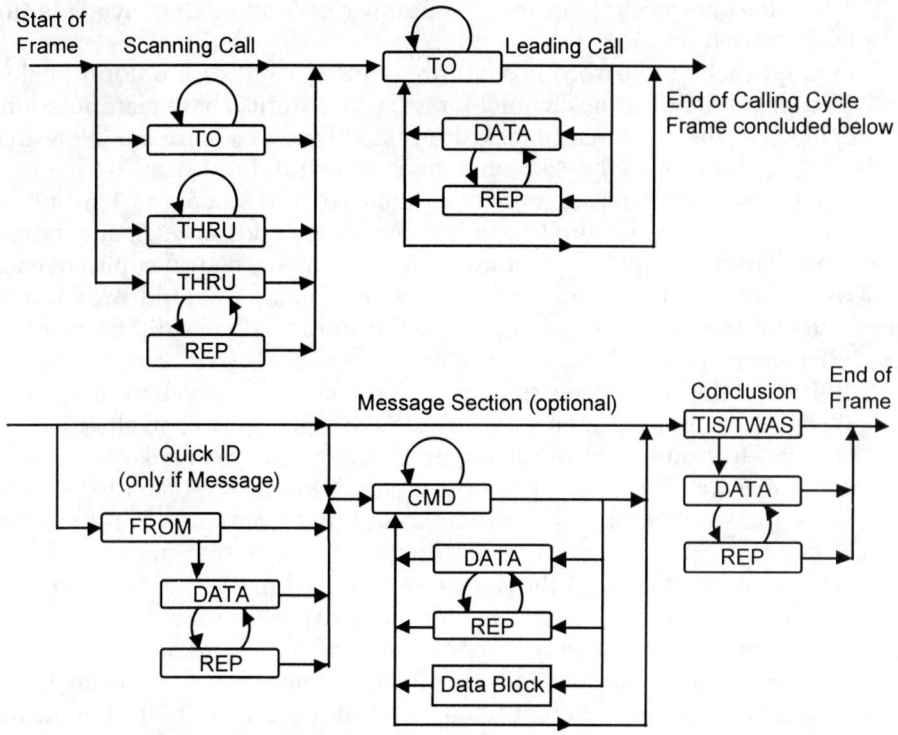

Figure 5.4 Allowed preamble (ALE word) sequences.

interference. In addition, strong interference can produce heterodyne tones within the passband of the ALE modem with sufficient power to overwhelm the ALE tones. This combination of channel disturbances is addressed by a matching combination of error dispersion and correction techniques.

5.4.1 Overview

Because HF channel conditions cannot be known a priori, each ALE word to be transmitted is expanded from 24 to 147 bits to help the receiver to overcome the error-producing effects of poor or marginal HF channels. First, the 24-bit ALE word is split into two 12-bit halves, and then each word half is encoded using the extended (24, 12, 3) Golay code. The two 24-bit *Golay words* are interleaved bit by bit, resulting in a doubling from 24 to 48 bits. A *stuff bit* (always a 0) is appended to these 48 bits so that the resulting 49-bit encoded ALE word does not occupy an integral number of 3-bit modem symbols (for reasons discussed later).

Finally, each 49-bit word is sent three times in succession, for a total of 147 bits transmitted over the channel for each ALE word. These manipulations are depicted in the functional block diagram shown in Figure 5.5. Note that A1 is the first bit sent to the ALE modem as shown in the figure.

The channel transmission rate of a standard ALE modem is 375 bps, so the triple-redundant ALE word transmission consumes 392 ms of channel time. The receiver performs majority voting among the three received copies of each ALE word, so error bursts up to about 130-ms duration should produce no errors after voting. Any errors that survive the voting will be split more or less evenly between the two Golay words by the de-interleaver, minimizing the probability that the error-correcting power of either Golay decoder is exceeded.

When an interfering tone overwhelms the signal power in the modem, it is possible to excise that interferer using a very narrowband notch filter. Unfortunately, the filter could also excise one of the ALE modem tones. The stuff bit ensures that groups of three adjacent bits do not appear in the same symbol (tone) in each repetition, which would result in the same tone being sent in all three repetitions. If the same bits appeared in the same tone in each repetition and that tone were excised, the majority voting would provide no benefit. However, by effectively moving the symbol boundaries through the ALE word for each repetition, in-band diversity results with each bit likely appearing in at least two different frequencies in the modem spectrum during the triple-redundant word. This "49th bit" improves the probability that majority voting can correctly recover all bits of the encoded ALE word despite excision of an interfering tone.

To summarize the ALE error-correction suite, the ALE standard consist of two pairs of cooperating techniques:

- Redundancy and majority voting try to overcome error bursts in marginal channels by spreading the ALE word over 392 ms in time. In-band diversity usually spreads the information of each bit over more than one tone in the ALE modem spectrum, improving the effectiveness of the majority voting.
- The second line of defense in eliminating the effects of errors is the Golay FEC coding, which can correct up to three errors in each 24 bits of output from the majority voter (i.e., 12.5%). Bit-by-bit interleaving within each encoded ALE word attempts to spread error bursts over two Golay words so that bursts of up to six errors per majority word can be corrected. Incidentally, counting the number of unanimous votes in the majority

Automatic Link Establishment Signal Structure 201

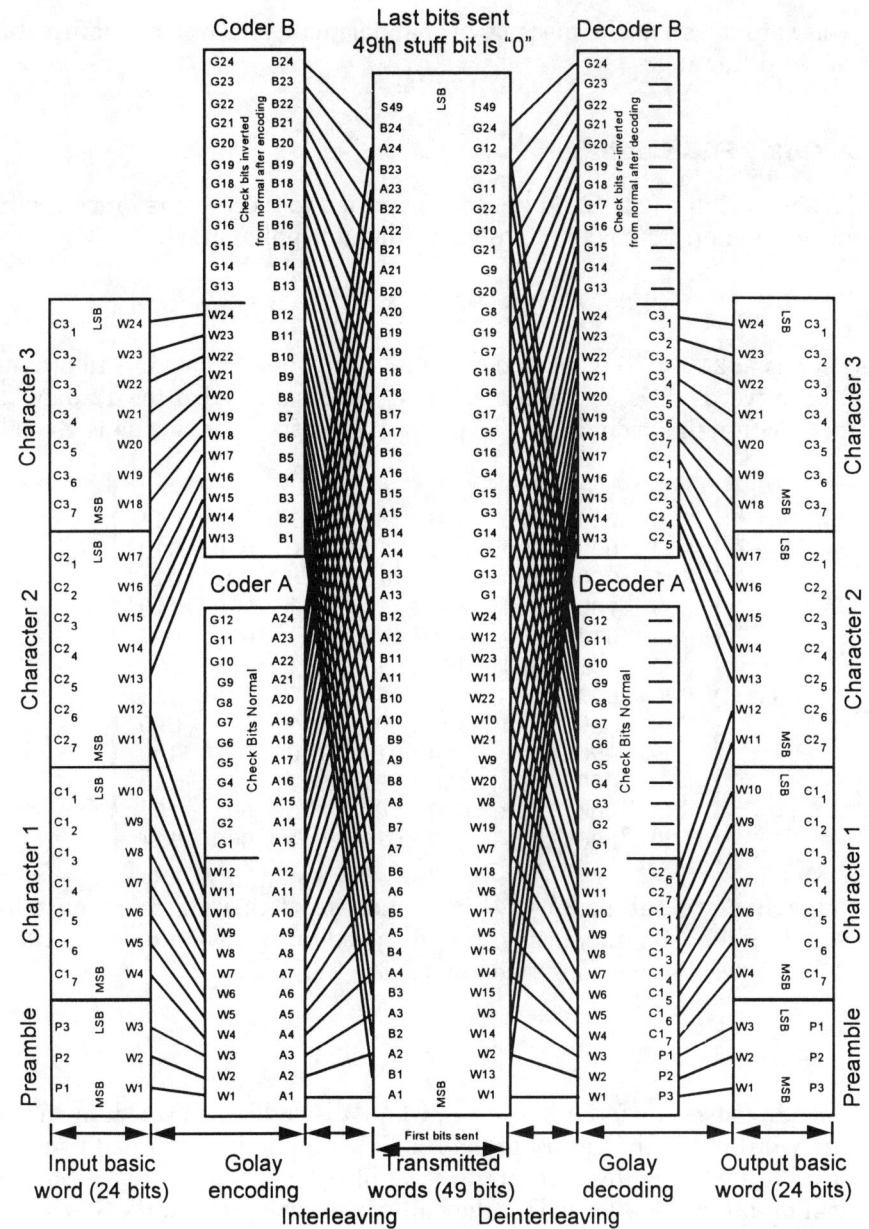

Figure 5.5 ALE word bit coding and interleaving.

voter provides a useful measure of channel quality, as well as confirmation of word framing.

5.4.2 Golay FEC Coding

The extended Golay (24, 12, 3) FEC code applied to ALE words for correction of transmission errors has the following generator polynomial:

$$g(x) = x^{11} + x^9 + x^7 + x^6 + x^5 + x + 1 \qquad (5.1)$$

This code is a (24, 12, 3) block code; that is, it encodes 12 data bits to produce 24-bit code words. It is also a systematic code, meaning that the 12 data bits are intact within the code word. The generator matrix for this code is given by

$$G = \begin{bmatrix}
100 & 000 & 000 & 000:101 & 011 & 100 & 011 \\
010 & 000 & 000 & 000:111 & 110 & 010 & 010 \\
001 & 000 & 000 & 000:110 & 100 & 101 & 011 \\
000 & 100 & 000 & 000:110 & 001 & 110 & 110 \\
000 & 010 & 000 & 000:110 & 011 & 011 & 001 \\
000 & 001 & 000 & 000:011 & 001 & 101 & 101 \\
000 & 000 & 100 & 000:001 & 100 & 110 & 111 \\
000 & 000 & 010 & 000:101 & 101 & 111 & 000 \\
000 & 000 & 001 & 000:010 & 110 & 111 & 100 \\
000 & 000 & 000 & 100:001 & 011 & 011 & 110 \\
000 & 000 & 000 & 010:101 & 110 & 001 & 101 \\
000 & 000 & 000 & 001:010 & 111 & 000 & 111
\end{bmatrix} \qquad (5.2)$$

The following two subsections describe the use of this matrix in encoding and decoding the ALE bit streams. A high-speed implementation of a suitable encoder and decoder was described by Johnson [4].

5.4.2.1 Encoding

ALE encoding uses the formula $\mathbf{x} = \mathbf{u}\,\mathbf{G}$, where the code word $\mathbf{x}$ is derived from the data word $\mathbf{u}$ and the generator matrix $\mathbf{G}$. Encoding is performed using the $\mathbf{G}$ matrix by summing (modulo 2) the rows of $\mathbf{G}$ for which the corresponding information bit is a "1". Note that the half of the generator matrix $\mathbf{G}$ is simply a 12 × 12 identity matrix $\mathbf{I}_{12}$. The right half of $\mathbf{G}$ generates the parity check-bit portion of the code word and is commonly called the $\mathbf{P}$ matrix. Thus, we can write $\mathbf{G} = [\mathbf{I}_{12}\ \mathbf{P}]$. Note that the parity check bits of the second Golay word are *inverted* before they are interleaved and sent as depicted in Figure 5.5.

5.4.2.2 Decoding

The introduction of errors in a communications channel is typically modeled as an error vector **e**, which is added (modulo 2) to the code word originally sent to produce a received word **y** = **e** + **x**. In block codes such as the Golay code used in the ALE standard, the decoder computes a syndrome **s** = **y H**^{T}, where **H** = [**P**^{T} **I**$_{12}$] is the parity check matrix derived from **G**, where the superscript "T" indicates matrix transposition. This vector-matrix multiplication simply computes the parity check bits for the received data bits and adds them (modulo 2) with the received parity check bits, giving a zero vector (**0** = [0 0 . . . 0]) result if a valid code word has been received.

Each correctable or detectable error vector **e** results in a *unique* syndrome vector **s**. Therefore when s is computed according to the preceding equation, this vector becomes an index into a look-up table to find the probable **e** vector. This **e** vector can then be added (modulo 2) to the y vector to give the corrected code word **x'**, such that **x'** = **x** when the error is correctable. If the **s** vector is not equal to the **0** vector, and **e** contains more 1s than the number of errors correctable by the Golay code, then an uncorrectable error has been detected.

5.4.2.3 Examples

To illustrate the ALE FEC process described in the preceding subsections, two examples of transmit coding are given in Figures 5.6(a) and (b). Note that the transmitted ALE words (TxALEwords) and Golay words are shown in octal representation, and that the parity check bits in the second Golay word are inverted before interleaving. Figure 5.6(a) shows the encoding process for "TO SAM" and (b) shows "TWAS AAA."

5.5 WORD SYNCHRONIZATION

The ALE system is inherently asynchronous and does not require any additional forms of system synchronization, although it is compatible with such techniques. However, the embedded timing and structure of the system provide "hooks" for achieving and maintaining word synchronization (*word sync*) during linking, orderwire, and anti-interference functions, as described later. Simulation studies have demonstrated that acquisition of correct word sync is a key factor in achieving the theoretical HF linking probabilities advertised for the ALE system.

5.5.1 Transmitter Word Phase

The ALE transmitting modulator accepts digital data from the encoder and provides modulated baseband audio to the transmitter. After the start of the

```
MIL-STD-188-141A FEC Layer Simulator

        0 DATA   1 THRU   2 TO    3 TWAS
        4 FROM   5 TIS    6 CMD   7 REP
    Select a preamble by number:  2
    Remainder of ALE word (three ASCII characters):  SAM
TxALEword = 25160315        Golay words = 25164546   03154021

Interleaved 49-bit word:
001 000 100 111 000 011 111 001 100 101 110 111 110 001 111 100 0

Triple redundant word tones in Hz (read across, then down):
  1000  750 2500 2000  750 1250 2000
  1000 2500 2250 1750 2000 1750 1000
  2000 2500  750 2500 1500 1250 2500
  1000 2000 2500 1750 1500 2000 1250
  2000  750 2000 1750  750 1500 1000
  1000 1750  750 2000 1750 1250 1000
                       (a)


MIL-STD-188-141A FEC Layer Simulator

        0 DATA   1 THRU   2 TO    3 TWAS
        4 FROM   5 TIS    6 CMD   7 REP
    Select a preamble by number:  3
    Remainder of ALE word (three ASCII characters):  AAA

TxALEword = 34060301        Golay words = 34063634   03017563

Interleaved 49-bit word:
001 010 100 101 000 000 101 001 001 010 101 100 001 011 110 000 0

Triple redundant word tones in Hz (read across, then down):
  1000 1500 2500 2250  750  750 2500
  1000 1000 1500 2250 2500 1000 1250
  1750  750  750 2250 1500 1500 2500
   750 1500 2500 2500 2250 1500 1750
   750 2250 2000  750  750 1500 2250
  1000 1500  750 1000 1500 1500 1500
  2250 1250  750 1500 2000 2500  750
                       (b)
```

Figure 5.6 Coded 49-bit word: (a) first example and (b) second example.

first transmission by a station, the ALE transmitting modulator must maintain a constant phase relationship (within the 10-ppm timing accuracy of the ALE modem) among all transmitted triple-redundant words until the final frame in the transmission is concluded. More specifically,

$$T_{(\text{later triple-redundant Word})} - T_{(\text{early triple-redundant Word})} = n \times T_{\text{rw}} \quad (5.3)$$

where $T_{(x)}$ is the event time of a given triple-redundant word within any frame, T_{rw} is the period of one triple-redundant word (392 ms), and n is any integer. This *word-phase tracking* is only required within a single continuous ALE transmission and not between separate ALE transmissions.

5.5.2 Synchronization at the Receiver

The internal timing references of the transmitting modulator and the receiver are initially independent and must therefore be synchronized. When a receiver arrives on a channel carrying ALE signaling, the receiving demodulator first achieves symbol synchronization with the transmitting modulator so that it can provide recovered digital data to the decoder. The decoder accepts this digital data and performs majority voting, de-interleaving, and FEC decoding. Before the ALE protocol state machine can be engaged, the decoder must achieve word sync. During the initial and continuing synchronization process, the following criteria determined by the decoder and ALE state machine are used to discriminate and read every ALE word: (1) Meet or exceed a threshold of unanimous votes in the two-thirds majority vote decoder; (2) successful Golay decode of both de-interleaved words (i.e., no uncorrectable errors); (3) acceptable preamble according to valid word sequences as shown in Figure 5.4, channel history, station status, expectations, and protocol; (4) acceptable bits in data field (e.g., all three characters must be from the Basic-38 subset, defined in a later section, while acquiring word sync); and (5) correct triple-redundant word phase. Note that the number of unanimous votes provides an easily adjustable BER-based signal quality discriminator. Although the remaining synchronization tests are sufficiently powerful to reject false sync in most cases, this simple scheme is essential for achieving full performance, especially when *linking protection* (see Chapter 7) is in use. A dynamically adjusted threshold is ideal, but a static threshold of 25 unanimous votes per word (of 48 votes) also works well in practice [5].

Dynamic adjustment of the Golay power is also important to achieving full performance. When acquiring word sync, the correct one/detect six mode is a good choice because it rejects many false word sync possibilities. After word sync is achieved, changing the FEC decoder to the correct three/detect four mode has been shown to minimize dropouts and the need for ALE frame retransmissions.

The acceptability of the received preamble is dependent on the recent signaling history of the stations heard, the state of the machine, the handshake(s) expected, and the protocol being used, if any. For example, a station that is awaiting calls would accept TO, if the accompanying address indicates an individual or net call (and possibly THRU or REP, if group call), as valid preambles for calls to it. The station would reject CMD as being irrelevant, because it missed the addresses at the beginning of the frame. TIS or TWAS would be tentatively accepted as a *sound* pending the necessary repetition of the sounding station address (see Chapter 6).

5.6 ADDRESSES

The ALE system employs a digital addressing structure based on the standard 24-bit (three-character) ALE word and the Basic-38 character subset. ALE stations have the capability and flexibility to link or network with single stations or with prearranged or "as-needed" groups of stations. In addition, efficient ALE operation usually requires that each station recognize and respond to multiple *self-addresses*. In fact, the standards require that ALE stations be able to store and use at least 20 self-addresses, with each self-address consisting of up to 15 characters representing any combination of individual and net calls.

Three ALE addressing methods are presented in this section: individual station, multiple station, and special modes. Protocols that use these addresses are discussed in Chapter 6. Note that certain alphanumeric address combinations may be interpreted to have special meanings for emergency or specific functions, such as SOS, MAYDAY, PANPAN, SECURITY, ALL, ANY, and NULL. Use of these combinations should be carefully controlled or otherwise restricted.

5.6.1 Character Sets

The ALE system provides and supports three hierarchical sets of characters, all of which are subsets of the ASCII characters.

5.6.1.1 Basic-38 Subset

The Basic-38 subset includes all capital alphabetic characters (A-Z) and all digits (0-9), plus two utility and wildcard symbols, @ and ?. The Basic-38 subset is used for all addressing functions as described later. A valid basic address must contain three alphanumeric characters (A-Z, 0-9) from the Basic-38 subset in any combination. In addition, the @ and ? symbols may be used for special functions, but only as described later. Digital discrimination of the Basic-38 subset should not be limited to examination of only the three MSBs, as a total of 48 bit combinations would satisfy this check. The additional 10 combinations included are invalid symbols, which, if improperly accepted, would reduce ALE linking performance.

5.6.1.2 Expanded-64 Subset

The Expanded-64 subset consists of all ASCII characters whose two MSBs are 01 or 10. This subset includes all capital alphabetic characters (A-Z), all digits (0-9), the utility symbols @ and ?, plus 26 other commonly used symbols. The Expanded-64 subset is used for the automatic message display (AMD) message function (see AMD discussion in Chapter 6).

5.6.1.2 Full-128 Set and Binary Data

The Full-128 set includes all characters, symbols, and functions available within the ASCII code. All seven-bit combinations are acceptable.

5.6.2 Individual Station Addresses

The address quantum in the ALE system is the single routing word, containing three characters, which forms the basic individual station address. This basic address word, used primarily for intranet and *time-slotted* group operations, may be extended to multiple words for increased address capacity and flexibility for internet and general use. An address that is assigned to a single station within a specific network is termed an *individual address*. If the address consists of one word (that is, no more than three characters), it is termed a *basic size* address; otherwise it is termed an *extended size* address.

5.6.2.1 Basic Address Word

The basic address word is composed of a routing preamble, such as TO or THRU, or possibly a REP word that follows TO or THRU, or a FROM, TIS, or TWAS. The preamble must be followed by three address characters from the Basic-38 subset, including the utility symbols @ and ?. The three characters in the basic individual address provide a maximum Basic-38 address capacity of 46,656 stations using only the 36 alphanumeric characters. This three-character single ALE word is considered the minimum address length. Use of one- or two-character addresses is strongly discouraged because these provide a Basic-38 address capacity of only 36 or 1,296 stations, respectively, with no significant advantages in speed, capacity, or reliability. As examples of proper usage, a minimum three-character call directed to "JIM" would be structured "TO JIM," and a shorter (discouraged) two-character call to "ED" would be structured "TO ED@" (see Sec. 5.6.4.1). Both addresses have the same size and ALE performance characteristics. In general, one-word addresses should be used only for abbreviated-address intranet and time-slotted response operations, with two-word (or longer) addresses used for internet and general intranet operations. All ALE stations should be assigned at least one single-word address for efficiency in slotted-response protocols.

5.6.2.2 Extended Address

Extended addresses provide address fields that are longer than one word (three characters), up to a maximum ALE-system limit of five words (15 characters). This 15-character capacity permits HF stations to be addressable within the Integrated Services Digital network (ISDN). Specifically, the ALE extended address word begins with an initial basic address word, such as TO or TIS. This initial word is directly followed by additional ALE words as necessary to contain the remaining address characters. These additional words employ much of the ALE-preamble sequence "DATA, REP, DATA, REP," a maximum of four additional ALE words. All address characters must be alphanumeric members of the Basic-38 subset.

Generally, all ALE stations should be assigned at least one two-word address for general use plus an additional address containing the station's assigned call sign. If a station has multiple call signs, the station should be assigned a corresponding ALE address for each sign. The recommended standard address size for intranet, Internet, and general non-ISDN use is two words. Any requirement to operate with address sizes larger than six characters must be a network management decision.

As examples of proper usage, a call to "EDWARD" would be "TO EDW DATA ARD," and a call to "MISSISSIPPI" would be "TO MIS DATA SIS REP SIP DATA PI@." The ALE addressing scheme is compatible with 32-bit Internet "dotted-quad" addresses by dropping the dots, which are not included in the Basic-38 subset. For example, the address 128.123.9.44 becomes "TO 128 DATA 123 REP 009 DATA 044." A more compact approach would also suppress implicit zeros, that is, 128.123.9.44 becomes "TO 128 DATA 123 REP 944," but this address is also an alias for "128.123.94.4" and others. This scheme is therefore strongly discouraged.

5.6.3 Multiple Station Addresses

5.6.3.1 Nets

A net is organized and managed with significant prior knowledge of the member stations, including their identities, capabilities, and, in most cases, their locations and necessary connectivities. These connectivities are the requirements for communications among net stations, which are usually determined to meet the functional objectives of the organization or organizations employing HF radio. In practice, this a priori knowledge facilitates optimization of net timing, addressing, and interchanges.

The purpose of a net call is to establish contact rapidly and efficiently with multiple prearranged stations (simultaneously if possible) by the use of a single net address. In this context, the net address is a common address assigned to all net members in addition to their individual ALE addresses.

5.6.3.2 Groups

Unlike a net, a group is not prearranged. In many cases, little or nothing is known about the stations except their individual addresses and scanned channels. Despite this lack of coordinating information, the ALE group addressing mechanism provides a means to create a new group where none previously existed. This mechanism uses a standardized protocol compatible with the other ALE protocols.

The purpose of a group call is to establish contact with multiple nonprearranged stations (simultaneously if possible) rapidly and efficiently by the use of a compact combination of their own individual addresses. A group address is formed from a sequence of the actual individual station addresses of the called stations, in the manner directed by the specific standard protocol.

5.6.4 Special Addressing Modes

The special addressing modes, which use the utility symbols @ (1000000) and ? (0111111) include the following: (1) stuffing, (2) Allcalls, (3) Anycalls, (4) wildcards, (5) self-address, and (6) null address. Each of these purposes is discussed in the following subsections.

5.6.4.1 Stuffing

The ALE basic address structure is based on single words that, of themselves, provide multiples of three characters. The quantity of available addresses within the system and the flexibility of assigning addresses are significantly increased by the use of address-character stuffing. This technique allows address lengths that are *not* multiples of three characters to be compatible with the standard address fields by "stuffing" the empty trailing positions with the utility symbol @. Either @ or @@ may be used to stuff address words with one character or two characters, respectively. These so-called "stuff-1" and "stuff-2" words are used only in the last word of an address and therefore should appear only in the *leading call*, designated T_{lc} of the calling cycle, designated T_{cc}. (These terms are explained in a later section.) As an example of proper usage, a call to the address "MIAMI" would be "TO MIA DATA MI@."

5.6.4.2 Allcalls

An *Allcall* is a general broadcast that does not request responses and does not designate any specific address. This essential function is useful for emergencies (e.g., "HELP!"), broadcast data exchanges, and channel quality and connectivity tracking. The *Global Allcall* special address is defined to be @?@. The Allcall protocol is discussed in Chapter 6.

As a variation of the Allcall, the calling station can organize (or divide) the available, though unknown, receiving stations into logical subsets using a *Selective Allcall* address. A Selective Allcall is identical in structure, function, and protocol to the Global Allcall, except that it specifies the last *single* character of the addresses of the desired subgroup of receiving stations. This specification limits the number of matching addresses for receiving stations to 1/36 of all possible addresses.

By replacing the ? symbol with an alphanumeric character, the Selective Allcall special address pattern becomes "TO @A@" or "THRU @A@" and "REP @B@," if more than one subset is desired. In this notation, the characters A (and B, if applicable) represent any of the 36 alphanumeric characters in the Basic-38 subset. These characters specify which character, or characters, must be positioned at the end of station addresses for them to stop scanning and listen to the transmitting station.

5.6.4.3 Anycalls

An *Anycall* is a general broadcast that requests responses without designating any specific station addresses. It is useful for emergencies, reconstitution of systems, and creation of new networks. An ALE station may use the Anycall to generate responses from essentially unspecified stations, and it thereby can identify new stations and connectivities.

The *Global Anycall* special address pattern is @@?, repeated as necessary for the entire calling cycle (T_{cc}). As discussed in Chapter 6, the Anycall protocol provokes slotted responses from the stations reached. If too many responses are received, or if the calling station must organize the available but unspecified responders into logical subsets, a Selective Anycall protocol is used. The Selective Anycall is a selective general broadcast that is identical in structure, function, and protocol to the Global Anycall, except that it specifies the last single character of the addresses of the desired subset of receiving stations as for the Selective Allcall.

By replacing the ? symbol with an alphanumeric, the Global Anycall becomes a Selective Anycall whose special address pattern is "TO @@A" in the calling cycle (T_{cc}). In a group Selective Anycall, "THRU @@A" and "REP @@B," and so on, are used alternately in the scanning call, designated "T_{sc}," and then "TO @@A" and "REP @@B," and so on, are used in the leading call (T_{lc}), where A and B are defined as earlier for the Selective Allcall.

If narrower acceptance and response criteria are required, the *Double Selective Anycall* may be used. The Double Selective Anycall is an operator-selected general broadcast that is identical to a Selective Anycall, except that its special address (using "@AB" format) specifies the last *two* characters in the addresses of the desired subset of receiving stations solicited to initiate a response (with a reduction in available addresses).

5.6.4.4 Wildcards

The special *wildcard* character is the ? symbol (0111111 in binary). A caller may use the wildcard character to address multiple stations with a single wildcard address. As described in Chapter 6, responses to a call containing an address with one or more wildcard characters are generated in pseudorandom slots to avoid collisions. Receivers must accept the wildcard character as a substitute for any alphanumeric character in their self-addresses in the same position or positions in which it appears in a wildcard address. That is, each wildcard character may substitute for any of 36 characters (A-Z, 0-9) in the Basic-38 character subset. The total length of the calling (wildcard) address, and any address that it matches, must be identical. Each wildcard character may substitute for only a single character at the corresponding position in a station address.

5.6.4.5 Self-Address

For self-test, maintenance, and other purposes, stations should be capable of using and responding to their own self-addresses in calls.

5.6.4.6 Null Address

For test, maintenance, buffer times, and other purposes, stations may use a *null address*, which is not directed to, accepted by, or responded to by any station. The *null address special address* pattern is @@@. The null address should only occur in the calling cycle (T_{CC}). Null addresses may be mixed with other addresses in a group call, in which case they should appear only in the leading call (T_{lc}) but not in the scanning call (T_{sc}). If a null address appears in a group call, no station is designated to respond in the associated slot; therefore, it remains empty. In this case, null addresses may be used as a buffer for tune-ups, or for overflow from the ALE station responding in the previous slot.

5.7 ALE FRAME STRUCTURE

Figure 5.7 contains examples of ALE frame construction. These examples include single-word and multiple-word samples of both single or multiple called station addresses for nonscanning (single-channel) and scanning (multiple-channel) use in individual, net, or group calls. Message sections are not included in these examples.

5.7.1 The Calling Cycle

As discussed earlier, the initial section of all frames (except sounds) is termed a *calling cycle*, and is denoted by T_{CC}. The calling cycle is divided into two

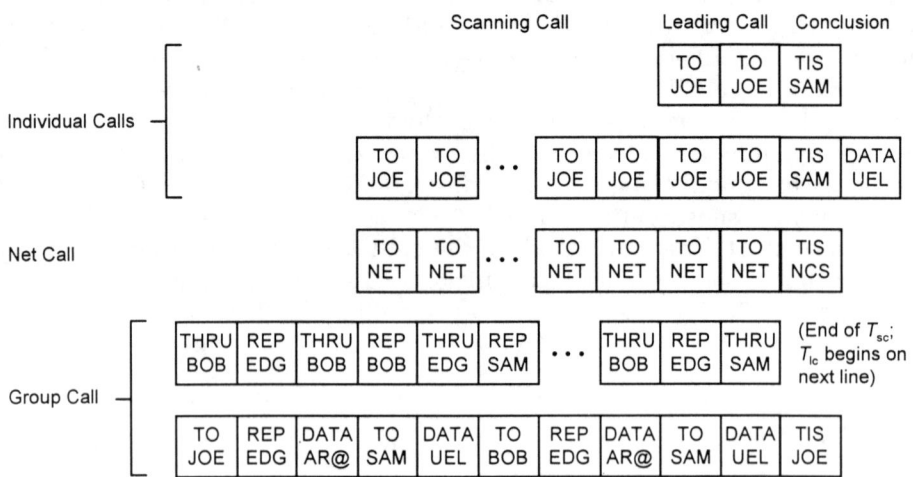

Figure 5.7 Examples of ALE frames.

parts: a scanning call, designated "T_{sc}," and a leading call, designated "T_{lc}." These terms and symbols are used interchangeably throughout the remainder of the book. The scanning call portion is required only when the called station, or stations, are scanning multiple channels. When all stations are known to be on a single channel, scanning call is optional and a call may begin with T_{lc}. Response and acknowledgment frames always omit the scanning call.

In an individual or net call, described in Chapter 6, the scanning call contains a repeated TO word, which carries the first word of the individual or net address. In a group call, the scanning call contains THRU words, which contain each *unique* first word, or words, of the called stations' addresses. In the case of more than one unique first word, THRU alternates with REP, as shown in Figure 5.7. The set of different address first words (T_{lc}) in all types of calls is repeated as necessary for a scanning call to fill the *scan period*, designated "T_s." Note that in the case of an odd number of unique first address words in a group call, an address that appeared in a THRU word in one cycle will appear in a REP word in the next cycle, as shown in Figure 5.3. The leading call is composed of TO (and possibly DATA and REP) words containing the *whole* address or addresses for the called station or stations. The entire list of whole addresses appears twice in the leading call.

There is no unique *flag word* or *sync word* for frame synchronization. Therefore, stations must be able to acquire and begin to read an ALE signal at any point after the start of a frame. Detection of the end of the calling cycle (T_{cc}) is based on the sequence of ALE word preambles in conjunction with the rules listed earlier. Receipt of a FROM, CMD, TIS, or TWAS clearly indicates that the leading call has ended. While processing the leading call, receipt of a TO, THRU, DATA, or REP word that violates the permissible sequence rules usually results in a receiver aborting the frame (and returning to scan mode if no link has been established).

The ALE standard requires that the transmitter have reached at least 90% of the selected RF power within 2.5 ms after the first tone transmission following call initiation. However, systems are permitted to transmit repeated duplicates of the scanning call word, or words, to be sent as the transmitter rises to full power. The transmitter may even use the ALE signal, instead of a tuning tone, while tuning, and then start the frame "officially" once the transmitter is at full power.

5.7.2 Message Section

The second and optional section of all frames, except sounds, is termed a *message*. Except for the so-called "Quick-ID," it is composed of CMD (and possibly REP and DATA) words from the end of the calling cycle (T_{cc}) until start of call conclusion as depicted in Figure 5.4.

The optional Quick-ID is composed of FROM (and possibly REP and DATA) words, containing the whole address of the sending station. The Quick-ID should only be used once between the end of the leading call and the start of the message section CMD sequences. The Quick-ID enables prompt caller identification and should be used if excessive length of a message section is a concern. It is never used without one or more following messages.

5.7.3 Conclusion Section

The final section of a frame is termed the *conclusion*. It contains either a TIS or a TWAS word, but not both, extended as necessary with DATA and REP words to carry the whole address of the sending station.

5.7.4 Sounds

Sounds, an exception to the above frame structure, start immediately with a TIS (or TWAS) word. A sound carries the whole address of the sounding station, as described in the sounding protocol presented in Chapter 6.

5.8 CONCLUSIONS

This chapter has presented the ALE signal structure, including the component preambles of transmitted ALE frames TO, TIS, TWAS, DATA, REP, THRU, CMD, and FROM, and the allowable sequences of the concomitant ALE words. Successful transfer of these frames, despite the HF channel anomalies described in Chapter 2, is accomplished with a robust signaling scheme composed of triple-redundant words, bit interleaving, and Golay FEC coding. The combination of these techniques provides inherent bit synchronization without the need for separate synchronization circuits in hardware implementations. The allowable structures imposed by the ALE addressing standards support a diverse variety of station-to-station, station-to-net, and station-to-group applications.

The structure of the ALE signals described in this chapter provides the fundamental link-layer requirements to meet the ALE standards. Although the ALE modem, along with the ALE signal structure presented in this chapter, may not be optimal for the HF channel, they nonetheless achieve significant HF linking probabilities in practice. These standards are available to all manufacturers with no licensing fee, having evolved from careful technical cooperation of the government and members of the HF Industries Association (HFIA). The resulting interoperability of HF automation systems built by different manufacturers therefore becomes a major attribute of this ALE signal structure. Because the global coverage of HF radio can potentially support propagation between any two stations on earth, this interoperability greatly increases the value of HF radio as a standard backup, or emergency, communication system.

References

[1] Department of the Army, Information Systems Engineering Command, *MIL-STD-188-141A: Interoperability and Performance Standards for Medium and High Frequency Radio Equipment*, Philadelphia, PA: Naval Publications and Forms Center, Attn. NPODS, Notice 2, Sep. 1993. Available on the Internet at URL ftp://tracebase.nmsu.edu/pub/hf/pubs/mil—std—188—141a.
[2] Department of Commerce, *Federal Standard 1045 (FED-STD-1045), Telecommunications: HF Radio Automatic Link Establishment*, Washington, DC: General Services Administration, Office of Information Resources Management, Jan. 1990. Available from GSA Specification Section.
[3] Watterson, C., et al. "Experimental Confirmation of an HF Channel Model," *IEEE Trans. Comm. Technol.*, Vol. COM-18, No. 6, 1970, pp. 792–803.
[4] Johnson, E. E., *An Efficient Golay Codec for MIL-STD-188-141A and FED-STD-104,5* Technical Report NMSU-ECE-91-001, New Mexico State University, Feb. 1991.
[5] Johnson, E. E., et al., *A Software Simulator for High Frequency Radio Automatic Link Establishment*, NTIA Technical Memorandum 94-162, Boulder, CO: National Telecommunications and Information Administration, May 1994.

Automatic Link Establishment Protocols 6

The robust signal structure described in Chapter 5 provides a basis for a suite of protocols that have been defined for evaluating and selecting HF channels, establishing links, and exchanging messages using the automatic link establishment (ALE) waveform and frame structure. These protocols are the subject of this chapter and are highlighted in the hierarchical structure of HF systems presented in Figure 6.1. Discussion of the data link protocol used with the high-speed HF data modems described in Chapter 4 is provided in Chapter 8.

6.1 OVERVIEW OF ALE OPERATIONS

As noted in the introduction to Chapter 5, ALE technology is often patterned after the procedures developed over many years by skilled human operators. One of the important and robust attributes of ALE operations that mimics the skilled operator is that the ALE protocols are asynchronous. In other words, there is no overall network synchronization with designated time slots assigned to each station.

Although the synchronous alternative can be more efficient for large congested networks, the asynchronous approach described in this chapter was designed to support reconstitution of a long-haul communications infrastructure in the event that the preexisting infrastructure is disabled. In such a scenario, network synchronization is a luxury that cannot be assumed. As a result of this design requirement, the ALE scheme documented in the U.S. military and U.S. military federal standards is also ideal for quickly establishing new networks, even interconnecting stations in different networks built by different vendors. It therefore adapts as easily to a massive loss of infrastructure as it does to normal changes in HF connectivity.

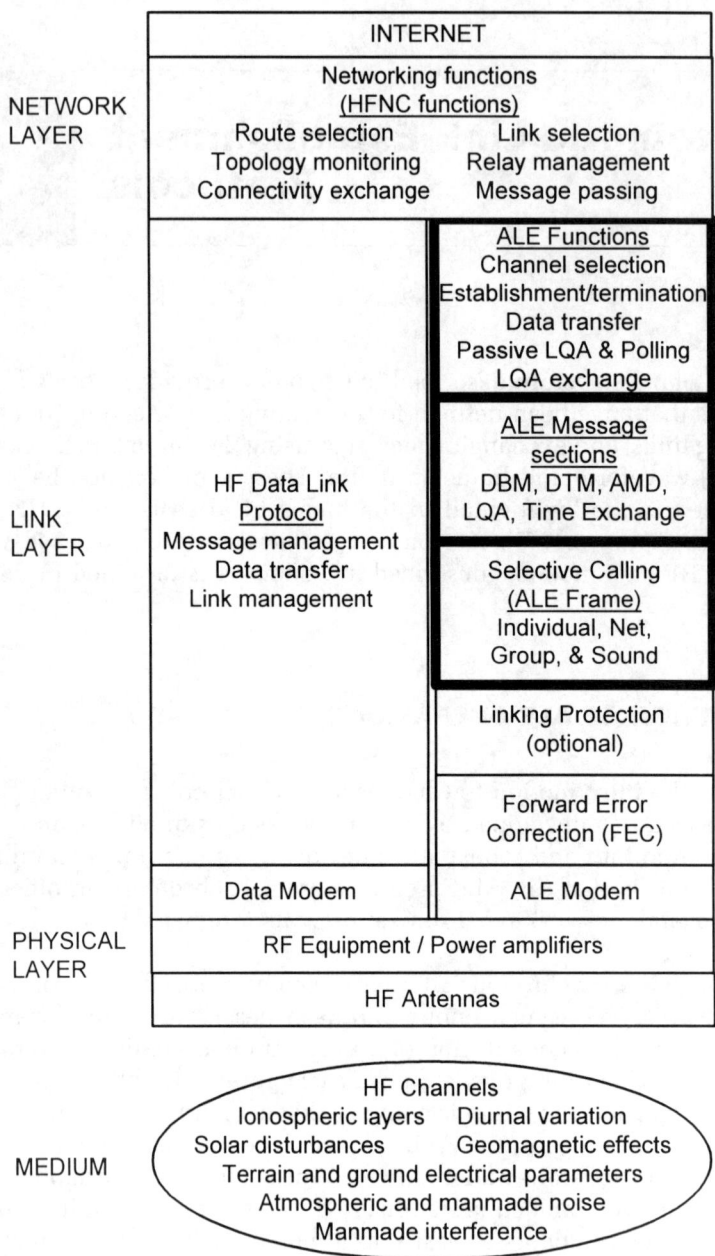

Figure 6.1 ALE functions, message sections, and selective calling protocols in the hierarchical-layered view of HF communications.

6.1.1 ALE Operational Rules

The ALE system incorporates the basic operational rules listed in Table 6.1. These operating rules constitute an expert system based on the procedures of skilled radio operators. Some of these rules may not be applicable in certain applications. For example, Rule 2, "always listening," is not possible while transmitting with a transceiver or when using a common antenna with a separate transmitter and receiver.

6.1.2 Listen Before Transmit

To comply with Rule 5 from Table 6.1, ALE controllers must listen to a channel before transmitting on that channel. Such a pause before transmission is implicit in all of the ALE protocols described in this chapter. Recognizing the presence of ALE signaling on a channel is simple for an ALE controller; however, detection of voice or other waveforms is not as simple. The recommended solution to this dilemma is to build ALE controllers with an input port to which other modems, voice processors, and so on, can provide a signal when they recognize their particular waveform. This approach ensures that ALE is a "good citizen" in the HF spectrum.

6.2 CHANNEL EVALUATION AND SELECTION

One of the most challenging aspects of communication over sky-wave channels is the dynamic nature of the ionospheric propagation medium as described in

Table 6.1
ALE Operational Rules (listed in order of decreasing precedence)

Rule No.	Rule Description
1	Independent ALE receiving capability in parallel with any others (critical).
2	Always listening for ALE signals (critical).
3	Always will respond (unless deliberately inhibited).
4	Always scanning (if not otherwise in use).
5	Will not interfere with active ALE channel (unless forced by operator).
6	Always will exchange LQA with other stations when requested (unless inhibited), and always measures the signal quality of others.
7	Will respond in the appropriate time slot to calls requiring slotted responses.
8	Always seek (unless inhibited) and maintain track of their connectivities with others.
9	Linking ALE stations employ highest mutual level of capability.
10	Minimize transmitting and receiving time on channel.
11	Automatically minimize power used (if capable).

Chapter 2. Changes in this medium and the resulting signal received over a sky-wave path are exhibited within nearly every time epoch, including (1) multipath effects on the scale of milliseconds, (2) fading on the scale of seconds to minutes, (3) diurnal variation on the scale of hours, (4) ionospheric disturbances that can persist for days, and (5) variation in sunspot activity on the scale of years. Multipath effects introduce intersymbol interference in modems and must be countered at that physical layer, as described in Chapter 4. Deep interleaving can overcome error bursts induced from signal fades, but the latency required to overcome long fades through interleaving can be unacceptable in many applications. Therefore, for fluctuations in propagation on time scales greater than a few seconds, the locus of "adaptivity" usually shifts to an ARQ protocol (see Chapter 8), or selection of better frequencies, alternate routes, or even alternate media (see Chapter 9).

The usability of a potential ionospheric channel is a function of a vast array of factors, some of which are quite predictable. For example, D-layer absorption and F-layer refraction both increase during the early hours of the day in middle and low latitudes. Therefore, higher frequencies are more likely to propagate than lower frequencies when the ray propagation path is sunlit, while lower frequencies prevail during darkness. Prediction programs, such as IONCAP described in Chapter 2, have been developed to reliably predict propagating frequencies at any time of day and year. However, other factors that affect propagation cannot be predicted with similar confidence, such as sporadic E-layer propagation and interference from other users on an otherwise usable channel. Therefore, channel quality prediction must generally be a supplement to channel evaluation.

6.2.1 Channel Evaluation

The term *channel evaluation* is subject to a wide variety of interpretations, ranging from whether or not the signal-to-noise ratio (SNR) exceeds a threshold, to a detailed statistical characterization of the channel in terms of parameters of a channel model. Indeed, a debate rages over whether detailed evaluation of channels is useful compared to simple "go/no go" evaluation. Questions also arise over whether channel evaluation using one waveform has much predictive value for a system that uses another waveform. In keeping with its robust self-sufficient philosophy, the ALE standards include mechanisms for channel evaluation using the ALE waveform. ALE controllers are required to maintain a database of such measurements, and to use such data in selecting channels. The results obtained by external channel evaluation schemes such as ionospheric sounders can, in principle at least, be incorporated into this channel quality database as well.

6.2.1.1 Scanning

Normally, stations that are not yet linked or otherwise transmitting will be continuously *scanning* an assigned set of frequencies listening for ALE signaling. With a well-chosen frequency scan set, at least one frequency will likely propagate between the station and any other station trying to communicate with it. By continuously scanning its assigned frequencies, an ALE station maximizes the probability that an incoming call will be received. This channel scanning behavior is one of the key automated techniques that enables reliable communication over HF without the presence of skilled operators.

When a station seeks to establish communications with a scanning receiver, its calling transmission must be of sufficient duration that the receiver will encounter that transmission at least once during its scan. As noted in Chapter 1, this minimum scanning call time, T_s, is equal to the number of channels scanned multiplied by an estimated pause on each channel of twice the duration of the ALE redundant word, T_{rw}. The 2 T_{rw} factor is used because a scanning receiver will pause for up to 2 T_{rw} on a channel carrying ALE signaling while it attempts to acquire ALE word synchronization (word sync).

6.2.1.2 Sounding

Sounding supports the empirical testing of selected channels and propagation paths by providing a very brief, beacon-like, self-identifying broadcast with the ALE waveform. Other stations that receive the sound can record the apparent connectivity, that is, the identity and availability of the sounding station. These data, collected from the unilateral sounding event, permit the scanning station to select known working channels for possible later use. The receiving station can also evaluate distortions in the received waveform for a more detailed picture of the link quality. Since sounding uses the standard ALE signaling, any ALE station can receive sounding signals.

The sounding signal is a unilateral one-way transmission that is performed at periodic intervals on unoccupied channels. The implementation is simple. A timer is added to the ALE controller to periodically initiate sounding signals after verifying that the channel is clear. Sounding is not an interactive bilateral technique, such as polling. However, the identification of unilateral connectivity with a station by correctly receiving its sounding signal indicates a high probability of bilateral connectivity. Moreover, connectivity may be acquired passively at the scanning station's receiver, that is, without transmissions from the scanning station. If propagation changes slowly, a long interval between soundings of perhaps one to two hours may be sufficient. However, if dynamic propagation conditions force constant change in usable frequencies, and the connectivity information is critical, sounding may be conducted more frequently (see later discussion).

The structure of the sound uses a modification of the standard ALE frame that includes only the conclusion, which identifies the transmitting station as shown in Figure 6.2. The whole address of the sounding station is sent at least twice to ensure reception by a scanning receiver. The use of TIS or TWAS in the sound indicates whether potential callers are encouraged or ignored, respectively. Normally, TWAS is used, and the sound is termed a call rejection sound. The *call rejection* sound is designed to provide usable channel data to scanning receivers without inviting responding transmissions that would collide with each other or interfere with the sounding process.

Early implementations of this sounding scheme revealed that ALE controllers frequently recorded hearing sounds from stations with rather bizarre addresses. (In those days, hearing *any* ALE signaling over the air was a surprise, because few implementations were extant.) A quick calculation revealed the source of these unexpected sounds. At that time, a single ALE word with a TIS or TWAS preamble and three Basic-38 characters was considered an acceptable sound. The probability of random noise bits "looking" like a sound was unacceptably high. In this case, 25% of all possible preambles are acceptable and nearly $(38/128)^3$ of the possible three-character combinations are valid one-word addresses, yielding a rate of acceptable random ALE words of 1 in 153. At that rate, a steady stream of bogus sounds quickly flooded the channel quality database, pushing out records of sounds from the desired stations.

The solution adopted to solve this problem was quite simple: An acceptable sound must contain the same sounding station address, using the same preamble(s), twice in immediate succession. This scheme reduced the probability of random noise appearing to be a sound by a factor of at least 2^{24}. Although this solved the bogus sounds problem, it was no longer possible for a scanning receiver to arrive at a channel at the conclusion of an ALE frame and count the single repetition of the sending station's address as a sound. This capability had been originally envisioned as a useful feature of the "unrepeated" sound design. However, the probability of acquiring sounds in this way was so small that the value lost was considered inconsequential.

The decision of whether or not to transmit a single- or multiple-channel sound is based on whether or not the *intended* recipients are scanning. If not,

TWAS SAM	DATA UEL	TWAS SAM	DATA UEL	...	(repeat as needed for scanning sound)

Figure 6.2 Structure of an ALE sound.

a single-channel sound containing at least two repetitions of the sounding station address is sent. Additional repetitions of the address increase channel occupancy, but also increase the probability of the recipients' correct acquisition of word sync and subsequent reading of the sound.

The multiple-channel (scanning) sound is identical to the single-channel sound except that additional repetitions of the sounding station's address are broadcast. The total duration of the sound must be sufficient to be encountered by all scanning receivers.

Sounding Rate

The required rate of sounding is determined by (1) the level of activity of the intended recipients and (2) the maximum acceptable probability that the most recently received sounds from some stations are older than some maximum acceptable age.

If P_b is the probability that a receiver is busy when a sound is sent, and assuming that busy periods occur independently of sounds, the probability that the receiver misses N consecutive sounds is $(P_b)^N$ for $N \geq 1$. If R_s sounds are sent per hour on a particular channel, the probability that the receiver in question has not received a sound from the sending station on that channel for A hours is $P_0 = (P_b)^{R_s A}$, where $R_s A \geq 1$. Thus, if A_s is the maximum acceptable age (in hours) for a sound, and P_0^* is the maximum acceptable probability that a sound is older than A_s, then at least R_s sounds must be sent per hour, where $R_s^* \geq \max[\log P_0^*/A_s \log P_b, 1/A_s]$ sounds/hr/channel. A_s and P_0^* are chosen by the network manager to satisfy network performance goals, while P_b is measured from actual network operation. Specifically, P_b is the fraction of time during a measurement period that the observed radio is not able to receive sounds. For a radio that is considered "available" when scanning, P_b is therefore the fraction of time that the radio is not scanning. Note that P_b includes busy time due to both message traffic and overhead, where overhead includes such ALE functions as sounding and polling.

For example, if polling is *not* used to fill in gaps in sounding data, then the station manager may be willing to accept only a 10% chance that each entry in the LQA table is more than one hour old, that is, $P_0^* = 0.10$. In this case, R_s is a function of P_b as illustrated in Table 6.2. Unfortunately, the equation for R_s^* in MIL-STD-187-721 does not reflect the $N \geq 1$ constraint on P^{Nb}. As a result, the sounding rate R_s for $P_b = 0.01$ is computed as 0.5 sounds per hour per channel. This result is obviously nonsense, because with a sounding interval of two hours, the chance that a sound has been received in the past hour is certainly less than 90%.

Table 6.2

Example Sounding Interval Calculation for $P_0^* = 0.10$

P_b, Probability Station Is Busy	R_s^*, Minimum Sounds per Channel per Hour	Maximum Equally Spaced Sounding Interval (min)
0.01	1.0	60
0.10	1.0	60
0.32	2.0	30
0.46	3.0	20
0.56	4.0	15
0.68	6.0	10

6.2.1.3 Link Quality Analysis

Link quality analysis (LQA) is the automatic propagation measurement of over one or more HF channels between stations. The resultant LQA data are used to *score* these channels and to support selection of a suitable channel for subsequent calling and communication. The LQA measuring technique is also used for the continuous monitoring of link quality during communications that use ALE signaling.

Pseudo Bit Error Ratio (PBER)

An ALE receiver essentially performs a "pass/fail" LQA test on every received signal by its critical examination of proper coding, structure, and format. In addition, there is an inherent PBER measurement capability within the ALE FEC-sublayer functions that provides more LQA resolution than is available from the absolute pass/fail approach. The analysis of ALE 8-ary FSK modulation on CCIR Poor HF channels showed that a roughly proportional measure of BER may be obtained by counting the number of nonunanimous (two-thirds) votes out of 48 in the majority vote decoder. The resulting PBER values range from 0 (no two-thirds votes) to 48 (no unanimous votes). This PBER measurement is performed on each redundant triplet (3 T_w) word received and properly decoded as a valid majority word. Therefore, the best (lowest) PBER value occurs when the majority vote decoder is properly aligned with the incoming signal during ALE word sync acquisition. Except for errors, all three-word inputs will then be occupied with identical redundant words. Because PBER may vary during an ALE transmission, a linearly averaged PBER is used. This measurement includes the measurements of all "good" words that were properly aligned. Unreadable words are assigned the worst PBER value ("48") and included in the average.

Signal-Plus-Noise-Plus-Distortion to Noise-Plus-Distortion Ratio (SINAD)

An optional signal-to-noise and distortion measurement defined for the LQA suite is a SINAD measurement $[(S + N + D)/(N + D)]$ averaged over the duration of the received ALE signal. If implemented, SINAD measurements are performed on all ALE signals.

Multipath

Multipath (MP) measurements have not yet been standardized. A bit field has nevertheless been reserved for reporting MP values in the LQA CMD word (see Sec. 6.5.1).

6.2.1.4 Polling

Polling is defined in MIL-STD-187-721 as the activity of actively requesting and measuring channel characteristics using LQA techniques. Polling protocols are used to acquire current bilateral link quality data by *handshaking* with one or more other stations, directly measuring the transmissions received, and exchanging these measurements with the other station or stations. Polling is used to actively acquire current bilateral LQA data for stations and channels for which recent LQA data are unavailable in the LQA table.

A simple, but time-consuming, approach to polling involves (1) calling other stations on each channel of interest to perform LQA during their responses and (2) requesting these called stations to respond with LQA determined from their reception of the polling call. More efficient protocols than this approach are described in the standard but will not be addressed in this book.

6.2.2 Channel Selection

Channel selection is the automatic identification of a suitable channel for initiating calls or broadcasts to one or more stations. This selection uses the information automatically collected and stored within the LQA data table of the calling ALE controller, including pseudo-PBER, SINAD, and MP. This information is used to speed link establishment and optimize the choice of quality channels for HF communications.

When attempting to establish a link, the sequence of channels to be tried is derived from information in the LQA data memory. Unless otherwise directed by the operator or controller, the channel or channels with the best scores are tried first until all LQA-scored channels have been tried. If LQA data are unavailable or exhausted, then the station will use all remaining available channels until success is achieved or all channels have been tried.

The standard levies no requirements on the algorithm used by the ALE controller to rank-order channels based on LQA data. Different manufacturers may employ unique innovative techniques that combine channel measurements, the ages of those measurements, the intended use of the link, channel prediction data, and so on, to perform this channel ranking. The standard requires only that LQA scores be displayed to the operator such that increasing numerical values correspond to increased channel quality.

6.3 LINK ESTABLISHMENT

6.3.1 Overview

A three-way handshake is sufficient to establish a link between a calling station and a single responding station. The fundamental protocol exchange for link establishment using this three-way handshake was first introduced in the overview of selective calling in Chapter 1. With the addition of slotted responses as described in Section 6.3.6.2, the same call/response/acknowledgment sequence can also link a single calling station to multiple responding stations. If a message (or CMD sequence) is to be sent in any of the ALE frames, it is inserted between the leading call and the frame conclusion. Such messages and CMD sequences can perform many functions, such as reporting LQA information, modifying the ALE protocol and timing, or conveying operator-to-operator orderwire messages (see Sec. 6.5).

A scanning calling station will send ALE calls on its scanned channels in the order dictated by its channel-selection algorithm. This station will link on the first channel that supports a handshake with the called station or stations. After "scan stop," "unmute," and "operator alert," the operators (or controllers) use the link as necessary. If a channel is rejected as unsuitable, the stations may restart the scan sequence to seek a better channel. This process is performed by muting, or resetting, their stations and reinitiating scanning calling. This reinitiation is usually performed by the original caller. If the calling station has LQA memory and scoring capability, it should lower the ranking of rejected or failed channels prior to *calling restart* on the best available remaining channel. If the station has a fixed calling channel sequence, it should restart the scanning call on the next channel that would have been tried previously had a link not been established.

During the scanning calling cycle, a caller may encounter occupied channels and skip them to avoid interfering with other traffic. After all available channels have been tried without success, the caller may use previously occupied channels if the interference has subsided. Once a calling station has failed to establish a link on any prearranged scan-set channels, it should immediately return to normal receive scanning. The ALE controller should also alert the operator and networking controller, if present, that the calling attempt was unsuccessful.

In the following discussion of ALE operations, we refer to an ALE controller as being in one of three conceptual *states*, (available, linking, or linked) although these states are not clearly defined in the standards.

6.3.2 Available State

A station is in the *available state* when it does not currently have a link with any other station and is not in the process of establishing a link. A station that is programmed for multichannel scanning operation will be scanning when it is in the available state. Single-channel stations will, of course, remain tuned to their assigned channel regardless of their state.

6.3.3 Linking State

A station enters the *linking state* from the available state when it sends or receives an ALE call frame. Scanning stations stop scanning when they enter the linking state. A station returns to the available state if the linking attempt does not complete successfully. Upon successful completion of a three-way handshake, stations in the linking state enter the linked state.

6.3.4 Linked State

A station is considered to be in the *linked state* if it has successfully completed link establishment with one or more stations and at least one link to which it is a party has not been terminated. While in the linked state, a *wait-for-activity* timer will be running (if not disabled by the operator). Stations programmed to scan will not be scanning while in the linked state. After link establishment, communication among linked stations is normally carried by additional three-way handshakes, but stations remain in the linked state during these handshakes. The ALE state diagram is depicted in Figure 6.3.

6.3.5 One-to-One Calling

The protocol for establishing a link between two individual stations consists of three ALE frames: a call, a response, and an acknowledgment (ACK). The

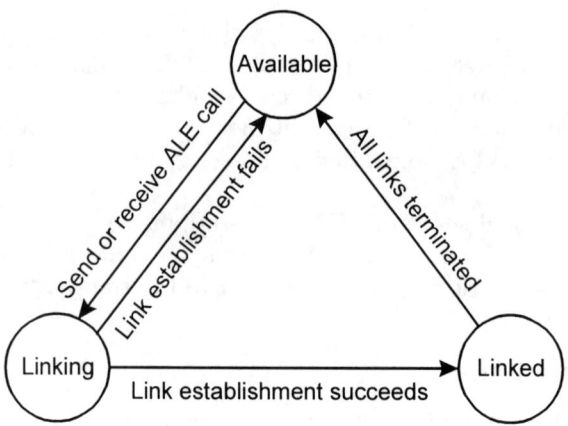

Figure 6.3 ALE state diagram.

sequence of events, and the timeouts involved, are discussed in the following paragraphs using a calling station named SAM and a called station named JOE. A hierarchical specification of the protocol is depicted in Figures 6.4 through 6.7. A typical three-way scanning call handshake requires between 9 and 14 sec to complete.

6.3.5.1 Sending an Individual Call

The caller SAM begins the ALE protocol after selecting a channel, listening to avoid disturbing active channels, and tuning its antenna coupler for transmission at the chosen frequency. If station JOE is known to be listening on the chosen channel (i.e., not scanning), station SAM will transmit a single-channel call. The single-channel call contains only a leading call and a conclusion as shown in the upper frame of Figure 6.8. Note that the entire called station address is used in the leading call section of any ALE calling cycle. If station JOE is scanning, station SAM will send a longer calling cycle. This calling cycle will precede the leading call with a scanning call of sufficient length to capture JOE's receiver as shown in the lower frame of Figure 6.8. Note that only the first word of the destination address is used in the scanning call section of the calling cycle, thus accelerating scanning and speeding the linking process.

To summarize the individual call (see Fig. 6.8), station SAM calls station JOE by transmitting a calling cycle containing JOE's address ("TO JOE"), followed by a conclusion containing its own address ("TIS SAM"). Station SAM then waits for a preset time to start to receive station JOE's response. In the single-channel case, the *wait-for-response* time, T_{wr}, includes anticipated round-trip propagation delay and station JOE's *turnaround* time. In the multi-

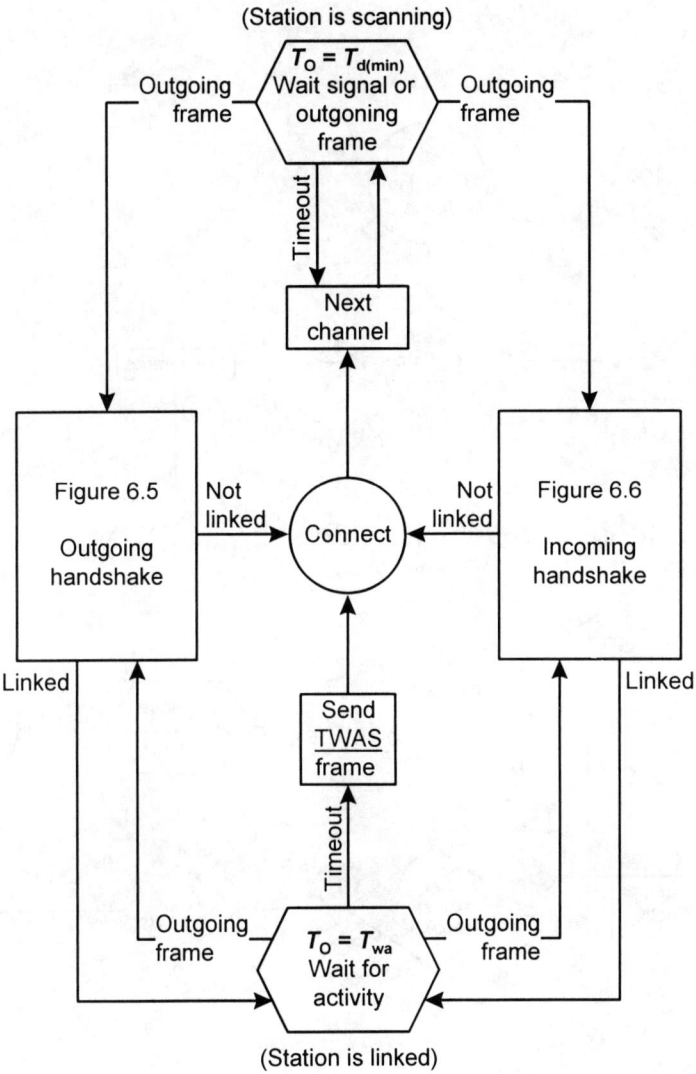

Figure 6.4 Top functional level of ALE (scanning).

channel case, station SAM waits for a *wait-for-response-and-tune* time, designated T_{wrt}, which also includes time for station JOE to "tune up" on the chosen channel.

If the expected reply from station JOE does not start to arrive within the preset wait-for-response time (T_{wr}), or within the wait-for-response-and-tune time (T_{wrt}), then the handshake terminates and the linking attempt on the

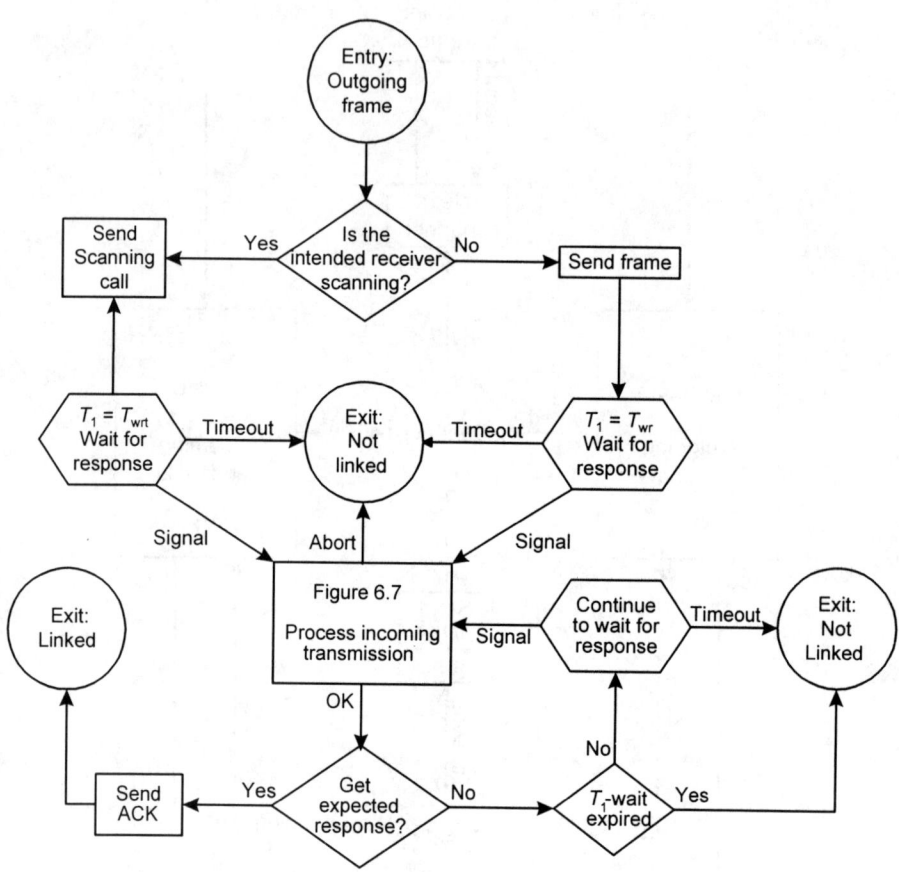

Figure 6.5 Outgoing handshake.

Automatic Link Establishment Protocols 231

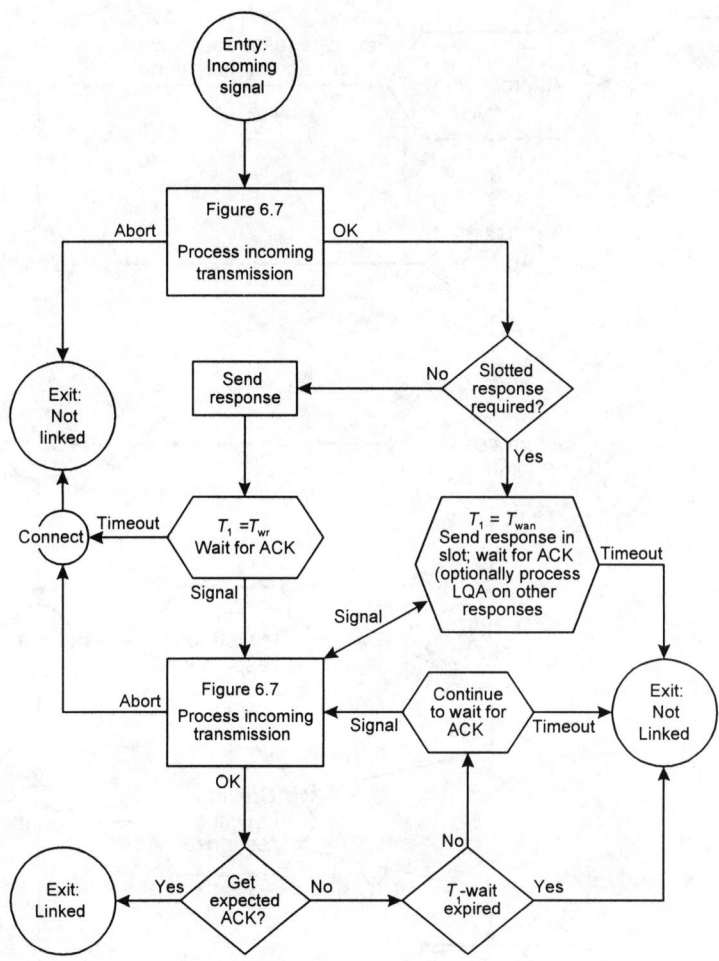

Figure 6.6 Incoming handshake.

channel is considered to have failed. At this point, the linking attempt will normally restart on a channel not yet tried as shown in Figure 6.4. Otherwise, the operator at station SAM or the station's networking controller is notified of the failed linking attempt.

6.3.5.2 Receiving an Individual Call

Assume that station JOE arrives on a given channel sometime during its *scan period* (T_s), at some point in station SAM's somewhat longer scanning calling

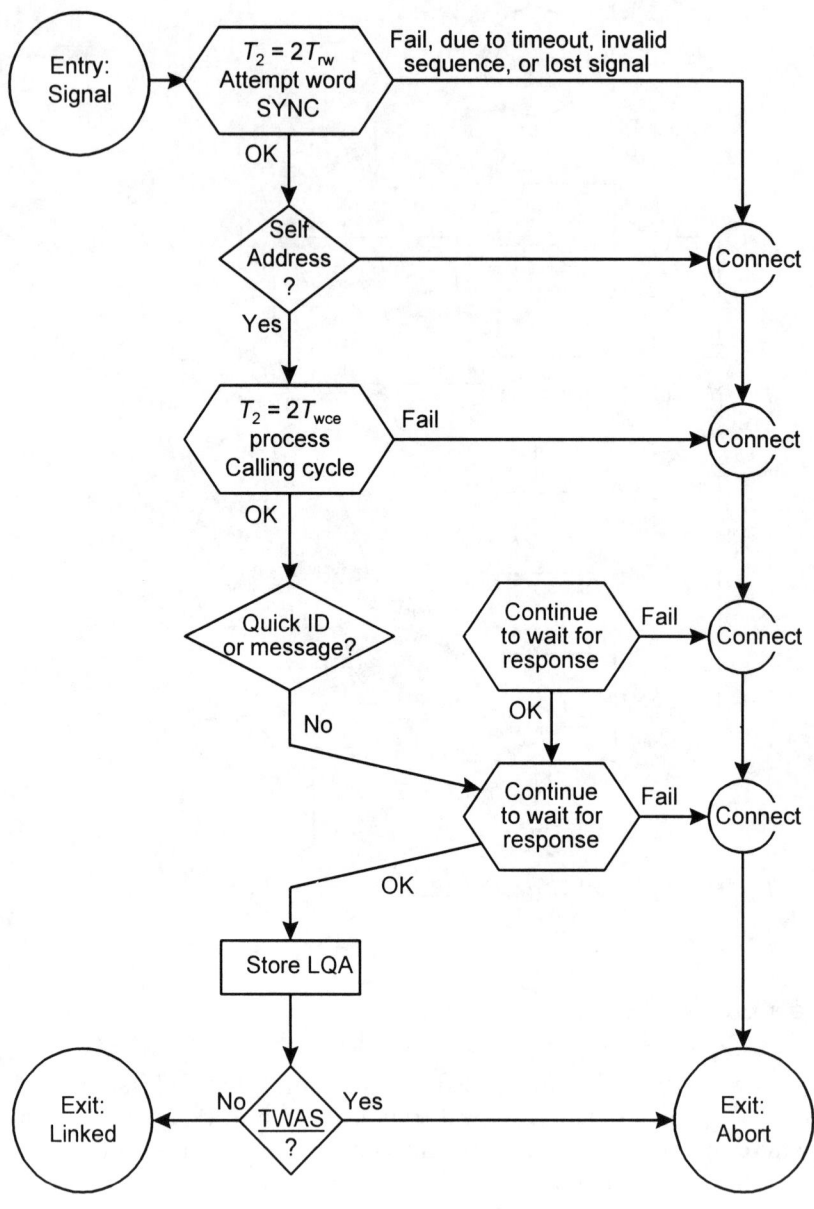

Figure 6.7 Incoming transmission.

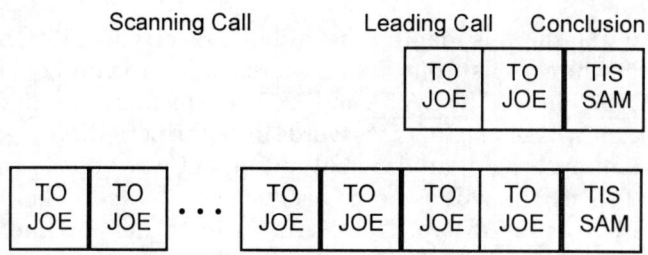

Figure 6.8 Individual ALE calls.

time (T_{sc}). Station JOE then will attempt to detect ALE signaling within a *dwell* time, $T_{d(min)}$, as indicated in Figure 6.4. If no ALE signaling, non-ALE signaling, or otherwise unrecognized interference is detected within period Td(min), station JOE will resume scanning. If ALE signaling is detected, station JOE will attempt to achieve ALE word sync within a preset time $T_{drw} = 2\ T_{rw}$ (see Figs. 6.6 and 6.7). If station JOE does not read appropriate ALE words within the time T_{drw}, it will leave the channel and resume scanning.

If station JOE reads "TO JOE", or an acceptable equivalent according to ALE protocols, it will (1) stop scan, (2) plan to reply (i.e., send a response frame), and (3) wait a preset time T_{wce} for station SAM's calling cycle to end and the message or conclusion to begin (see Fig. 6.7). Meanwhile, station JOE will continually read incoming ALE signaling from station SAM to identify additional information. This information could include the type of call and additional station addresses, if any. Station JOE also attempts to detect invalid ALE word sequences. If an invalid sequence is found, station JOE will automatically reject the call and immediately resume scanning, unless directed otherwise by the operator or controller.

If a Quick-ID or message section starts within period T_{wce}, station JOE will wait and attempt to read the message within a preset time limit given by $T_{m(max)}$. Station JOE will resume scan if a Quick-ID, message, or conclusion is not detected within period T_{wce} or a conclusion does not begin within period $T_{m(max)}$. If an invalid message sequence is read, station JOE will immediately resume scan. If a frame conclusion begins, such as "TIS SAM", station JOE will wait and attempt to read the calling station's address (i.e., SAM's address) within a new preset time period $T_{x(max)}$. If an unacceptable conclusion address sequence is read, station JOE will immediately resume scanning. If an acceptable conclusion sequence with TIS is read, station JOE will prepare to respond to station SAM while identifying the entire address. If TWAS is read instead, station JOE will resume scanning immediately after reading the entire address without responding to station SAM.

All receiving stations identify the end of a received ALE signal using the following procedures. First, the station searches for a valid conclusion, such as TIS or TWAS, followed by DATA and REP for a maximum of five ALE words total. The station will search for these words until the period $T_{x(max)}$ is exceeded. The conclusion must maintain constant redundant word phase with previous ALE words. The station will examine each successive redundant word phase following the TIS (or TWAS) word for the first of up to four possible nonreadable or otherwise invalid words. The end of the received transmission is detected by (1) failure to receive an error-free word, (2) detection of an improper word sequence, and (3) detection of the last REP word plus the last-word-wait delay time, $T_{lww} = T_{rw}$. The maximal acceptable terminator sequence is TIS (or TWAS), DATA, REP, DATA, REP. If all of the preceding criteria for an acceptable ALE call frame are satisfied, and if station JOE is not otherwise directed by the operator or ALE controller, then station JOE will immediately initiate an ALE response.

6.3.5.3 Response

On receipt of SAM's call and after recognition of both station JOE's and SAM's addresses, station JOE's ALE controller will do the following (see Fig. 6.7): (1) Tune to the transmit frequency, (2) send the response shown in Figure 6.9, and (3) start its wait-for-response timer (T_{wr}). Note that T_{wr} is used rather than the longer wait-for-response-and-tune timeout (T_{wrt}) because station SAM has already tuned to the appropriate frequency. After transmitting its call, station SAM attempts to detect ALE signals and read a response frame from station JOE within an appropriate time period, either T_{wr} or T_{wrt} (see Fig. 6.5).

Station SAM will automatically terminate the protocol and resume scanning calling if it does not receive the following: (1) an appropriate response calling cycle (i.e., "TO SAM") starting within the timeout period, and (2) the appropriate conclusion word (i.e., "TIS JOE") starting within period T_{lc}, or $T_{lc} + T_{m(max)}$ if a message is included. If station SAM receives the proper conclusion, "TIS JOE," starting within period T_{lc} (or within period $T_{lc} + T_{m(max)}$ if a message is included) then station SAM will plan to reply with an ACK. If "TWAS JOE" is received instead of "TIS JOE," however, station SAM's linking attempt is terminated. Otherwise, if an invalid ALE word sequence occurs or

TO SAM	TO SAM	TIS JOE

Figure 6.9 ALE response frame.

the end of the conclusion frame is not detected within period T_{lww} (plus the additional multiples of T_{rw} if an extended address), station SAM will terminate the protocol and resume scanning calling.

If the events following the flowchart in Figure 6.7 exit via "OK," then station SAM alerts its operator that a correct response has been received. If not otherwise directed by the operator or controller, station SAM's ALE controller will (1) initiate the ALE acknowledgment (using "TIS SAM"), (2) enter the linked state with JOE, and (3) unmute station SAM's speaker. A *wait-for-activity* timer is set for period T_{wa}, typically 30 seconds, that causes the link to be dropped if it remains unused for extended periods as possible in Figure 6.4.

6.3.5.4 Acknowledgment

The ACK from station SAM to station JOE is shown in Figure 6.10. If the expected ACK from SAM does not start to arrive within the preset wait-for-response time (T_{wr}) (see Fig. 6.7), then (1) the handshake is terminated, (2) station JOE does not enter the linked state, and (3) station JOE's operator or controller is notified. However, the linking attempt may be reinitiated by the operator or controller at any time. In rare cases when the ACK to station JOE is lost, station SAM will be in the linked state (without JOE). In these cases, station SAM eventually returns to the scanning or available state after the wait-for-activity timer, set to T_{wa}, expires.

Note that station SAM's ACK sent to station JOE appears identical to station SAM's individual call to station JOE. Station JOE does not respond to station SAM's ACK, however. Such an endless "ping-pong" handshake is avoided because station JOE *expects* to receive station SAM's ACK within the narrow time window T_{wr} immediately following JOE's response. If station SAM's ACK arrives after a time T_{wr}, then station JOE treats it as a new, or second, individual call. If this late ACK employs the TIS word, then station JOE generates a new response.

6.3.5.5 Link Termination

Termination of an ALE link after a successful linking handshake is accomplished by sending a frame concluded with TWAS to any linked station to be

TO	TO	TIS
JOE	JOE	SAM

Figure 6.10 ALE acknowledgment frame.

terminated. For example, if station SAM sends "TO JOE, TO JOE, TWAS SAM," then the link between stations SAM and JOE will terminate. Unless station JOE retains links with other stations on the channel, its ALE controller will immediately mute station JOE's speaker and return to an available state. Likewise, station SAM will also immediately mute its speaker and return to the available state in the absence of other linked stations on the channel.

Manual Termination

Station operators may reset the stations (mute the speakers) at any time, thus restoring them to the ALE available state. This *manual reset* should cause a link-terminating transmission using the TWAS word as previously described.

Automatic Termination

If an operator or networking controller does not key the *push-to-talk* (PTT), or if the station is not "used" within the preset wait-for-activity time limit, T_{wa}, the station will (1) automatically mute the speakers, (2) terminate the linked state with any linked stations, and (3) return to the ALE available state. The wait-for-activity timer is mandatory, but it may be disabled by the operator or net manager. This timed reset is not required to generate a link terminate transmission using the TWAS word. Nevertheless, it is strongly recommended that a termination be sent to reset the other linked station or stations and immediately "free" them into the available state.

In general, link termination during a handshake or protocol by the use of TWAS (or a timer) causes the receiving (or timed-out) station to immediately (1) end the handshake or protocol and (2) terminate the linked state (with that station). Unless the station remains linked with another station, it will also (1) remute the speakers and (2) return to the available state.

6.3.5.6 Nonlinking ALE Transmissions

Normally, station SAM's use of the TIS word in a call compels station JOE's response. On the other hand, the substitution of a TWAS word suppresses station JOE's response and terminates the ALE protocol. Consider a case in which station JOE received the individual call using "TWAS SAM," possibly including a message. Station JOE receives station SAM's call and realizes that it was intended as a one-way broadcast for station JOE. In this case, station JOE would not respond unless otherwise compelled by the message, the operator, or a networking controller. Likewise, if station JOE sends a TWAS word in a response containing a TIS, then station SAM will neither alert the operator or networking controller nor send an ACK, thus terminating the protocol.

Any station, such as JOE, may terminate the linking handshake under several circumstances. First, station JOE may be unavailable because it is currently engaged in other traffic or has detected that the channel is busy locally. Alternatively, station JOE may be compelled by prearrangement or other adopted convention to respond without having an obligation to continue. Such nonbinding scenarios include mandatory roll calls and optional chat mode exchanges.

6.3.5.7 Collision Detection

During reception of an ALE signal, a channel *dropout* may occur due to such factors as interference or fading. This loss of continuity is indicated by failure to detect acceptable ALE words at every redundant word time interval, T_{rw}. If a channel dropout during the initial received call continues without reception of other acceptable words for more than 3 T_{rw} (1,176 ms) after the last acceptable word was received, the receiving station will abandon its attempt to link and, if operating in a multichannel mode, will immediately return to normal receive scanning.

In all cases, all included words within each individual ALE signal must be received at consistent and uniform redundant word phase (T_{rw}), irrespective of channel dropouts. This redundant word phase restriction is necessary to distinguish ALE words to be considered acceptable and valid by the ALE controller at the receiving station. This stipulation tacitly assumes that the propagation delay incurred on the link will remain approximately constant during reception of the entire ALE signal. Any variation of this word phase indicates interference or a collision and such variation should be rejected as not part of the signal. Nevertheless, stations should be able to read interfering ALE signals because they may contain useful (or even critical) information. This capability is the basis for the ALE operational rule that a station be "always listening." Finally, consider a station that receives a complete individual call including TO with the whole address, but does not receive the expected conclusion with the TIS or TWAS words. In this case, the station may broadcast a single-channel call Acceptance sound to reinitiate the calling station (see Sec. 6.2.1.2).

6.3.6 One-to-Many Calling

A critical requirement for HF systems is the capability to rapidly initiate links and interoperate with multiple stations. Linking among multiple stations is significantly more difficult than linking between two individual stations. In particular, the number of required links can increase geometrically as stations are added, and yet all of the available channels still retain their individual diverse propagation and interference characteristics. In general, full connectivity cannot always be achieved on any single frequency for one or more reasons, including (1) differing ionospheric conditions along the propagation paths between stations, (2) different interference environments at each station, and (3) differences in RF equipment, particularly antenna patterns, at each station. The complexities for fixed stations are magnified for mobile stations. The changing relative positions and absolute locations of these mobile stations can cause significantly greater time variability in channel performance than experienced between fixed stations.

There are three topologies from which any network can be constructed. Each topology requires significantly different quantities of links L determined by the number of stations N, which are included as shown in Table 6.3. In this context, a point-to-point path is considered to be a single link between two stations (repeaters are discussed in Chapter 9). A star topology involves several stations in a *one-to-many* configuration and requires one fewer links than the number of stations. A multipoint topology involves several stations in a *many-to-many* configuration. The multipoint topology requires a quantity of links that grows as N^2. In this section, two different one-to-many protocols are described, the *star net* and *star group*.

6.3.6.1 Star Net Calling

A star net is a prearranged collection of stations designed to link and interoperate primarily with a single hub station. The purpose of a star net call is to rapidly and efficiently establish contact with a prearranged collection of stations by the use of a single net address (see Sec. 6.4.3) which is common to all net members. In most cases, this hub station also operates as the net control station (NCS). A star net is usually organized and managed with significant prior knowledge of the member stations. This prior knowledge includes the number of stations, their individual identities, respective capabilities, and in most cases,

Table 6.3
Number of Links Versus Network Topology

Topology	Number of Links
Point-to-point	$L = 1$
Star	$L = (N - 1)$
Multipoint	$L = N(N - 1)/2$

their locations and necessary connectivities. These connectivities may include sets of stations who must interoperate to perform some desired function. A star net call is identical to an individual call, except that the called station address is a net address, as shown in Figure 6.11.

6.3.6.2 Slotted Responses

Because a net call addresses multiple stations, the simple three-way handshake used for individual links cannot be used. If it were used, the responses from the called stations would collide with each other. Instead, a time-division multiple-access (TDMA) scheme is used. In this scheme, each member station of the net is assigned a unique time slot for its response. Although the ALE scheme is considered to be asynchronous in general, participating stations in a link establishment handshake are necessarily locked into a fairly precise timing structure. This synchronous *overlay* on the ALE protocol is especially necessary during slotted-response protocols. In these protocols, any timing "slop" allowed in each slot is multiplied by the number of slots and could result in unacceptably long latencies for link establishment.

At the end of the net-call frame, the following events take place:

1. The calling station sets a wait-for-response-and-tune timer (WRTT) that will trigger its ACK after the last response slot time has expired. The period of this timeout is denoted T_{wrn}.
2. The called stations set their own WRTT values that bound their wait times for an ACK. The wait time is set to $T_{wan} = T_{wrn} + 2T_{rw}$ in order to allow time to acquire word sync during the leading call of the ACK.
3. Each called station sets a *slot-wait-time* timer, T_{swt}, that will trigger its response. If the calling station is a member of the net, it will not respond so its corresponding time slot will be empty.
4. The called stations tune on the appropriate channel and prepare to respond during the slot time, called *slot 0*, immediately following the end of the net call frame.

From this point, the ALE protocol literally runs like clockwork. As each station's slot wait timer expires, it releases its response and awaits the expiration of its

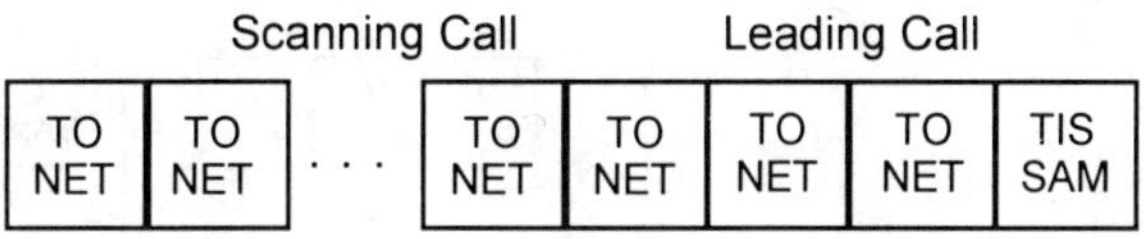

Scanning Call			Leading Call				
TO NET	TO NET	...	TO NET	TO NET	TO NET	TO NET	TIS SAM

Figure 6.11 ALE net call.

WRTT. Should that timer expire before an ACK begins, the station assumes that the linking attempt has failed and returns to scan.

As illustrated in Figure 6.12, slotted-response frames are identical to responses in the one-to-one calling protocol. Each consists of a leading call, composed of two repetitions of the complete address of the calling station in TO words (extended as necessary with DATA and REP words), followed by a conclusion using TIS to accept the call or TWAS to reject it.

When the calling and responding stations use one-word addresses as shown in Figure 6.12, each slot consumes 3 T_{rw}, or 9 T_w. However, the slots are somewhat larger than 9 T_w to allow some extra time for long-distance propagation and for signal detection and processing delays within the ALE controllers. With one-way propagation times for long-range links on the order of tens of milliseconds, it is clear that propagation time must be considered in ALE timing.

In particular, an allowance for round-trip timing must be included in the slot width margin. This allowance is necessary because slot timings computed by a distant net member will not begin until it receives the end of the call *at its location*. As a result, the distant net member will have slots delayed by the one-way propagation time relative to the calling station. Its responses will likewise be delayed in transmit by a one-way propagation time. These signal delays result in timing skew among responses from net members of as much as the round-trip propagation time to the most distant station. The standards therefore require a 14 T_w minimum slot size, which allows response frames with single-word addresses to propagate to and from the other side of the globe, and accounts for the delays in commonly available HF equipment.

When net member addresses used in net calls exceed one ALE word, member responses and their allocated slots are expanded by one T_{rw} per additional ALE word required. Slots are extended by *two* T_{rw} for each additional ALE word in the calling station address, because that address appears twice in each response. Similarly, responses and slots are expanded to accommodate message or LQA CMD words included in member responses. As each slot is expanded, all subsequent slots are delayed commensurably.

The correct setting for the slot wait timer, when one or more slots vary from the minimize size, is given as a function for a selected slot number, s_n, by

Figure 6.12 ALE slotted responses.

$$T_{swt}(s_n) = s_n[5T_w + 2T_{ac} + d_{\text{LQA}} T_{rw} + d_m T_m] + T_{ac} + \sum_{m=1}^{m=s_n-1} T_{am} \quad (6.1)$$

where T_{ac} is the address length of the calling station given by an integer multiple of T_{rw}; T_{am} is the address length of the called net member programmed to respond in slot m. Note that the length of slot 0 is determined by using the address length of the calling station; d_{LQA} is equal to 1 if and only if an LQA response was requested in the call and is set to 0 otherwise; d_m is equal to 1 if and only if a message section was requested in the call and is set to 0 otherwise; and the $\sum_{m=1}^{m=s_n-1} T_{am}$ is the sum of called addresses in all previous slots except slot 0. Note that the length of the message section, T_m, is assumed to be the same for all slots. Given this result, the wait-for-net-reply timeout value, T_{wrn}, for the calling station is given by $T_{wrn} = T_{swt}(N_s)$ where N_s is the total number of slots, that is, the number of net member stations plus one to account for slot 0. The expression for the called-station-ACK timer then becomes $T_{wan} = T_{wrn} + 2T_{rw}$.

6.3.6.3 Star Group Calling

When a net is programmed, member stations are assigned a special address that refers to them collectively as well as a specific indexed slot time in which to respond. The group calling feature of the ALE standards extends the power of one-to-many calling to ad hoc collections of stations that have not been preprogrammed as a net. Nothing need be known about the stations except their individual addresses and scanned frequencies. Because a group is not set up in advance, however, stations must be able to derive a collective address and individual response slots on the fly. The group calling capability is particularly valuable during emergency situations in which stations that normally do not interact are compelled to form a net to support crisis warning, response, and recovery.

Naming a Group

As defined in Section 5.4.3, a group address is produced by combining the individual addresses of the stations that are to form the group. During a scanning call, only the first ALE word or words of addresses are sent, just as for individual or net calls. However, the group member individual addresses may have different first words, and all of them must be sent to capture all of the group members. Thus, the group calling protocol must have a mechanism for telling scanning stations that arrive on the calling channel during the scanning period, T_{sc}, to persist in the search for their respective addresses. In particular, the first word received by a scanning station during its scan period may not be the first word of one of its self-addresses. In this case, the station will continue to monitor the call because a subsequent word might match one of its address words. This persistent search for self-address words is requested by using THRU instead of TO words during the scanning period of a group call as depicted in Figure 6.13.

When group member addresses share a common first word, that word is sent only once during the scan period. Furthermore, only five unique first words may be sent in rotation during this scan period. The complete addresses of the prospective group members are sent during the leading call period of duration T_{lc} using TO words as usual. A total of up to 12 address words are allowed in the standard for the full addresses of group members, so T_{lc} in a group call may reach a maximum value of 24 T_{rw}.

Derived Slots

The second preprogrammed parameter used in linking with a net that is lacking in a group call is slot numbers. Slot numbers are derived for group call responses from the order in which individual addresses appear in the call. The last received address is assigned the first slot time, the next-to-last address is assigned the second slot time, and so on. Although this procedure may seem odd at first, it actually simplifies implementations and increases the robustness of the protocol. The sequence of events needed to derive slot times proceeds using the following steps:

1. Consider each station as it pauses on a channel carrying the scanning call portion of a group call. Each station will necessarily receive either a THRU or a REP word preamble. If the address word in this first received word matches the first word of one of its individual addresses, it will stay to receive the leading call during period T_{lc}. Otherwise, the station will continue to read first address words until it finds either (a) a match with the first word of a self-address, (b) a repetition of a word it has already

| THRU | REP | THRU | REP | THRU | REP | ... | THRU | REP | THRU | (End of T_{sc}; T_{lc} |
| BOB | EDG | SAM | BOB | EDG | SAM | | BOB | EDG | SAM | begins on next line) |

| TO | REP | DATA | TO | DATA | TO | REP | DATA | TO | DATA | TIS |
| BOB | EDG | AR@ | SAM | UEL | BOB | EDG | AR@ | SAM | UEL | JOE |

Figure 6.13 ALE group call.

seen, or (c) five unique words. In the latter two cases, the station concludes that it has not been called and returns to scan.
2. When the leading call period, T_{lc}, starts, a station potentially named during the scanning call period, T_{sc}, monitors its receiver output for its complete address. If its complete address is found, a slot counter is set to 1 and incremented for each subsequent station address. If that address is found again (as it should be, because the address list is repeated during the leading call), the counter is then reset to 1 and incremented for each subsequent address as before. The number of words in each subsequent address is also noted for use in computing T_{swt} from the above expression.
3. At the end of the leading call period, T_{lc}, each station named in the group call has computed its slot number and recorded the sizes of the addresses in the slots that will precede it.
4. The message section (if any) and the frame conclusion are processed as usual. At the end of the frame, the calling and called stations set timeouts as described for a net call and proceed through slotted responses in an identical fashion.
5. The ACK of responses from a group call may be addressed to any subset of the members originally called. This standard feature is necessary because not all of the called stations may have received the group call. Group members that responded but were not named in the calling station's ACK will simply return to scanning on expiration of their respective WRTT.

Now consider a station that begins to monitor a channel carrying the group call after the calling station's scanning calling period has concluded. This station may still correctly derive its slot data from other addresses received after its address during the leading call period, T_{lc}. This derivation will not be possible if (1) the station was assigned to respond during slot 1 and (2) it arrives on channel and receives only the last repetition of its address at the end of the leading call period. This station would be unaware that a group call was under way and would respond in slot 0. The calling station might still receive this response if the tuning performed by other group members does not corrupt more than the first T_{rw} of that response.

Slotted-Response Timing

The formulas for star group slotted responses (T_{swt}, T_{wrn}, T_{wan}) are the same as those for star net slotted responses.

In the example group call shown in Figure 6.13, station SAMUEL will respond in slot 1, with $T_{swt} = 14\ T_w$. (Note that in the one-word address JOE causes slot 0 to be 14 T_w in duration.) Next, station EDGAR will respond in slot 2 with $T_{swt} = (14 + 17)T_w = 31T_w$. Slot 1 is 17 T_w because of station SAMUEL's two-word address. Station BOB will respond in slot 3 with $T_{swt} = 48\ T_w$. As a result, station JOE will send an ACK after $(48 + 14)T_w = 62T_w$.

Consider a called station that does not identify the maximum length of addresses in the called group and therefore cannot compute the correct T_{wan} value. In this case, the station should use a default value for T_{wan} given by the longest possible group call of 12 one-word addresses. Based on the slotted response formula $T_{wan} = 107T_w + 27T_{ac} + 13d_{LQA}\ T_{rw} + 13d_m T_m$. In the case of no message field ($d_m = 0$) and a one-word address caller, the maximum value for $T_{wrn} = 188T_w \approx 25$ sec with no LQA requested ($d_{LQA} = 0$), or $T_{wrn} = 277\ T_w \approx 30$ sec with LQA.

Long Responses

Consider a situation in which the caller wishes a called station to respond with a longer transmission than fits within its assigned slot time. For example, the caller may request the called station to respond with a special message. In this case, the calling station may insert the NULL address in the previous adjacent position *in the leading call only* of a group calling cycle. This NULL address will provide a blank slot for *overflow*, immediately following that responder's slot time. This overflow slot, typically assigned the minimum width because the NULL address is a one-word address, provides an additional data word capacity of more than 4 T_{rw} in duration for the special message or other desired special response.

Multiple Self-Addresses

As another special case, consider a station called multiple times in a single group call by different addresses. In this case, the station should properly respond to at least one address. Note that the calling station may be unaware that the called station has multiple addresses. In some cases, it would be confusing or inappropriate to respond to one address, but not another address, despite their reference to the same station. Redundant calling address conflicts can be resolved after successful linking, however, if it results in operational problems.

6.3.6.4 Allcall Protocol

Recall from Chapter 5, Section 5.4.4.2, that an Allcall requests all stations receiving this calling protocol to stop scanning and listen, but not respond.

The Allcall special address structures are based on @?@ or some combination of @*@ where the asterisk (*) is one of the 36 uppercase alphanumeric characters. These structures should be the exclusive member, or members, of both T_{sc} and T_{lc} periods of the calling cycle during the initial call. They should not be used in any other address field or any other part of the handshake. In addition, the Global Allcall address should appear only in TO words. Selective Allcalls with more than one selective Allcall address, however, use the THRU word during the scanning calling period to form a Group Allcall.

Unless inhibited or otherwise directed by the operator or controller, all stations receiving an Allcall must temporarily stop their scan for a preset time $T_{cc(max)}$. If a message section or terminator section does not arrive within a time $T_{cc(max)}$, the station automatically resumes scanning. If a Quick-ID arrives beginning with a FROM word after the calling cycle, the pause for the message section is extended for no more than five words (i.e., 5 T_{rw}). If a CMD word does not arrive, the called station will then resume scanning. If a message arrives following receipt of a CMD word, the station pauses for a preset limited time, $T_{m(max)}$, to read the message. If the frame conclusion does not arrive within period $T_{m(max)}$, the station automatically resumes scanning. If a conclusion is received using a TIS or TWAS word, the station will pause for a preset limited time, T_{xx}, to read the caller's (transmitter's) address. If the end of the signal does not arrive within T_{xx}, the station automatically resumes scanning.

If an Allcall frame is successfully received and concluded with a TIS word, then the called station will stop scanning, alert the operator, and unmute its speaker to receive a message. If there is no activity for a preset time period, T_{wa}, the called station automatically mutes its speaker and returns to scan. If a station receiving an Allcall desires to link with the calling station, it initiates a handshake within the pause after TIS is received. If an Allcall is successfully received with a TWAS word, the called station will automatically resume scanning and not respond unless otherwise directed by the operator or controller. Note that the Allcall address should not be used in any handshakes other than the calling cycle of the initial Allcall. Finally, stations should have the capability to disable receipt of the Allcall to minimize adverse effects resulting from overuse or abuse of Allcalls. Otherwise, the Allcall is normally enabled.

6.3.6.5 Anycall Protocol

The only significant difference between an Anycall and an Allcall is that the Anycall solicits responses while the Allcall inhibits these responses. Use of the Anycall special address structures is identical to that for the Allcall special address structures. Unless inhibited or otherwise directed by the operator or controller, each station receiving an Anycall temporarily ceases scanning and examines the call using the procedure for Allcalls. The Anycall procedure therefore includes the $T_{cc(max)}$, $T_{m(max)}$, and $T_{x(max)}$ limits. If the Anycall is successfully received, the station will automatically generate a slotted response similar to that for a star net (scanning) call protocol as modified below.

Because an unknown collection of stations is expected to respond to an Anycall, neither preprogrammed nor derived slot data are available. Instead, the Anycall slotted responses use 17 standardized slots, composed of slot 0 plus 16 others, each of duration $20T_w = 2613.33$ ms. If the calling station requests LQA, the response slots increase in duration by 3 T_w to include the LQA.

Each responding station individually and *randomly* selects a slot (slot 1 through 16) for its response. In this protocol, collisions are expected and tolerated, so the station sending the Anycall will attempt to read the best response in each slot. Responses are standard star net responses consisting of TO words with the caller's station address and TIS words with the responding station's address. In addition, LQA CMD words are also included if requested. Responders must use a self-address no longer than five words minus twice the caller address length. For example, if the caller's address is two words, the responder must use a one-word self-address. They must not use the Anycall special address.

On receipt of the slotted responses, the calling station will transmit an acknowledgment (ACK) to any selected combination of responding stations using an individual or group address. The caller must not use the Anycall special address in the ACK. The caller then selects the conclusion of its ACK to either (1) maintain the link for additional interoperation and traffic with the responders using the TIS word or (2) return all stations to scan using the TWAS word, depending on the caller's original purpose in issuing the Anycall.

The responders that receive ACKs will alert their controllers or operators, unmute their speakers, and either pause for traffic if the TIS word is received or drop the link immediately if the caller's ACK conclusion included the TWAS word. If there is no activity for the preset time interval, T_{wa}, then the stations automatically mute their speakers and return to scanning. Any responding stations not included in the ACK also immediately depart and resume scanning. To minimize possible adverse effects resulting from overuse or abuse of Anycalls, stations should have the capability to disable receipt of the Anycall. Normally, however, the Anycall feature should be enabled.

6.3.6.6 Wildcard Calling Protocol

Wildcard addresses should be the exclusive members of a calling cycle in a call. They may not be used in any other address sequence in the ALE standard frame or handshake. The following rules govern responses to wildcard calls: (1) Calls to wildcard addresses that conclude with a TWAS word should cause no responses, and are handled similarly to Allcalls; and (2) calls to wildcard addresses that conclude with a TIS word should cause responses in pseudo-randomly selected slots, and should use the protocol otherwise identical to the Anycall. As in both the Allcall and Anycall protocols, the operator or controller should be able to disable acceptance of wildcard cards, but it should normally be enabled.

6.4 ALE DATA STRUCTURES

The ALE functions described in Section 6.3 are controlled by a large collection of programmable parameters. Conceptually, these operating data are organized into the four tables and a group of scalar variables described in the following sections. Most of these data may be examined and modified by distant control stations using the network management technology described in Chapter 9.

6.4.1 Channel Table

The *channel table* stores programmed operating parameters for each assigned channel, including (1) associated transmitting and receiving frequencies, (2) desired modulation (e.g., USB, LSB, ISB, AM, etc.), (3) transmitting power, (4) antenna type to be used with azimuth angles for steerable beams, and (5) sounding data. In addition, channels can be grouped into scan sets and assigned a corresponding index. In this way, the station can automatically select between different scan sets by merely changing the scan-set index.

6.4.2 LQA Table

The *LQA table* stores the results of channel evaluation for all active (channel, station) pairs. Since channel conditions are time variable, each measurement is time-tagged to facilitate age-based discrimination of stored LQA data. Simply put, older LQA measurements are generally weighted less significantly than more recent measurements during automatic channel evaluation for link establishment. In addition to time tagging, LQA data are indexed by the identity of the distant station. Since channel conditions may not be reciprocal on an HF link, LQA data are required in each link direction.

6.4.3 Self-Address Table

The *self-address table* lists all of the addresses to which a station will respond. In practice, an address may be restricted to use only on certain channels. In this case, these channels will be stored with the associated address. In the case of a net address, the corresponding individual address or addresses and assigned slot data are also stored.

6.4.4 Other Address Table

The ALE controller also maintains the *other address table*, which contains lists of addresses for other stations and nets that might be called. Channel restrictions may be listed for each address, such as no unauthorized transmissions on designated net controller broadcast frequencies. In practice, this table is dynamically updated when sounds are received from new stations. The address records for callable net addresses will include the addresses of net members and their assigned slots. ACK timeouts (T_{wrn}) may also be preprogrammed for net addresses.

6.4.5 Scalar Variables

Table 6.4 contains a list of the ALE operating parameters defined in the standards for remote ALE control using the HF management information base. Some of these parameters have been defined in previous sections while the definition of the remaining parameters is evident from their descriptive name.

6.5 ALE ORDERWIRE FUNCTIONS

In addition to automatically establishing links, ALE-equipped stations have the capability to transfer information within the message section of ALE frames. This section describes the protocols used for these messages, including data transfer, system control, error checking, networking, and special-purpose func-

Table 6.4
Remote Control Parameters

Timing Parameters	Switches	Operations
Scan rate (channels/s)	Accept Anycall	Scan set
Maximum channels scanned	Accept Allcall	Linked stations
Maximum tune time (T_{tx})	Accept AMD	
Turnaround time (T_{ta})	Accept DTM	
Activity timeout (T_{wa})	Accept DBM	
Listen time (T_{wt})	Request LQA	
	Auto power adjust	

tions. Table 6.5 provides a summary of the orderwire functions. Some of these orderwire functions are specified in the ALE standards, MIL-STD-188-141A [1] and FED-STD-1045A [2]. Other functions are from the planning standard MIL-STD-187-721C [3] and represent advanced ALE functions.

6.5.1 Link Quality Analysis

6.5.1.1 Overview

The LQA CMD word is designed to support the exchange of current LQA information among ALE stations. For the cost of one extra ALE word in each transmission, station A can report to station B the link quality just measured on the immediately preceding transmission from B to A. This real-time channel evaluation and reporting mechanism is one of the most powerful features of this generation of ALE technology, and it comes at a low incremental cost when piggybacked on other ALE transmissions. Support for the LQA CMD word is mandatory in the ALE standards.

The LQA CMD word is constructed as shown in Figure 6.14. The first character is "a," which identifies the LQA function "analysis." It carries three types of analysis information, BER, SINAD, and MP, which are separately generated by either ALE or other analysis capability. When the control bit KA1 (W11) is set to 1, the receiving station is requested to respond with an LQA report in its next frame of the handshake. If KA1 is set to 0, a report is not required.

6.5.1.2 Bit Error Ratio

A pseudo-BER measurement is empirically derived by all ALE stations during reception of an ALE signal as described in Section 6.2.1.3 and is quantified in a five-bit field, BE5 through BE1. The contents of this field give the average number of nonunanimous redundant word votes per T_{rw} interval counted during the immediately preceding transmission from the station to whom this frame is addressed. The number 30 represents thirty or more nonunanimous votes, and the number 31 (all 1s) indicates that no report of BER is available.

6.5.1.3 Signal-Plus-Noise-Plus-Distortion to Noise-Plus-Distortion Ratio (SINAD)

If SINAD is analytically derived from the analog ALE signaling by suitably equipped stations, then it is communicated as a five-bit field, SN5 through SN1. The quantitative range for the SINAD report is 0 to 30 dB in 1-dB steps.

Table 6.5
Summary of CMD Functions

First Character		Second Character		Function
Any of the Extended-64 character set				Automatic message display
"	1100000			Advanced LWA
a	1100001			LQA
b	1100010			Data block analysis
c	1100011			Channels
d	1100100			Data text message
f	1100110			Frequency
m	1101101			Mode selection commands
		a	1100001	Analog port Selection
		c	1100011	Crypto negotiation
		d	1100100	Data port selection
		n	1101110	Modem negotiation
		q	1110001	Digital squelch
n	1101110			Noise report
p	1110000			Power control
r	1110010			LQA report
t	1110100			Scheduling commands
		a	1100001	Adjust slot width
		b	1100010	Station busy
		c	1100011	Channel busy
		d	1100100	Set dwell time
		h	1101000	Halt and wait
		l	1101100	Contact later
		m	1101101	Meet me
		n	1101110	Poll operator (default NAK)
		o	1101111	Request operator ACK
		p	1110000	Schedule periodic function
		q	1110001	Quiet contact
		r	1110010	Respond and wait
		s	1110011	Set sounding interval
		t	1110100	Tune and wait
		w	1110111	Set Slot Width
		x	1111000	Do not respond
		y	1111001	Year and date
		z	1111010	Zulu time
v	1110110	c	1100011	Capabilities
		s	1110011	Version
x	1111000			Cyclic redundancy check (CRC)
y	1111001			(16-bit CRC overflows into
z	1111010			the two least-significant bits of
{	1111011			the first character)
|	1111100			User-unique functions
~	1111110			Time exchange

3	7	1	3	5	5
CMD	1100001 ('a': LQA)	KA1	MP3 … MP1	SN5 … SN1	BE5 … BE1

Figure 6.14 LQA CMD word format.

Thus, the number "00000" corresponds to 0 dB or less, "11110" to 30 dB or more, while the number "11111" indicates that no measured value is available.

Three bits, MP3 through MP1, are reserved for the reporting of estimated multipath values in milliseconds. The original intention at the 1987 "cantina meeting" of the Technical Advisory Committee was to report multipath in 1-ms steps, but this convention has not yet been standardized at this printing. Until standardized, these bits must be set to "111" to indicate that no measured value is available.

6.5.2 LQA Report

The LQA reporting protocol is optional as defined in MIL-STD-187-721C. It is used to exchange previously measured LQA data rather than to report current channel measurements. By exchanging unilateral measurements from passive LQA (e.g., measurements of sounds), ALE-equipped stations can quickly accumulate bilateral data for many channels and stations.

6.5.2.1 LQA Report CMD

A group of stations can exchange the unilateral LQA data derived from measurements performed on traffic and sounds received on several channels. The purpose of this exchange is to provide each station with bilateral LQA data for those channels in support of channel selection algorithms. Thus, the LQA report protocol "closes the loop" with passive LQA. In this regard, the LQA report protocol is used by station A to report LQA data to station B that were collected by station A during transmissions received from station B.

LQA reports are embedded in the message section of ALE frames. The first word of an LQA report message is an LQA report CMD word formatted as shown in Figure 6.15. This word is immediately followed by LQA reports for the number of channels specified in the *Chan* field in the CMD word. These subsequent reports are carried either using DATA and REP words, or in a DBM data block, as specified by control bit KR5. The individual LQA reports have one of three formats as specified by control bits KR4, KR3, and KR2, namely, (1) data only, consisting of 16 bits per report; (2) channel and data, containing 16

3	7	2	5	2	5
CMD	1110010 ('r': LQA report)	0 0	Control	0 0	Chan

Figure 6.15 LQA report CMD format.

bits of data and 7-bit channel designators; or (3) frequency and data amounting to 36 bits, as shown in Figure 6.16. The control bits and their respective meanings to the ALE controller are given in Table 6.6. The data portion of the report has the same format in each case. In the standardized encoding scheme, 3 bits specify the age of the data, 3 bits are retained for multipath data, 5 bits are used to report SINAD, and 5 bits quantify the BER measurement. When a channel or frequency designator is included in each report, the designator field immediately precedes its corresponding data field and is formatted as described later in Section 6.5.2.3.

The BER field in LQA reports is encoded as in the LQA CMD words. The SINAD measurement is encoded as an integer in decibels measured from 0 to 30, with the code "31" (11111) reserved to indicate that no SINAD report is available. Multipath is similarly encoded as an integer in milliseconds, 0-6, with the code "7" (111) reserved to indicate no multipath report. The age of each report is encoded as shown in Table 6.7.

When DATA and REP ALE words are used to carry LQA reports, the reports are packed bit by bit into the 21-bit data fields. Note that these bits are packed without aligning each report with the start of a new ALE word. When the reports do not fill the final ALE word, the remaining bits after the last report are filled with 0s. The DBM data block is likewise filled with packed reports, with the length field set to indicate the length of reports.

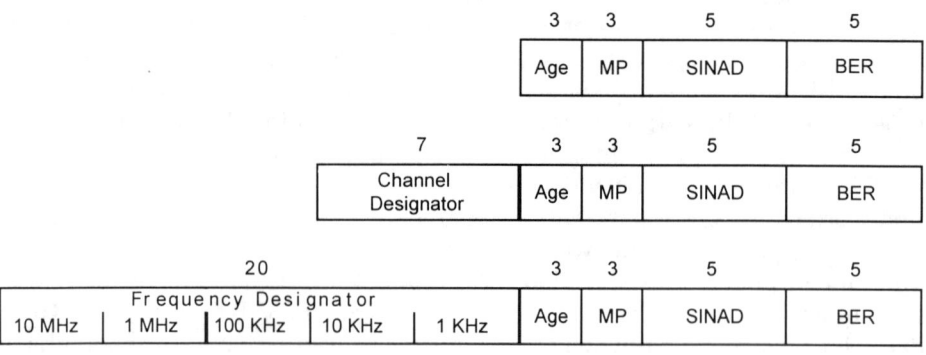

Figure 6.16 LQA report formats.

Table 6.6
Control Bit Assignments (LQA Report CMD Word)

Control Bit		Meaning
KR5 (msb)	0	Reports are in $\overline{\text{DTM}}$ message
	1	Reports are in $\overline{\text{DBM}}$ message
KR4, 3, 2	000	Data only, no channel or frequency designators
	001	Data and channel designator with negative offset
	010	Data and channel designator with absolute channel no.
	011	Data and channel designator with a positive offset
	100	Data and frequency designator with a negative offset
	101	Data and frequency designator with absolute frequencies
	110	Data and frequency designator with a positive offset
	111	Reserved
KR1 (lsb)		Reserved

Table 6.7
Age Field Encoding

Age Field	Age of Reported Data
0	0-15 min
1	15-30 min
2	30-60 min
3	1-2 hr
4	2-4 hr
5	4-23 hr
6	23-25 hr
7	Older than 25 hr or unknown

6.5.2.2 LQA Report Request

LQA reports may be sent either on request or by prearranged schedule. A station may request an LQA report by sending an LQA report request formatted as shown in Figure 6.17. The bits of the control field are used to request the format of LQA reports. These bits are formatted as was shown in Table 6.6. The age field specifies the maximum age acceptable for available LQA reports and is

3	7	2	5	4	3
CMD	1110010 ('r': LQA report)	1 0	Control	0 0 0 0	Age

Figure 6.17 LQA report request CMD format.

formatted as shown in Table 6.7. However, if the age field is set to 6, then all channels with LQA measurements of any age are requested. If the age field is set to 7, then LQA measurements for all channels common to the two stations are requested, including those for which no data are available.

6.5.2.3 Channel and Frequency Designators

Three different encodings are used when two or more stations need to explicitly refer to channels or frequencies other than those in use for a link:

1. A frequency is designated using a binary coded decimal (BCD) string. The standard frequency designator is a five-digit string (20 bits), in which the first digit is the 10-MHz digit, followed by 1-MHz, 100-kHz, 10-kHz, and 1-kHz digits.
2. A frequency designator is normally used to indicate an absolute frequency. When a bit in the command associated with a frequency designator indicates that a frequency offset is specified instead, the command will also contain a bit to select either a positive or a negative frequency offset.
3. A channel differs from a frequency in that a channel is a logical entity that implies not only a frequency, or two frequencies for a full-duplex channel, but also various operating mode characteristics as described in Section 6.4.1. Channels may be specified either absolutely or relatively. In either case, a seven-bit binary integer interpreted as an unsigned integer in the range 0 through 127 is used to index individual channels. Bits in the associated command word indicate whether the channel designator represents an absolute channel number or, alternatively, a positive or negative offset from the current channel index. These encodings are defined in MIL-STD-187-721C for use with the advanced ALE functions.

6.5.3 Noise Report

The sounding protocols necessarily result in only unilateral link quality assessments at the receiving station. The noise report CMD supports an alternative broadcast technique that permits receiving stations to predict the approximate *bilateral* link quality for the channel used by the station sending the noise report. The utility of this optional technique is apparent in networks for which most stations rarely transmit, but which nevertheless require a high linking probability on the first attempt with a sounding or broadcasting station.

A station receiving any ALE transmission can measure unilateral link quality on the channel from the sounding station. If a noise report CMD is included in a broadcast transmission, then the recipient can compare the noise level at the broadcast station to its own local noise level and estimate the bilateral link quality. Note that noise report CMDs are not sent in sounds, but in the message section of ALE frames. This noise report CMD word is an advanced ALE function described in MIL-STD-187-721.

A noise report is contained in a single CMD word as shown in Figure 6.18. The noise report gives the mean and maximum noise power measured on the channel in the past 60 min. Units for the maximum and mean bit fields in decibels are relative to 0.1 mV in a 3-kHz noise bandwidth. If the local noise measurement to be reported is less than 0 dB on this relative scale, then a "0" is sent. For measured noise ratios of 0 to 126 dB, the ratio in decibels is rounded to an integer before transmission. For noise ratios greater than 126 dB, a value of "126" is sent. The code value "127" (all 1s) is sent when no report is available for a field. In practice, a station receiving noise reports from a distant station can make a better selection of channels for linking attempts with that station. This selection is based on knowledge of both propagation characteristics and the interference/noise environment reported on several channels by the distant station.

6.5.4 Advanced Link Quality Analysis

Advanced link quality analysis (ALQA), an extension to the standard LQA mechanism, is an optional technique defined in MIL-STD-187-721. ALQA techniques differ from the basic LQA techniques in two respects: (1) Measurements are histogrammed rather than averaged to provide a more detailed characterization of recent channel behavior, and (2) the quality of a channel is reported as the fraction of measurements that exceeded a threshold, rather than an average of the measurements. ALQA measures are divided into two types, channel

3	7	7	7
CMD	1101011 ('n': noise report)	Max	Mean

Figure 6.18 Noise report format.

quality measures (CQMs) and link performance measures (LPMs). CQMs evaluate the performance of individual channels between pairs of stations, whereas LPMs combine measures of channel performance and traffic load to derive a single quantitative measure of link performance for use by network controller routing algorithms.

6.5.4.1 Channel Quality Measures

The following CQMs have been standardized for ALQA:

- SINAD measurements are derived from the ALE modem on a baud-by-baud time scale.
- Articulation index (AI) estimates are also derived from the ALE modem using a nonuniform weighting of SINAD values from each ALE tone. The weighting approximates the importance of each spectral band to the intelligibility of speech.
- PBER is derived from the majority vote decoder as described for LQA. For ALQA use, however, the PBER values are histogrammed on a word-by-word basis over several epochs, including all past transmissions since power-up, rather than contributing to a running average for a single transmission.
- Spectral distortion due to fading and multipath effects causes degradation in the BER performance of the ALE modem as compared to its performance over Gaussian noise channels. The severity of spectral distortion is gauged by comparing instantaneous samples of BER and SINAD measured on received ALE words to theoretical AWGN channel performance.
- The mean error-free interval (EFI) is the average duration of contiguous error-free ALE words expected on the link. Because ALE transmissions are not continuous, indirect calculation of this parameter must be performed from available measurements. In this case, ALE word errors detected by the Golay decoder and ALE protocol are accumulated to estimate word-error and word-error-burst probabilities. These probabilities are then used to estimate EFI.

All implementations of ALQA support SINAD and PBER. AI, SD, and EFI are optional as are the LPMs, AVQ and ADC defined next.

6.5.4.2 Link Performance Measures

The following LPMs have been standardized for ALQA:

- Achievable voice quality (AVQ) samples combine SINAD or AI measurements with channel occupancy to produce a histogram of the voice quality

of the best channel available during each scan of the channels performed by the ALE controller.
- Data link performance is gauged using available data capacity (ADC) samples. ADC samples measure the effective data rate available to each station, including both the effects of retransmissions due to channel errors and contention for transmission facilities from other traffic.

These LPMs provide single-valued performance measures of voice and data link performance, respectively, for use by the HF network controller described in Chapter 9.

Further details of ALQA operation can be found in MIL-STD-187-721C and in a recent paper [4].

6.5.5 Channel and Frequency Commands

Channel and *frequency* CMD words are used to modify the effect of other CMD words. As a result, the function to be performed will occur on a specified channel or frequency rather than on the default or current channel. This optional functionality is specified in MIL-STD-187-721. The *frequency select* CMD word is formatted as shown in Figure 6.19. A *frequency designator* (see Sec. 6.5.2.3) is sent in a DATA word immediately following the frequency select CMD word. Bit W4 of this DATA word is set to 0 as shown in the figure. The 100- and 10-Hz fields in the frequency select CMD word contain BCD digits that extend the precision of the standard frequency designator. These digits are normally set to 0, except when it is necessary to specify a frequency that is not an even multiple of 1 kHz, such as when many narrowband modem channels are allocated within a single 3-kHz voice channel.

Setting the control field to "000000" specifies a frequency absolutely; whereas setting this same field to "1000000" indicates a positive frequency offset, and "110000" indicates a negative frequency offset. Other encodings

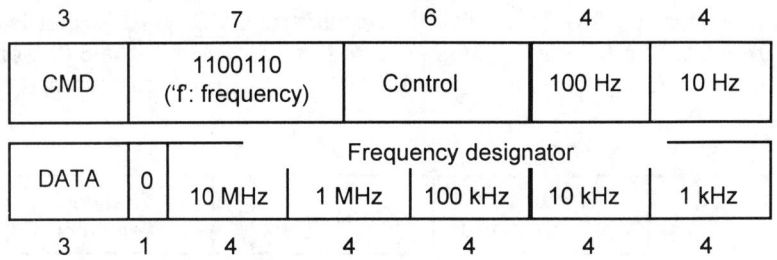

Figure 6.19 Frequency select CMD format.

were included in Harrison's original scheme [5] but were not included in the standards.

A station receiving a frequency select CMD word should in general make whatever response is required by an active protocol on the indicated frequency. For example, transmitting and receiving frequencies to be used on a full-duplex link may be negotiated independently as follows:

- Station A selects a frequency expected to propagate to the distant station B, the prospective responder, and places a call to station B on that frequency. Station A embeds a frequency select CMD word in the call to request that station B respond on a frequency exhibiting the best propagation from station B to station A. Most likely, these data would come from data in the LQA table at station A.
- If station B hears the call, it will respond on the second frequency. At this point, station B may ask station A to switch to a better transmitting frequency by embedding a frequency select CMD word in its response, possibly also based on LQA data.
- Station A sends an ACK on the frequency chosen by the responder, using the original frequency by default, and the *FDX independent* link is established.

An equivalent channel select CMD word could be formed as shown in Figure 6.20. This word has a seven-bit channel designator in the least significant bit positions and uses the same control field encodings as for the frequency select CMD word. However, the ALE standards do not include such a channel select $\overline{\text{CMD}}$ word.

6.5.6 Power Control

The *power control* orderwire command is used to advise linking or linked communicants that they should raise or lower their transmitting power for optimum system performance. The power control CMD word format is shown in Figure 6.21. The KP control bits are used as shown in Table 6.8. Normally, a station receiving a power control request (KP3 = 1) approximates the requested

3	7	6	1	7
CMD	1100011 ('c': channel)	Control	0	Channel Designator

Figure 6.20 Hypothetical channel select CMD format.

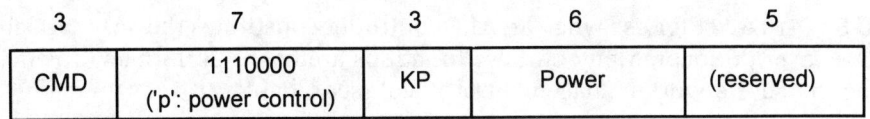

3	7	3	6	5
CMD	1110000 ('p': power control)	KP	Power	(reserved)

Figure 6.21 Power control CMD format.

Table 6.8
Power Control Command Bits (KP1-3)

Bit	Value	Meaning
KP3 (msb)	1	Request to adjust power
	0	Report of current power level
KP2	1	Relative power (dB)
	0	Absolute power (dBW)
KP1 (lsb)	1	Relative power (dB) is positive
	0	Relative power (dB) is negative

effect as closely as possible. Once completed, the station responds with a power control report (KP3 = 0) to specify (and to demonstrate) the result of its power adjustment.

6.5.7 Mode Control

6.5.7.1 Overview

Many of the advanced features of an ALE controller are *modal* in the sense that when a particular option setting is selected, that selection remains in effect until changed or reset by some protocol event. The mode control CMD is used to select many of these operating modes as described in the following paragraphs. The CMD word is formatted as shown in Figure 6.22. The first character is "m" to identify the mode control CMD word, the second character identifies the type of mode selection being made, and the remaining bits specify the new setting for that mode. The optional mode control CMD words are included in

3	7	7	7
CMD	1101101 ('m': mode control)	Mode ID	Mode selection

Figure 6.22 Mode control CMD format.

MIL-STD-187-721 for use when an ALE controller constitutes the only available higher level protocol. Many of these functions are better performed, however, using the HF network management protocol (see Chapter 9).

6.5.7.2 Modem Negotiation and Handoff

An ALE data link can be used to *negotiate* a modem to be used for data traffic by exchanging *modem negotiation* messages. Each such message contains a modem selection CMD word formatted as shown in Figure 6.23. The leading CMD word may be followed by one or more DATA words as discussed next.

Modem negotiation employs messages in the following protocol:

1. The station initiating the negotiation will send a modem selection CMD word containing the code of the modem it wants to use.
2. The responding station, or stations, may either accept this modem selection or suggest alternatives. A station accepting a suggested modem sends a modem selection CMD word containing the code of that modem.
3. A station may extend the negotiation by sending a modem selection CMD word containing all 1s in the modem code field, followed by one or more DATA words containing the codes of one or more suggested modems. Modem codes are listed in order of preference in the DATA word or words. Unused positions in the DATA words are filled with the all 1s code.
4. The negotiation is concluded when the most recent modem negotiation message from all participating stations contained an identical modem selection CMD word with the same modem code (not all 1s). When this occurs, the station that initiated the negotiation will normally begin sending traffic using the selected modem.

6.5.7.3 Crypto Negotiation and Handoff

An ALE data link can also be used to negotiate the particular encryption device to be used for voice or data traffic by exchanging *crypto negotiation* messages. The crypto selection CMD word is formatted as shown in Figure 6.24. Crypto

3	7	7	7
CMD	1101101 ('m': mode control)	1101110 ('n': mode negotiate)	Modem code

Figure 6.23 Modem selection CMD format.

3	7	7	7
CMD	1101101 ('m': mode control)	1100011 ('c': crypto select)	Crypto code

Figure 6.24 Crypto selection CMD format.

negotiation employs crypto negotiation messages in the protocol described earlier for modem negotiation.

6.5.7.4 Analog Port Selection

The *analog port selection* CMD word is used to enable and disable audio inputs and outputs individually at a station. The analog port selection CMD word is formatted as shown in Figure 6.25. The bits of the analog port field in the CMD word are assigned as indicated in Table 6.9. A bit set to 1 enables the corresponding analog port; when set to 0, the corresponding analog port is disabled. The analog ports controlled by these standardized bits are components of the radio that is receiving the command. The other bits may be employed to control any other analog ports at a station. Multiple inputs and outputs may be simultaneously enabled by this command. If the equipment at a station

3	7	7	7
CMD	1101101 ('m': mode control)	1100001 ('a': analog port select)	Analog Port Bits

Figure 6.25 Analog port selection CMD format.

Table 6.9
Analog Port Selection Bits

Bit	Analog Port Assignment
VP7 (msb)	Operator microphone (input)
VP6	Line-level input
VP5	Local significance
VP4	Local significance
VP3	Local significance
VP2	Line-level output
VP1 (lsb)	Operator speaker/headset (output)

cannot fully implement a command, then the equipment should approximate the requested effect as closely as possible.

6.5.7.5 Data Port Selection

The *data port selection* CMD word shown in Figure 6.26 is used to specify the destination for the immediately following DTM or DBM message. By default, any DTM or DBM message that arrives without an immediately preceding data port selection CMD is assumed to carry a message for the station operator. In this case, the message is routed to an appropriate data port, such as an operator display or a printer. Note that the data port selected by a data port selection CMD word persists only until the end of the DTM or DBM message that immediately follows that CMD word. Thus, a data port selection CMD word has no effect if it is not immediately followed by a DTM or DBM message.

The station operator data port, which is the default, is explicitly specified by preceding a message with a data port selection CMD word with port number 0. A message destined for an attached network controller must be preceded by a data port selection CMD with a port number of 1. Similarly, a data fill message destined for the ALE controller must be preceded by a data port selection CMD word with a port number of 2. Other port numbers from 3 through 15 have station-specific meanings.

6.5.7.6 Digital LINCOMPEX Zeroization

The *digital LINCOMPEX zeroization* CMD word depicted in Figure 6.27 may be used to zeroize a digital LINCOMPEX system. The third character position is set to "1111111."

3	7	7	3	4
CMD	1101101 ('m': mode control)	1100100 ('d': data port select)	0 0 0	Port #

Figure 6.26 Data port selection CMD format.

3	7	7	7
CMD	1101101 ('m': mode control)	1111010 ('z': LINCOMPEX zeroize)	Subcommand

Figure 6.27 Digital LINCOMPEX zeroization CMD format.

6.5.7.7 Digital Squelch

The *digital squelch* CMD word depicted in Figure 6.28 is used for remote control of a radio's audio output. The second character position is set to "q" in order to indicate a "C" command. The third character position (i.e., "subcommand" in the figure) is set to "1111110" to mute the speaker of a distant radio or set to "0000000" to unmute. A receiving ALE controller that cannot mute the radio speaker may respond with a digital squelch CMD word with a subcommand of "1111111." In other cases, no response is necessary from the receiving ALE controller.

6.5.8 Scheduling Commands

Table 6.10 lists the groups of ALE scheduling functions. The interested reader should refer to the standards texts for details of the encoding and use of these commands. Each of these functions employs a scheduling CMD word with the generic format shown in Figure 6.29. The first character in every scheduling CMD word is "t," i.e., "1110100." The second character in the CMD word identifies the specific scheduling function to be performed. For all scheduling functions except the date and time group, the third character position contains a time code from Table 6.11. The time offset indicated in the time code is added to the time of receipt of the end of the transmission carrying the CMD word. This computed sum becomes the time T at which the specified function is to be performed as described in the relevant following paragraph. Note that the time code definitions specified in Table 6.11 are those specified in MIL-STD-187-721, which differs from the definitions in MIL-STD-188-141A. The 141A definitions support delays up to 29 hours. These delays were considered unreasonably long by the developers of MIL-STD-187-721.

In some cases noted under the Date/Time Option column in Table 6.10, the third character position may be set to "1111111." Setting this character to all 1s specifies that the function is to be performed at an absolute date and time rather than at a time offset from the end of the transmission. The desired date and time are specified in a DATA word. This DATA word immediately follows the CMD word and is formatted as shown in Figure 6.30. The month field indicates the desired month (1 to 12), the day field the desired day (1 to

3	7	7	7
CMD	1101101 ('m': mode control)	1110001 ('q': digital squelch)	Subcommand

Figure 6.28 Digital squelch CMD format.

Table 6.10
Scheduling Command Groups

Group	Scheduling Functions	Second Character	Data/Time Option
Future calls	Contact later	l	Yes
	Meet me	m	
	Quiet contact	q	
Wait on channel	Halt and wait	h	No
	Respond and wait	r	
	Tune and wait	t	
	Do not respond	x	
Congestion management	Station busy	b	No
	Channel busy	c	
	Set dwell time	d	
Slot width	Adjust slot width	a	No
	Set slot width	w	
Periodic functions	Schedule periodic function	p	Yes
	Set sounding interval	s	
Poll operator	Poll operator (default NAK)	n	Yes
	Request operator ACK	o	
Date and time	Year and date	y	Implicit
	Zulu time	z	

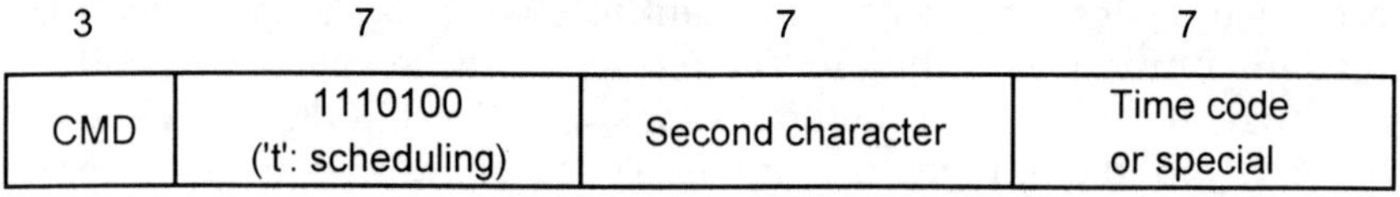

3	7	7	7
CMD	1110100 ('t': scheduling)	Second character	Time code or special

Figure 6.29 Generic scheduling CMD format.

Table 6.11
Time Codes for Scheduling

Time Code Bits	Encoding	Meaning	Range (approx.)
	00	Time unit is 1 T_w ($\approx$ 1/8 sec)	0-4 sec
TB7 (msb), TB6	01	Time unit is 8 T_w ($\approx$ 1 sec)	0-32 sec
	10	Time unit is 64 T_w ($\approx$ 8 sec)	0-4 min
	11	Time unit is 1024 T_w ($\approx$ 120 sec)	0-69 min
TB5-TB1 (lsb)	00000-11110	Time offset is indicated multiple of time unit	
	11111	Use absolute data and time from following DATA word	

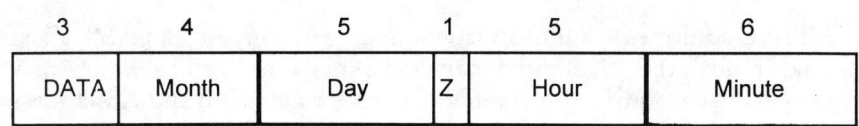

Figure 6.30 Date and time word format.

31, depending on the month), the hour field the desired hour (0 to 23), and the minute field the desired minute (0 to 59). The Z bit is set to 0 if the specified date and hour are Zulu time (UTC). Otherwise, the date and hour correspond to the local time zone of the sending station.

6.5.9 Automatic Message Display

The automatic message display (AMD) mode, which is mandatory in the ALE standards, was conceived to enable stations to communicate short orderwire messages or prearranged codes to one or more selected stations. This basic data transfer function exploits the communications, processing, controller, and operator interfaces already present in ALE stations and systems. The operators and controllers are able to send and receive simple ASCII text messages using only the existing station equipment. The Expanded-64 character subset is used for this purpose.

Every ALE station should have the capability to both send and receive AMD word messages from and to both the operator and the network controller. The station should also have the capability to display any received AMD messages directly to the operator upon arrival as well as alert the operator and network controller when an AMD message has arrived. The message display has a minimum required capacity of at least 20 characters, although a minimum of 40 characters is preferable. To provide recall of the most recently received message or messages, the AMD message storage capacity is at least 90 characters, along with the sending station address or addresses. A total of at least 400 characters is preferred. The operator or controller should be able to display any message in the AMD memory and identify the address of the originating station.

Receipt of the CMD AMD word warns the receiving station that an AMD message is about to be received. The controller alerts the operator and displays the message, unless these actions have been disabled. The operator and controller have the capability to disable the display and the alarm when their functions would be undesired. In addition, the ALE-equipped station can distinguish between multiple independent AMD messages and display them separately. This capability is independent of whether they were received in one or several transmissions.

All ALE-equipped stations must be capable of receiving an AMD message contained in any ALE frame, including calls, responses, and ACKs. An AMD message is constructed in the standard word format, with the AMD message inserted in the message section of the ALE frame. Within the AMD structure, the first word is a CMD AMD word containing the first three characters of the message. This first word is followed by a sequence of alternating DATA and REP words containing the remainder of the message. The CMD, DATA, and REP words must each contain only characters from the expanded ASCII 64 subset, which identifies them as a legitimate AMD transmission.

Each separate AMD message is kept intact, sent in a single frame, and transmitted in the exact sequence of the message itself. In many cases, one or two additional characters are required to fill the standard ALE word triplet in the last word sent. In this case, the ALE controller automatically "stuffs" these positions with the *space* character "0100000." The end of the AMD message is indicated by the start of the frame conclusion or by the receipt of another CMD word. Multiple AMD messages may be sent within a frame, but each message must start with a new CMD word.

Up to 30 AMD words may be sent within an AMD message. However, up to 29 other CMD words may precede the beginning of an AMD message. In such cases, the value of $T_{m(max)}$ (see Sec. 6.3.1.2) is expanded from a duration of 30 ALE redundant words to a maximum of 59 redundant words. The maximum AMD message remains 30 words, however, regardless of additional CMD words in the frame. If words are received that have the proper AMD format, but are within a portion of the message section under the control of another message protocol (such as DTM), the other protocol takes precedence and the words will be ignored by the station's AMD function.

The received message characters within the AMD structure are displayed verbatim as received. If a detectable information loss or error occurs (e.g., detected by the FEC sublayer), the station "flags" this event by substitution of a unique and distinct error indication. For example, this occurrence could be indicated when all display elements are activated to form a *block* in each character position. Note that when higher data integrity or reliability is required than provided by the ALE FEC capability, DTM or DBM protocols with CRC should be used instead of AMD messages.

6.5.10 Data Text Message Mode

6.5.10.1 Overview

The *data text message* (DTM) mode orderwire function enables stations to communicate text or binary messages. DTM orderwire messages may be unilateral or bilateral, that is, transmitted in one or both link directions, respectively, and may be broadcast or use ARQ. The DTM function provides a modem-like function for associated data terminals or other DTE devices connected to the station through the ALE controller DCE ports. The DTM data transfer function is a standard speed mode like AMD, but with improved robustness. In particular, this special orderwire message function enables utilization of the inherent redundancy and FEC techniques to work with weak HF signals and tolerate short noise bursts while detecting uncorrected errors. The DTM data blocks are fully buffered at each station, so the underlying DTM channel should appear transparent to the DTEs connected to the station. Routing of DTM messages upon arrival may be controlled using the data port selection CMD word previously described in Section 6.5.7.5.

There are four DTM modes: BASIC, EXTENDED, NULL, and ARQ. The DTM BASIC block can include a maximum of 651 bits, or 93 characters, with a data length resolution of one bit. The DTM EXTENDED blocks contain a maximum of 351 ALE words, consisting of 1,053 characters or 7,371 bits; the data length resolution of one ALE word may require padding. If the DTM structure transmission time exceeds the maximum limit for the message section (T_{max}), the DTM protocol overrides the T_m limit. The DTM NULL and ARQ are used for link management, error handling, and flow control. The characteristics of the DTM orderwire message modes are listed in Table 6.12.

The DTM structure is inserted within the message section of the standard ALE frame. A CMD DTM word precedes the DTM data block and specifies the mode, the block length, and four control bits. The message data are transferred in ALE words following the DTM CMD, using DATA and REP word preambles. A CMD CRC word immediately follows the data block words in the Basic and Extended modes only, carrying the error control frame check sequence (FCS). The control bits in the DTM CMD word are used to request data ACK, for flow control, to number alternate frames, and to identify message boundaries. Oddly,

Table 6.12
DTM Characteristics

DTM Mode	Basic	Extended	ARQ Null
Maximum data bits	651	7371	0
Cyclic redundancy check	16 bits	16 bits	None
Data capacity (ASCII characters)	0-93	3-1053	0
Capacity quantum (characters)	1	3	
Data capacity (bits)	1-651	21-7371	0
Capacity quantum (bits)	1	21	
Time diversity (in T_{rw})	3 fixed	3 fixed	0
Data transmission time	≈ 12 sec	≈ 2.29 sec	0

the bit used to mark message boundaries changes in the *first* frame rather than in the *last* frame of a series of ALE transmissions. This convention implies that the end of a message is not detected until the next frame arrives.

The DTM NULL CMD word is used (1) to interrupt ("break") the DTM and message flow, (2) to interrogate a station to confirm DTM capability before initiation of the DTM message transfer protocols, and (3) to terminate the DTM protocols while remaining linked. The protocol frame terminations for all involved stations during a DTM transfer use TIS words until all the DTM messages are successfully transferred (and all are acknowledged, if ARQ error control is in use). The only exceptions occur when (1) the protocol is a one-way broadcast or (2) a station is forced to abandon the exchange by the operator or controller. In either case, the termination will necessarily use the TWAS word.

6.5.10.2 Corrupted Frames

The ALE standards developers tacitly assumed that the reception of ALE frames and DTM data blocks would be subject to deep signal fades, interfering signals, noise bursts, and channel collisions. It was therefore mandated that once the receiving station had achieved word sync with the frame and DTM data block transmissions, it would maintain this synchronization. The ALE-equipped station can read and process any colliding ALE signals, even if these signals were significantly stronger than the intended DTM signal, without confusing them with the DTM signal ("basic ALE reception in parallel" and "always listening"). Therefore, useful information derived from readable collisions of ALE signals is not arbitrarily rejected or wasted.

Even when some words in a DTM data block are lost, the words not flagged by the ALE FEC sublayer as containing uncorrectable errors are retained for use in overlays as retransmissions of the previously corrupted frame are received. "Good" words from multiple generations of the frame may then be combined and the reconstructed frame checked for integrity using the CRC. Such a technique can substantially increase the throughput of DTM frames transmitted on marginal channels.

6.5.10.3 Asymptotic Throughput

Data throughput of the DTM mode is bounded by the ALE modem and the standard FEC sublayer. For extended DTM data blocks, the overhead of each frame becomes small in relation to the frame length (e.g., seconds compared to minutes), and we approach the following asymptotic throughput $X_{\max} = 375$ (bps) $\times (24/147) = 61$ bps. Of course, repeated frames sent to correct errors reduce this throughput by a factor equal to the average number of times each frame is transmitted.

6.5.11 Data Block Message Mode

6.5.11.1 Overview

The *data block message* (DBM) mode orderwire message mode seeks to improve throughput and robustness over DTM by eliminating the usual ALE triple redundancy in favor of deep interleaving. The control bits and protocol used for DBM are the same as for DTM.

As with DTM, there are four DBM modes: BASIC, EXTENDED, NULL, and ARQ. The DBM BASIC block is a fixed size, and contains a variable quantity of data from 0 to 572 bits, which is exactly specified to ensure integrity of the data during transfer. The DBM EXTENDED blocks are variable in size in integral multiples of the BASIC block, up to 262,820 bits of user data. Finally, the DBM NULL and ARQ variants are used for link management, error handling, and flow control. The characteristics of the DBM orderwire message modes are listed in Table 6.13.

The DBM structure is inserted within the message section of a standard ALE frame. A DBM CMD word precedes the DBM data block and contains the same types of information as a DTM CMD word. This DBM CMD word receives the usual ALE FEC processing. However, the data that follow the DBM word are treated quite differently.

The data to be transferred are broken into blocks of the size specified in the DBM CMD word. This size value is a multiple of 588 bits (49 × 12) less 16 bits for an embedded CRC FCS. The FCS is computed over the block of user data and occupies the final 16 bits of the data block. The data block (including

Table 6.13
DBM Characteristics

DBM Mode	Basic	Extended	ARQ Null
Maximum data bits	588	262,836	0
Cyclic redundancy check	16 bits	16 bits	None
Data capacity (ASCII characters)	0-81	81-37,377	0
Capacity quantum (characters)	1	3	
Data capacity (bits)	1-572	572-262,280	0
Capacity quantum (bits)	1	21	
Time diversity (in T_{rw})	49 fixed	3 fixed	0
Data transmission time	3.1 sec	3.1 sec-23 min	0

the FCS) is Golay encoded, 12 bits at a time, and the resulting 24-bit Golay words are written to an interleaver buffer. The capacity of this buffer is a multiple of 49 Golay words. Each Golay word constitutes a row in the interleaver, and the interleaver depth (ID) is the number of rows in the buffer.

When the buffer is full, the bits are read out column by column. In other words, the first bit of the first Golay word is read out, followed by the first bit of the second word, and so on until the first bit of the last Golay word is read. Then, the second bit of the first Golay word is read, and so on until the buffer is empty. This bit stream is sent over the channel using the ALE modem and is de-interleaved at the destination.

After error correction using the Golay code, each received DBM data block is examined using the CRC data integrity test embedded within the DBM structure and protocol. If a received data block passes the CRC test, the data are passed to the appropriate DCE port or normal output as directed by the operator or controller. If the data block is part of a larger message that was segmented before DBM transfer, it is reassembled in the ALE controller before being output.

DBM data blocks that fail the CRC data integrity test, or that contain detectable but uncorrectable errors, are tagged for further analysis, error control, or inspection by the operator or controller. If ARQ is required, the received, albeit unacceptable, data block is temporarily stored. Next, a DBM ARQ NAK is returned to the sending station, which will then retransmit an exact duplicate DBM data block. Upon receipt of the duplicate, the receiving station will again test the CRC. If the CRC is successful, the data block is passed through as previously described, the previously unacceptable data block deleted, and a DBM ARQ ACK returned. If the CRC fails again, both the duplicate and the previously stored data blocks may be used to correct errors and to create an "improved" data block as described earlier for DTM mode.

6.5.11.2 Use of the Basic-Size DBM Data Block

The DBM BASIC data block has a fixed size (ID = 49) and may be used to transfer any quantity of message data between 0 and 572 bits or 81 ASCII characters. The DBM BASIC CMD word reports the exact number of message data bits contained in the data block. It should be used as a single DBM for any message data within this range. It should also be used to transfer any message data in this size range that is an "overflow" from the larger size DBM EXTENDED data blocks (which should immediately precede the DBM BASIC block in the DBM sequence).

6.5.11.3 Use of the Extended-Size DBM Data Block

The DBM EXTENDED data blocks are variable in size and quantified in increments of 588 bits. These blocks should be used to maximize the advantages of

DBM deep interleaving, and higher speed transfer of data as compared to DTM or AMD techniques. The ID of the EXTENDED data block provides the largest data field size that can be totally filled by the message data to be transferred. Any *overflow* is contained in a message data segment sent within DBM EXTENDED or BASIC data blocks immediately following the first block's transmission.

6.5.11.4 Fragmentation

Under operator or controller direction, multiple DBM EXTENDED data blocks, with smaller than the maximum appropriate ID sizes, may be selected if they will optimize DBM data transfer throughput and reliability. However, these multiple data blocks will require that the message data be divided into multiple segments at the sending station and sent in the exact order of the segments in the message. The receiving stations must reassemble the segments into a complete received message. When binary bits are being transferred, the EXTENDED data field is filled exactly to the last bit. When ASCII characters are being transferred, the EXTENDED data field may have 0 to 6 stuff bits inserted. Individual ASCII characters are not be split between DBM data blocks. Finally, the receiving station reads the decoded data field on a 7-bit basis, discarding any remaining stuff bits.

6.5.11.5 Asymptotic Throughput

HF link data throughput using the DBM mode can be substantially higher than for the DTM or AMD modes. This increased throughput is due both to elimination of triple redundancy and the stuff bit, and to a reduced number of retransmissions resulting from the deep interleaving. For large extended DBM data blocks, the following asymptotic throughput in the error-free case provides an upper bound on link performance $X_{max} = 375$ (bps) $\times$ (588/1176) = 187.5 bps. Of course, repeated frames needed to correct errors reduce this throughput by a factor equal to the average number of times each frame is sent.

6.5.12 Cyclic Redundancy Check

A *cyclic redundancy check* (CRC) error-checking function is available to provide data integrity assurance for any form of message in an ALE frame. This CRC function is optional in the standards, but mandatory when used with the DTM or DBM modes. The sixteen-bit FCS and associated method[1] as specified by FED-

---

[1] The federal standard procedure (FED-STD-1003) specifies that the 16-bit shift register used to compute the CRC be initially filled with 1s (versus 0s), and that the remainder be inverted for use as the FCS.

STD-1003 are used for all ALE CRC operations. The FCS provides a probability of undetected error of 2^{-16} regardless of the number of bits checked. The generator polynomial is $X^{16} + X^{12} + X^5 + 1$ and the 16 FCS bits are designated

$$\text{(MSB) } X^{15}, X^{14}, X^{13}, X^{12}, \ldots, X^1, X^0 \text{ (LSB)} \tag{6.2}$$

The ALE CRC is employed in a stand-alone CRC CMD or within the DBM data field as described earlier. The stand-alone usage is described in this section.

The CMD CRC word is constructed as shown in Figure 6.31. The first character in this word is one of the following: x (1111000), y (1111001), z (1111010), or { (1111011). The two most significant FCS bits occupy positions C1-2 and C1-1 (W9 and W10) in the first character field, resulting in four identifying characters. The CMD CRC message normally appears at the end of the message section of a frame, but it may be inserted within the message section any number of times for any number of separately checked messages.

The CRC analysis is performed on all ALE words in the message section that precede the CMD CRC word bearing the FCS information. The CRC analysis is limited to the message section bounded by the beginning of the calling cycle, or the previous CMD CRC word, whichever is closest. The selected ALE words are then analyzed in their nonredundant and unencoded (or FEC decoded) basic 24-bit ALE word form in the bit sequence (MSB) W1, W2, W3, W4, ..., W24 (LSB). This analysis continues with the unencoded bits W1 through W24 from the next word received, followed by the bits of the next word, until the first CMD CRC message is found. Therefore, each CMD CRC message inserted and sent in the message section ensures the data integrity of all the bits in the preceding ALE words as well as their respective preambles.

If the ALE words in the calling cycle (e.g., TO) preceding the message section must be checked, then an optional calling cycle CMD CRC is used as the calling cycle terminator. This optional CMD is received first in the message section, so the calling cycle words can be analyzed in their simplest (T_c), nonredundant, and nonrotated form. If it is necessary to check the words in a conclusion (e.g., TIS or TWAS), an optional conclusion CRC should directly precede the conclusion portion of the call. This optional conclusion CRC is placed at the end of the message section and must be directly preceded by a separate CMD CRC. This preceding CMD CRC is used to check the message

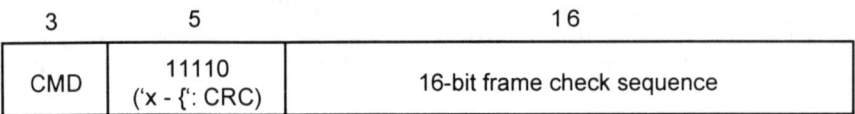

Figure 6.31 CRC CMD format.

section or calling cycle. ALE controllers should perform CRC analysis on all received ALE transmissions and are prepared to compare computed FCS values with any CMD CRC words received. If a CRC FCS comparison fails, an automatic (or operator initiated) retransmission request (ARQ) or other appropriate procedure, is used to correct the message as described previously.

6.5.13 User-Unique Functions

An interesting feature of the ALE system is the provision for *user-unique functions* (UUFs). UUFs are for special uses, as coordinated with specific users or manufacturers, of the ALE system in nonstandard or non-ALE ways. The UUF CMD word (Figure 6.32) uses the pipe character in the first character position of the word, while the remaining 14 bits are available to specify the specific unique function to be performed. Often, the first 7 bits of this field are used to specify a particular manufacturer or user community, and the final 7 bits identify one of the 128 possible UUFs for that community. Management of the UUF *name space* (i.e., assignment of these 14-bit codes) has been delegated to the NTIA Institute for Telecommunication Sciences in Boulder, Colorado.

Some UUFs simply invoke manufacturer-specific functions at the receiving station in response to the UUF CMD. Other UUFs include a data field that follows the UUF CMD word in a series of DATA and REP words. It is even possible to use a UUF CMD to embed non-ALE signaling within an ALE frame. An example of such a function is a user that needs to link using ALE, transmit an HF channel characterization (non-ALE) waveform during the ALE frame or handshake, and conclude the underlying protocol. UUF activity that uses non-ALE signaling should be conducted completely outside of the frame and should not interfere with the protocol timing. If non-ALE UUF activity (1) must be conducted within the message section, (2) will occupy time on the channel, and (3) is incompatible with the ALE system, that activity should be conducted immediately after the UUF CMD and it should be restricted to a limited time. A STAY CMD should precede the UUF instruction to indicate that time. The sending station must resume the previous redundant word phase when the frame resumes to ensure synchronization.

UUFs are used only among stations that are specifically addressed and included within the protocol. Furthermore, they must be addressed only to

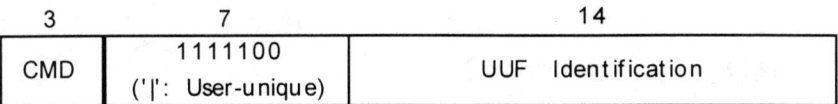

Figure 6.32 User-unique CMD word format.

stations specifically capable of participating in UUF activity. Links with all nonparticipating stations should be terminated. There are two exceptions to this termination rule for stations that are not capable of participating in the UUF but which nonetheless must be retained in the protocol until concluded:

1. The calling station could direct all stations (a) to stay linked for the duration of the UUF, (b) to read and use anything that they are capable of during that time, and (c) to resume acquisition and tracking of the ALE frame and protocol after the UUF ends. To accomplish this, the sending station should send a STAY CMD immediately before the UUF, and indicate the time period (T) for which the receiving stations should wait for resumption for the frame and protocol.
2. The sending station could also use any standard CMD function to direct the nonparticipating stations to wait or return later, or to do anything else that is appropriate and controllable through the standard orderwire functions.

In both of these cases, the scheduling function overrides functions such as the wait-for-activity timer (T_{wa}) that might otherwise interfere with the protocols or maintenance of the link.

The STAY command is similar to the scheduling commands described earlier, but it has two important differences from those commands: (1) The STAY command applies to the current transmission instead of future transmissions, and (2) the time T specified in the STAY command is encoded as a 12-bit count of redundant word times (T_{rw}) instead of the time codes used in the scheduling commands.

6.5.14 Time Exchange

The time exchange protocols developed for maintaining coarse synchronization in the ALE controller's real-time clocks are described in Chapter 7. Such synchronization is not used to align ALE system timing per se, but it is used to compensate for long-term drift that would interfere with the time-based encryption of ALE words by the linking protection mechanism.

6.6 FUTURE DEVELOPMENTS

As this book goes to press, the developers of military standards are working on revisions to both of the automated HF system standards.

References

[1] Department of the Army, Information Systems Engineering Command, *MIL-STD-188-141A: Interoperability and Performance Standards for Medium and High Frequency Radio Equipment*, Philadelphia, PA: Naval Publications and Forms Center, Attn. NPODS, Notice 2, Sep. 1993. Available on the Internet at URL ftp://tracebase.nmsu.edu/pub/hf/pubs/mil_std_188_141a.

[2] U.S. Department of Commerce, *Federal Standard 1045 (FED-STD-1045), Telecommunications: HF Radio Automatic Link Establishment*, Washington, DC: General Services Administration, Office of Information Resources Management, Jan. 1990. Available from GSA Specification Section.

[3] Department of the Army, Information Systems Engineering Command, *Military Standard 187-721C (MIL-STD-187-721C): Interface and Performance Standard for Automated Control Appliqué for HF Radio*, Philadelphia, PA: Naval Publications and Forms Center, Attn. NPODS, Nov. 1994, Available on the Internet at URL ftp://tracebase.nmsu.edu/pub/hf/pubs/mil_std_187_721c.

[4] Desourdis, R. I., and E. E. Johnson, "Advanced Link Quality Analysis for ALE HF Radio," *IEEE MILCOM '93 Conf. Proc.*, Boston, Oct. 1993.

[5] Harrison, G. L., and F. C. Leiner, Mitre working paper 86W0033335, Sep. 30, 1986.

Linking Protection[1]

7.1 INTRODUCTION

In many HF radio networks, it is necessary to prevent uninvited stations from surreptitiously entering the net or denying its use to legitimate members by means of deception. *Linking protection* (LP) is a means of preventing these uninvited stations, called *adversaries* in this context, from automatically establishing links with protected stations. The LP technique developed for use with second-generation ALE employs time-varying encryption of ALE transmissions to authenticate those transmissions. In this way, an adversary is inhibited from generating ALE transmissions that "make sense" when decrypted by recipients. Due to this encryption of ALE words, the contents of those words are made unintelligible to eavesdroppers, but this is merely an incidental benefit and not the principal function of LP. LP is an optional feature of ALE as depicted in the hierarchical structure shown in Figure 7.1.

A guiding principle in the design of the LP technique discussed in this chapter was that its use should be transparent to the existing forward error correction (FEC), ALE, and other protocols. In other words, the LP technique that is chosen must achieve and maintain cryptographic synchronization transparently to the ALE waveform, word structure, and protocols employed. In particular, scanning radios should be able to acquire crypto synchronization ("sync") at any point in the scanning call portion of a protected transmission (see Chapter 6), as long as this transmission was encrypted under the key in use by the receiving station. Thus, LP may not insert sync bits into the data stream, and must acquire crypto sync without the use of synchronization preambles that would impact the ALE protocol. This design requirement was necessary not only to ensure interoperability with the ALE protocol, but also to

---

[1]Updated from an earlier work by Johnson, E. E., and R. Moore, "Evaluation of HF ALE Linking Protection," *MILCOM'93 Conf. Proc.,* Bedford, MA: IEEE, Oct. 1993.

278   Advanced High-Frequency Radio Communications

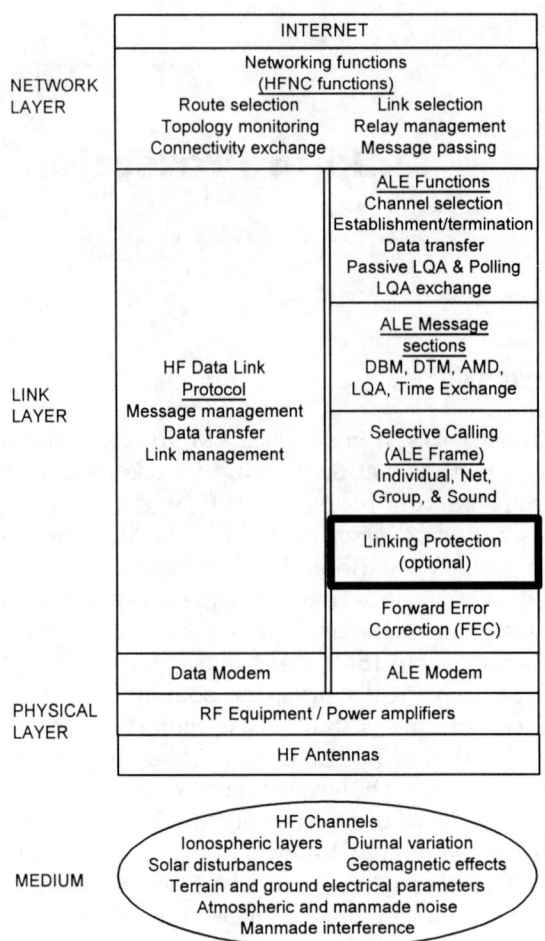

Figure 7.1 ALE modem and FEC in the hierarchical-layered view of HF communications.

minimize the additional processing required to add LP to existing ALE controllers.

A necessary characteristic of an encryption algorithm used for authentication is that small differences in a cipher-text word to be decrypted should change about *half* of the bits in the plain-text result. This rule of thumb results in error multiplication in received words, however, which is clearly undesirable given the noisy channels that must often be used for sky-wave HF communications. Thus, it is necessary to correct as many errors as possible in a received word *before decryption*. LP is therefore placed in the intermediate protection sublayer of the ALE data link layer as shown in Figure 7.2. In this hierarchical position, LP can make full use of the error-correcting power of the FEC sublayer while still intercepting unauthorized attempts to establish unauthorized HF links before they reach the ALE sublayer.

Data flow through an automated HF radio with LP is shown in Figure 7.3. The active entity within the protection sublayer is termed the *linking protection control module* (LPCM). The LPCM uses a scrambler to encrypt and decrypt the 24-bit ALE words exchanged by the ALE protocol modules. In conjunction

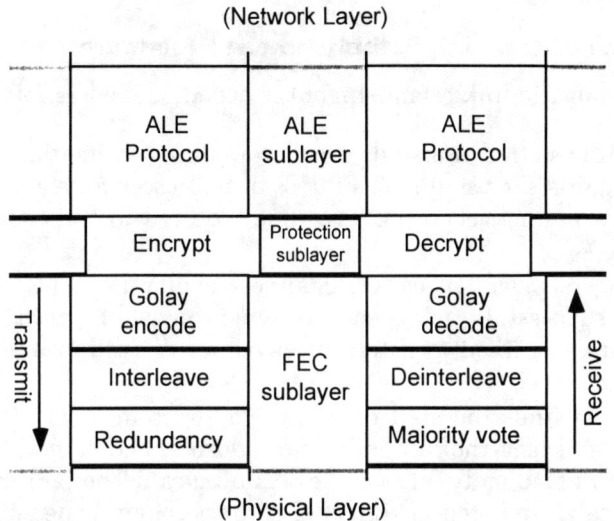

Figure 7.2 Sublayers of ALE data link layer.

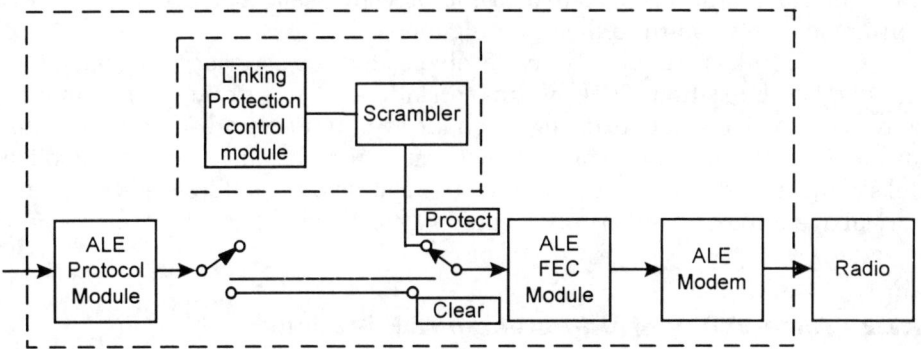

Figure 7.3 Block diagram of protected ALE system.

with the ALE protocol module, the LPCM determines the authenticity of each received ALE transmission. The inputs for the scrambler (hardware or software) include a private key variable, time- and frequency-dependent randomization data, and a plain-text ALE word. The scrambler outputs a 24-bit cipher-text word when in encrypt mode, with the positions of cipher-text and plain-text interchanged for decryption. Time is quantized for LP into *protection intervals* as discussed later.

7.2 LP TECHNIQUE

7.2.1 Review of the Link Establishment Procedure

To review, automatic link establishment is accomplished as follows:

1. The calling station transmits a call containing the initial portion of the called station's, or stations', address or addresses for a sufficient time to capture scanning receivers. The time required for this scanning call is denoted T_{sc}.
2. Scanning receivers pause on channels containing ALE signaling. If a receiver successfully achieves word synchronization (word sync), the ALE words are examined to determine whether the call is intended for that receiver.
3. If a station finds the start of its own address in a call, it will stay on channel and read the full address or addresses contained in the *leading call* that immediately follows the scanning call. The period defining the leading call is denoted T_{lc}. If its full address is found, the receiving station reads the rest of the call, including the caller's address, and completes a three-way handshake to establish a link.

If any of the conditions specified in the preceding list is not met, a receiver will immediately return to the scanning mode.

Once a link is established, the receivers alert the corresponding operators and display the station ID of the linked station. At this point, operators may converse in voice mode, exchange data, or even reprogram each other's automated HF communication controllers. Clearly, if an adversary could establish links with unsuspecting and unprotected stations, it could severely disrupt normal operations. This disruption could be particularly tragic in an emergency situation.

7.2.2 Vulnerability of Unprotected ALE Stations

As described, ALE stations are normally scanning their assigned channels for incoming calls. The timing of ALE scanning calls is based on the assumption that scanning stations will not dwell for a duration of more than $2T_{rw}$ on each channel. For this reason, a simple "attack" on an ALE network would consist of continuously broadcasting the scanning call section of a net call. Once each station in the victimized net scanned the channel carrying the bogus call, it would stop scanning and await the conclusion of the call, but this conclusion is never transmitted by the adversary. Of course, each receiving station in the net would eventually time out while awaiting call termination and return to the scanning mode. However, these net stations would then quickly scan through the other channels and once again be "captured" by the "spoofer." Thus, each station would spend most of its time "hung" on an unusable channel, driving the probability of establishing *legitimate* links nearly to zero.

If a station operator discovered the spoofer, the operator could remove the channel carrying the bogus call from the scan list. Although this removal defeats the spoofer in the short term, it also removes a usable HF channel and thus increases spectrum congestion. After a time, the adversary could switch channels and ultimately force the net to abandon all usable channels one by one. The net stations would be then be forced to manually check abandoned channels for the spoofer before restoring them to active service. As this example demonstrates, the adversary can achieve significant disruption of automated network operations. Furthermore, manual intervention to excise spoofed channels requires that operators be present at the net stations to take such evasive action. Unattended HF stations would be less likely to take action against such a spoofer.

This simple spoofing example requires only that the adversary know a single ALE address for the net. If ALE addresses are changed periodically, the spoofer could modify this attack to incorporate an audio tape recorder with a tape loop. The recorder could be switched to record mode on receipt of an ALE call from the victim network, and returned to playback mode when the call had been recorded. The attack then proceeds as described previously. From the perspective of an adversary, a net call would be the best type of ALE call to record, although an individual call would be useful as well. If several individual calls were received, they could be used to produce an endless group call with the addition of an ALE modem and a laptop computer to modify the TO preambles to THRU.

So far, the discussion has focused on the disruption of ALE. More aggressive attacks could seek to impersonate legitimate net members for voice communications, since the address of the calling station is displayed on the called radio and would therefore look authentic, or to gain access to the data systems of one or more stations in the victimized net. Although the mischief achievable through the ALE modem is unpleasant, it is limited to messages sent using the ALE functions AMD, DTM, or DBM, and the associated protocols. Some HF networks use the ALE modem only for initial link establishment, followed by transition to a modern high-speed data modem for the transmission of internet and intranet traffic or reprogramming station equipment via network management protocols. Thus, many networks will want to authenticate ALE calls to avoid a variety of scenarios involving both disruption and unauthorized access to potentially sensitive information.

7.2.3 Overview of the LP Procedure

Linking protection seeks to exclude disruption and deception from ALE operations. The method chosen to shield the receiving ALE module from such attacks is the time-varying encryption of ALE words. In this method, the sender encrypts ALE words using a secret key and a *seed* containing time of day and the frequency in use on a link. Note that using the time of day (TOD) to vary the encryption results thwarts an adversary that replays previously recorded transmissions. Likewise, the frequency of a transmission is used in the encryption to preclude *playback* of an intercepted transmission on a different frequency.

The receiver decrypts received words using its assigned key along with several seeds, each of which covers a narrow range of times of day. The TOD range covered by each seed is called the *protection interval* (PI). Unless the received word was encrypted using the same key and a TOD in that range, it will appear as noise to the receiver (see Figure 7.4) and will be accepted as a legitimate call with a probability of less than 10^{-7}. The randomization data, or seed, contains date, PI, ALE word, and frequency fields as shown in Figure 7.5.

- The date field contains a 4-bit field for the current month (1 to 12) and a 5-bit field for the current day (1 to 31).
- The PI field contains an 11-bit coarse time field used to count minutes since midnight and a 6-bit fine time field giving the residual seconds. The fine time field is always a multiple of the PI length and is aligned to PI

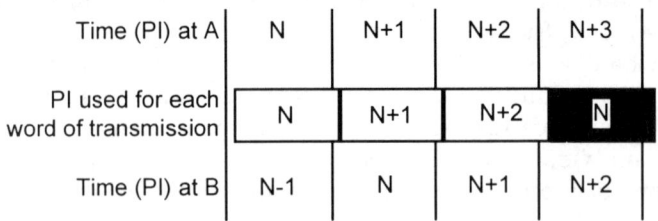

Figure 7.4 Transmission from A to B with an optimum PI.

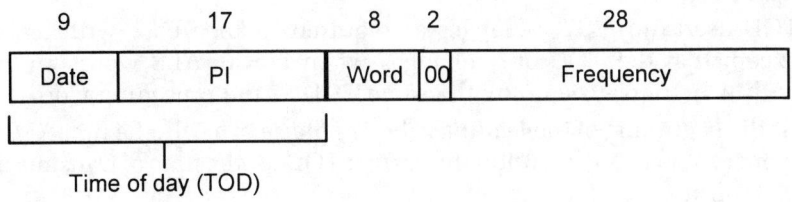

Figure 7.5 LP seed format.

boundaries. For example, a 2-sec PI requires that the fine time value is always even. If a 1-min PI is used, the fine time field is set to all 1s.
- The word field, which counts ALE words within a PI as described below.
- The frequency field, which contains the frequency of the current protected transmission in binary-coded decimal (BCD).

The PI number field in the randomization seed is a principal source of the time variation in the protection process and is updated at the beginning of each protection interval.

All stations must be synchronized to some degree because the receiving LPCM must use the same seed employed by the transmitting station to recover each plain-text ALE word. LPCMs receiving ALE words will usually attempt to decrypt received cipher-text under several seeds with adjacent PI numbers to discover the one currently in use at the transmitting LPCM. This approach reduces the probability that imperfect station synchronization will prevent link establishment between two stations employing this LP technique. To keep the number of valid seeds manageable, the range of local times among LPCMs in a network should be at most one PI. This recommendation, as well as other considerations in choosing an appropriate PI, are discussed at length after the following explanation of LP processing.

7.2.4 Transmit Processing

The LP module in a transmitting station encrypts each 24-bit ALE word to be sent using the current seed data and delivers the encrypted word to the FEC module. In the course of a transmission, the procedure described in step 3 causes the pseudo-time represented by the TOD fields to advance more slowly than real time. This discrepancy is corrected at the start of each new transmission.

The word field w in the seed is used in the LP procedure as follows:

1. During the scanning call phase of a call (i.e., during T_{sc}), the calling station alternates transmission of words encrypted using $w = 0$ and $w = 1$. The

TOD used during T_{sc} changes as required to keep pace with real time, except that the TOD only changes when $w = 0$. ALE words encrypted with $w = 1$ always employ the same TOD as the preceding word.

2. At the beginning of the leading call (T_{lc}) phase of a call, the first ALE word is encrypted using $w = 0$ and the correct TOD for the time of transmission of that word.
3. All succeeding ALE words of the call use word numbers incremented modulo w_{max}, where w_{max} is determined by the PI length. When w becomes 0, the TOD is incremented.
4. Responses and all succeeding transmissions begin with $w = 0$ and the current (corrected) TOD, with these fields incremented as described earlier for each succeeding ALE word.

7.2.5 Receive Processing

The receiving side of an LP module is responsible for achieving crypto synchronization with transmitting stations and for decrypting protected ALE words produced by the Golay decoder. When a scanning receiver arrives on a channel carrying valid ALE tones and timing, the FEC sublayer (majority voter, deinterleaver, and Golay decoder) will process the output of the ALE modem. When an acceptable candidate word has been received, the FEC sublayer will alert the LP receiving module. Such an alert occurs on average once every 8 ms when the Golay decoders are correcting three errors, or once every 78 ms when correcting one error per Golay word. The receive LP module must then decipher the candidate word, and determine whether word sync has been achieved by checking for acceptable preamble and data bits.

This task is complicated by the possibility that the received word may have been encrypted using a slightly different TOD than the current TOD at the receiver. Thus, the receiving LP module must decrypt each candidate word under several different TODs. A further complication in this situation is the possibility that more than one of these decrypted words may satisfy the preamble and data field checks. In this case, all TODs that produced valid words are used to decrypt the next word, which is then checked for compliance with the ALE protocol in the context of the word decrypted under the predecessor TOD. The probability that more than one TOD will remain viable after this second word is checked in this way is vanishingly small.

While attempting to achieve word sync, the receiver typically attempts to decode received words under six different combinations: the current TOD and two adjacent TODs (future and past), and both $w = 0$ and $w = 1$. After achieving word sync, the number of valid combinations is significantly reduced by the LP protocol. Stations using a PI of 2 sec or less do not accept more than one transmission encrypted using a given TOD, and therefore need not check combinations using that TOD. This feature further reduces the number of valid combinations that must be tried when achieving word sync.

The state diagrams shown in Figures 7.6 and 7.7 succinctly summarize the operation of the sending and receiving LP modules, respectively. Note, however, that the procedure used to acquire word and TOD synchronization at the receiver is not shown. In these figures, the states of the LP process are labeled using the current PI number (denoted N) and the current word number used in the seed to the encryption algorithm. The phrase "Increment N" denotes a PI transition.

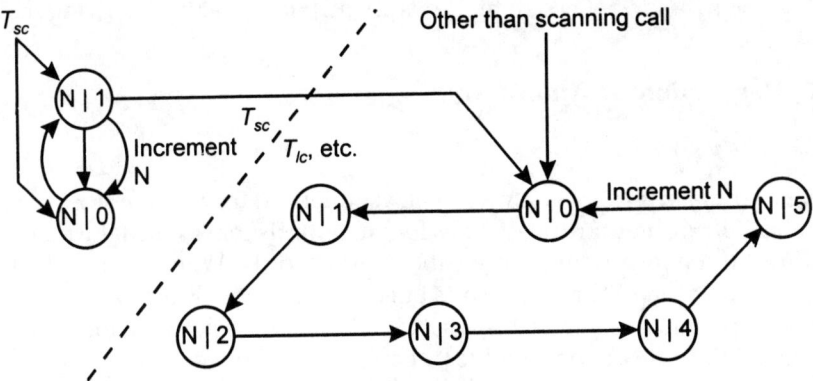

Figure 7.6 Transmitting state diagram (2-sec PI).

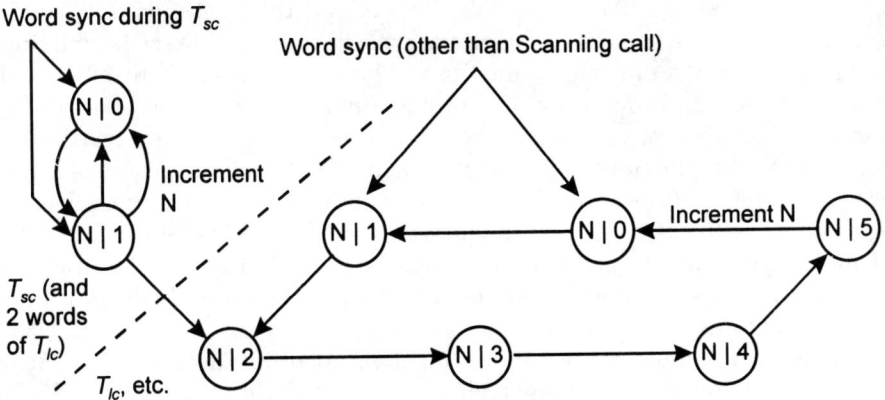

Figure 7.7 Receiving state diagram (2-sec PI).

The correct PI and word number to be used at the transmitting station are always determined unambiguously by the state diagram in Figure 7.6. However, there are several sources of ambiguity at the receiver that can cause linking attempts to fail when they should succeed. These sources of ambiguity are (1) uncertainty of the TOD the transmitter, (2) unknown word number for ALE word sync acquisition during T_{sc}, (3) unknown locations of PI transitions during T_{sc} in the received ALE word stream, and (4) unknown location of the transition from T_{sc} to T_{lc}. The first two uncertainties require decryption of received words under a range of PI/word number combinations during word sync acquisition. In all cases, the usual word sync tests must be used to resolve ambiguity, due to the lack of synchronization codes in the ALE word stream. Techniques that may be used to resolve these ambiguities are discussed in the following sections.

7.2.6 Resolution of Ambiguity

7.2.6.1 Word Sync

During unprotected (non-LP) word sync acquisition, the receiver's FEC module examines the received stream of tones for bit patterns that exceed the unanimous vote threshold and produce correctable Golay words. When a candidate word is produced by the FEC module, it is checked by the ALE protocol module for acceptable preamble and ASCII subset and, if these checks pass, for compliance with the ALE protocol. When all tests concur that the received word is acceptable, word sync is assumed, and the FEC module settles into checking and returning one word every T_{rw} thereafter until otherwise notified by the ALE protocol module.

From Figure 7.2, the LP function is an interposed sublayer between the FEC sublayer and the ALE protocol. Thus, when the FEC module returns a candidate word, the LP sublayer must decrypt this word using seeds containing all valid PI/word number combinations. The LP sublayer then delivers the results of the decryption to the ALE protocol module where the final series of tests is applied. In most cases, at most one seed will produce a word that is acceptable to the ALE module. When this occurs, TOD synchronization between the transmitter and receiver will be achieved at the same time as word synchronization. On rare occasions, a candidate word from the FEC module will produce acceptable ALE words under two or more seeds. In this case, the resulting ambiguity must be resolved before the LP function can properly decrypt subsequent ALE words.

The word number sequencing in the LP protocol was designed specifically to assist in the resolution of this ambiguity. For individual and net calls with single-word addresses (a common case), the next ALE word received following word sync acquisition must be identical to the first received word. This result occurs when these ALE words are decrypted under word numbers alternating between 0 and 1, possibly with a PI change, whether word sync is achieved during the T_{sc} or T_{lc} calling periods. In this case, the word sync function can resolve seed ambiguity by simply waiting for the next received ALE word. This word can then be decrypted under the appropriate combinations to determine which PI/word number combination was the correct one for the first (word sync) word. This next received ALE word is then retained to be returned when requested by the receiving ALE protocol software.

7.2.6.2 Transitions During T_{sc}

Note that in Figure 7.6, three arrows emerge from the $N/1$ state during T_{sc}, corresponding to simple alternation to word number 0; transition to the next PI and word 0; and a transition to $N/2$, indicating that T_{lc} began two words before the current word. These transitions are identified by evaluating incoming words under all possibilities and selecting the most probable word when more than one word passes the tests. In the following example, this technique is implemented by trying seeds in order of decreasing probability (i.e., no change, then PI change, then start of T_{lc}) and accepting the first seed that produces an acceptable ALE word.

7.2.7 Linking Protection Processing Example

The LP procedure is illustrated in the following example of a scanning call followed by a response and acknowledgment. Each line represents one call to the scrambler and includes the seed fields (month/day minute:second/word), the input to the scrambler (as the argument to the encrypt or decrypt function call), and the output of the scrambler. Decrypted words at the receiver are shown to the right in the figures. A PI duration of 2 sec was in use. The following ALE words, shown in octal, are used in the example:

```
25160315 TO SAM
54523705 TIS JOE
24523705 TO JOE
55160315 TIS SAM
```

First, the transmitting LPCM (JOE) encrypts and sends several words before the receiver (SAM) scans the active channel. Note the word number alternation:

```
seed( 9/ 5 180: 0/0)  encrypt(25160315) -> 06355273
seed( 9/ 5 180: 0/1)  encrypt(25160315) -> 01405770
seed( 9/ 5 180: 0/0)  encrypt(25160315) -> 06355273
seed( 9/ 5 180: 0/1)  encrypt(25160315) -> 01405770
```

```
seed( 9/ 5 180: 0/0) encrypt(25160315) -> 06355273
```

The receiver SAM then arrives on the active channel, detects ALE signaling, and achieves possible word synchronization using the FEC sublayer tests. The final attempt at synchronization uses decryption of the prospective word under the six possible seeds (i.e., word number 0 or 1, with current, next, and previous PI numbers):

```
seed( 9/ 5 180:   0/0) decrypt(06355273) -> 25160315 TO SAM
seed( 9/ 5 180:   0/1) decrypt(06355273) -> 41671704 FROM gD
seed( 9/ 5 180:   2/0) decrypt(06355273) -> 67316636 CMD v;
seed( 9/ 5 180:   2/1) decrypt(06355273) -> 66751750 CMD o'h
seed( 9/ 5 179:58/0) decrypt(06355273) -> 51111235 TIS %
seed( 9/ 5 179:58/1) decrypt(06355273) -> 37101254 TWAS r,
```

Only one of the decrypted ALE words has a valid preamble and three ASCII-38 characters, so station SAM assumes that it has achieved word and crypto sync. Station JOE has meanwhile encrypted and sent the next word in the scanning call. Of course, this is the same ALE word but with a different word number:

```
seed( 9/ 5 180: 0/1) encrypt(25160315) -> 01405770
```

SAM now tests synchronization of the new received word by decrypting with the only logical successor to the seed that worked for the preceding word as follows:

```
seed( 9/ 5 180: 0/1) decrypt(01405770) -> 25160315 TO SAM
```

The resulting decrypted word is identical to the first word, so synchronization is confirmed. Transmitting station JOE and receiving station SAM now stay synchronized for the remainder of the call. Note that station SAM always assumes that the seconds field (PI number) remains unchanged until it is forced to increment this field:

```
seed( 9/ 5 180: 2/0) encrypt(25160315) -> 47762707
seed( 9/ 5 180: 0/0)   decrypt(47762707) -> 07754077 DATA 0?
seed( 9/ 5 180: 2/0)   decrypt(47762707) -> 25160315 TO SAM
seed( 9/ 5 180: 2/1) encrypt(25160315) -> 76113624
seed( 9/ 5 180: 2/1)   decrypt(76113624) -> 25160315 TO SAM
seed( 9/ 5 180: 2/0) encrypt(25160315) -> 47762707
seed( 9/ 5 180: 2/0)   decrypt(47762707) -> 25160315 TO SAM
seed( 9/ 5 180: 2/1) encrypt(25160315) -> 76113624
seed( 9/ 5 180: 2/1)   decrypt(76113624) -> 25160315 TO SAM
seed( 9/ 5 180: 2/0) encrypt(25160315) -> 47762707
seed( 9/ 5 180: 2/0)   decrypt(47762707) -> 25160315 TO SAM
seed( 9/ 5 180: 2/1) encrypt(25160315) -> 76113624
seed( 9/ 5 180: 2/1)   decrypt(76113624) -> 25160315 TO SAM
```

```
seed( 9/ 5 180: 4/0)   encrypt(25160315) -> 77661202
seed( 9/ 5 180: 2/0)    decrypt(77661202) -> 75461117 REP YDO
seed( 9/ 5 180: 4/0)    decrypt(77661202) -> 25160315 TO SAM
seed( 9/ 5 180: 4/1)   encrypt(25160315) -> 31345675
seed( 9/ 5 180: 4/1)    decrypt(31345675) -> 25160315 TO SAM
seed( 9/ 5 180: 4/0)   encrypt(25160315) -> 77661202
seed( 9/ 5 180: 4/0)    decrypt(77661202) -> 25160315 TO SAM
seed( 9/ 5 180: 4/1)   encrypt(25160315) -> 31345675
seed( 9/ 5 180: 4/1)    decrypt(31345675) -> 25160315 TO SAM
seed( 9/ 5 180: 6/0)   encrypt(25160315) -> 07336353
seed( 9/ 5 180: 4/0)    decrypt(07336353) -> 07373375 DATA wm}
seed( 9/ 5 180: 6/0)    decrypt(07336353) -> 25160315 TO SAM
seed( 9/ 5 180: 6/1)   encrypt(25160315) -> 17357624
seed( 9/ 5 180: 6/1)    decrypt(17357624) -> 25160315 TO SAM
seed( 9/ 5 180: 6/0)   encrypt(25160315) -> 07336353
seed( 9/ 5 180: 6/0)    decrypt(07336353) -> 25160315 TO SAM
seed( 9/ 5 180: 6/1)   encrypt(25160315) -> 17357624
seed( 9/ 5 180: 6/1)    decrypt(17357624) -> 25160315 TO SAM
```

Station JOE now begins the leading call (T_{lc}) although station SAM will not detect this leading call until it is forced to try a seed containing $w = 2$ exactly two ALE words later:

```
seed( 9/ 5 180: 6/0)   encrypt(25160315) -> 07336353
seed( 9/ 5 180: 6/0)    decrypt(07336353) -> 25160315 TO SAM
seed( 9/ 5 180: 6/1)   encrypt(25160315) -> 17357624
seed( 9/ 5 180: 6/1)    decrypt(17357624) -> 25160315 TO SAM
seed( 9/ 5 180: 6/2)   encrypt(54523705) -> 25147235
seed( 9/ 5 180: 6/0)    decrypt(25147235) -> 70127642 REP _"
seed( 9/ 5 180: 8/0)    decrypt(25147235) -> 67101005 CMD r
seed( 9/ 5 180: 6/2)    decrypt(25147235) -> 54523705 TIS JOE
```

This example call is now complete. For the response, station SAM will be the transmitting station and station JOE the receiving station. Station SAM encrypts and sends the ALE word "TO JOE":

```
seed( 9/ 5 180: 8/0)   encrypt(24523705) -> 11576752
```

Station JOE must now synchronize:

```
seed( 9/ 5 180: 8/0)    decrypt(11576752) -> 24523705 TO JOE
seed( 9/ 5 180: 8/1)    decrypt(11576752) -> 71310355 REP !m
seed( 9/ 5 180:10/0)    decrypt(11576752) -> 14724173 THRU NP{
seed( 9/ 5 180:10/1)    decrypt(11576752) -> 02022046 DATA H&
seed( 9/ 5 180: 6/0)    decrypt(11576752) -> 45563336 FROM [M^
seed( 9/ 5 180: 6/1)    decrypt(11576752) -> 25567654 TO [_,
```

Although acceptable preambles are frequent in the decrypted ALE words shown here, the requirement for three ASCII-38 characters is quite effective in rejecting spurious words. The remainder of the response presents no ambiguity in PI or word number as seen from steps 3 and 4 in the LP procedure:

```
seed( 9/ 5 180: 8/1)  encrypt(24523705) -> 21616614
seed( 9/ 5 180: 8/1)   decrypt(21616614) -> 24523705 TO JOE
seed( 9/ 5 180: 8/2)  encrypt(55160315) -> 35713131
seed( 9/ 5 180: 8/2)   decrypt(35713131) -> 55160315 TIS SAM
```

Having received a good response, station JOE encrypts and sends the acknowledgment:

```
seed( 9/ 5 180:10/0)  encrypt(25160315) -> 34172163
```

Station SAM then synchronizes:

```
seed( 9/ 5 180:10/0) decrypt(34172163) -> 25160315 TO SAM
seed( 9/ 5 180:10/1) decrypt(34172163) -> 76611716 REP 1'N
seed( 9/ 5 180:12/0) decrypt(34172163) -> 22110461 TO ""1
seed( 9/ 5 180:12/1) decrypt(34172163) -> 17264052 THRU uP*
seed( 9/ 5 180: 8/0) decrypt(34172163) -> 27767261 TO ]1
seed( 9/ 5 180: 8/1) decrypt(34172163) -> 27750370 TO !x
```

and, as noted for the response, the transmission proceeds without further ambiguity:

```
seed( 9/ 5 180:10/1)  encrypt(25160315) -> 66745516
seed( 9/ 5 180:10/1)   decrypt(66745516) -> 25160315 TO SAM
seed( 9/ 5 180:10/2)  encrypt(54523705) -> 03620500
seed( 9/ 5 180:10/2)   decrypt(03620500) -> 54523705 TIS JOE
```

As might be expected, the ambiguities that must be resolved by the LP receiving module can result in degraded linking performance unless the decision trees are constructed optimally. This issue is analyzed following a discussion of optimal PIs.

7.3 PROTECTION INTERVAL ANALYSIS

Because this LP mechanism is time based, all stations must be synchronized to some extent. As noted earlier, the range of LPCM local times within a network should not significantly exceed the length of a PI if the number of TODs checked by receivers is to remain manageable. Because the effectiveness of LP increases as the length of the PI decreases [1], a tightly synchronized system is desirable from a security standpoint. However, the cost of implementing and operating the protected system increases with the degree of synchronization required, so the choice of a PI length must balance security with the costs to produce and use such a protected system.

Encryption of the transmissions used to establish links among HF stations clearly provides protection from casual analysis of those transmissions. When the encryption algorithm employed has sufficient nonlinearity to provide authentication of those transmissions, such encryption can provide protection against unauthorized link establishment as well, by forcing the adversary to guess bit combinations that will decrypt to acceptable transmissions. Even with the relatively short 24-bit words used in the standard ALE protocol, the necessity to guess the encrypted versions of two words for two different PIs reduces the probability of success for such a guessing attack to 2^{-48}, or an expected time until success of millions of years.

To protect against playback of previous transmissions, the encryption process must be time varying so that old transmissions decrypt to noise. The period of this variation is closely related to the degree of synchronization required among the stations. Furthermore, this period must be chosen to strike a balance between protection, which increases with shorter periods, and feasibility, which increases with longer periods. If the playback threat is considered negligible, then a PI of 1 min provides significant protection with minimal operational complexity. Full protection against playback attacks, however, requires a PI of 3 sec or less combined with rejection of transmissions that duplicate a TOD value [1]. (The need to reject duplicate TOD values can be eliminated entirely if the PI is set to 392 ms.) To achieve sufficiently "tight" synchronization, the standard LP technique employs a two-stage synchronization procedure in which operators load time accurate to within ±30 seconds to place radios in *coarse sync*. Once coarse sync is established, the LPCMs automatically acquire fine sync and are ready for fully protected operation.

7.4 PERFORMANCE EVALUATION OF LINKING PROTECTION

In this section, results obtained from a detailed ALE simulator are compared to results of nonprotected simulations and of measurements performed using early implementations of LP. The metric used in these comparisons is the *probability of successful link establishment,* determined as a function of parametric channel conditions. The conditions include a range of signal-to-noise ratios (SNRs) for three standard channels (ITU-R Reports 549-2 [2] and 520-1 [3]): (1) a pure additive white Gaussian noise (AWGN) channel, defined as exhibiting no fading or multipath effects; (2) the CCIR Good channel, described as possessing a 0.1-Hz fading bandwidth with 0.5-ms multipath delay; and (3) the CCIR Poor channel, with a 1-Hz fading bandwidth and 2-ms multipath delay. The fading bandwidth and multipath spread of HF channels were discussed in Chapter 2.

7.4.1 The Simulator[2]

The simulator used to evaluate the performance of LP algorithms is written in the C language and structured consistently with the protocols to be simulated as shown in Figure 7.8. Each simulator shown was validated by comparison with the corresponding measured results. The HF channel simulator implements the well-known Watterson model [4]. The remaining simulators implement the protocols as specified in the standards, albeit with a few minor limitations in the ALE protocols that can be simulated. In particular, only single-word addresses and individual or net calls are simulated. This limitation results in simpler protocol processing in the simulator than in fielded ALE controllers.

7.4.2 Simulations for Protected Versus Nonprotected ALE Links

7.4.2.1 Overview

Each plot presented in Figures 7.9(a), (b), and (c) compares the predicted linking probabilities for three cases [(1) ALE with no LP, sometimes called AL-0 LP; (2) AL-1 LP, that is, a 60-sec PI; and (3) AL-2 LP, that is, a 2-sec PI] for AWGN, CCIR *Good*, and CCIR *Poor* channels, respectively. Other experimental condi-

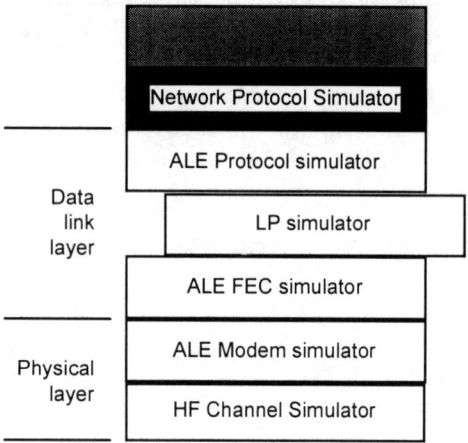

Figure 7.8 Simulator stack.

---

[2]This software was developed by Johnson Research (Las Cruces, New Mexico), primarily under the sponsorship of the Institute for Telecommunication Sciences, National Telecommunications and Information Administration, U.S. Department of Commerce, Boulder, Colorado, following some initial work sponsored by the U.S. Army Information Systems Engineering Command, Ft. Huachuca, Arizona.

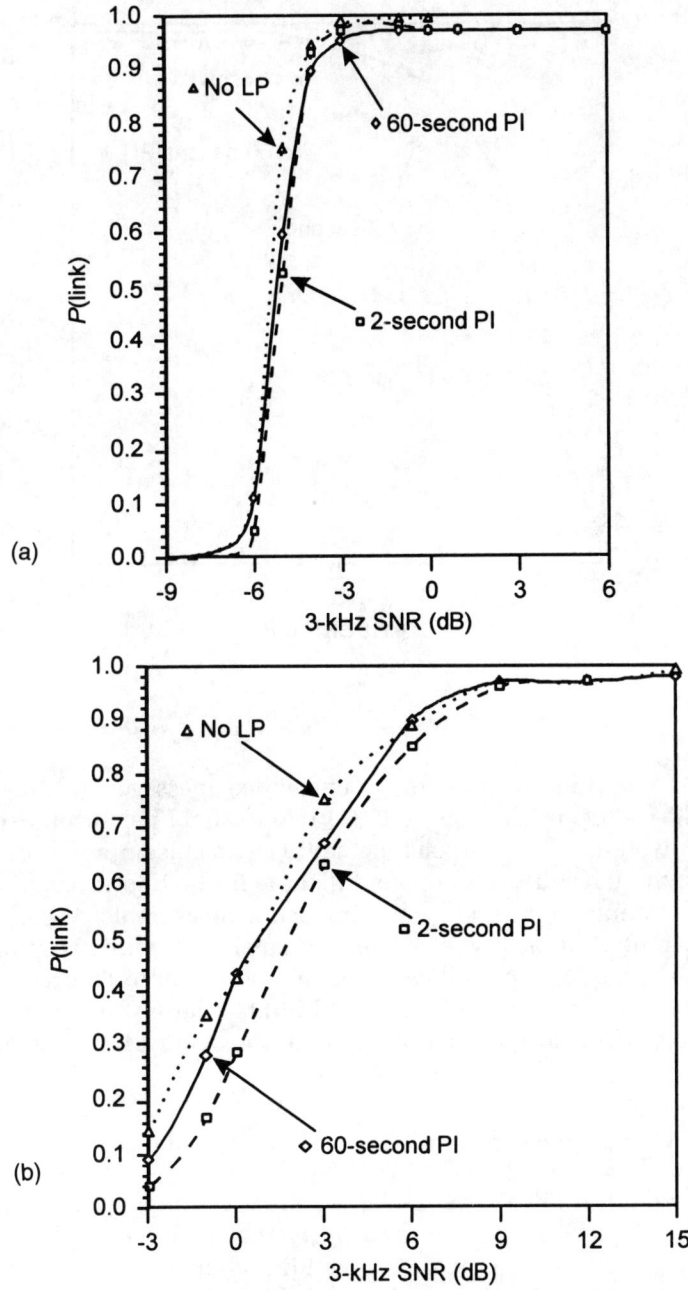

Figure 7.9 Effect of LP on *P*(link): (a) Gaussian channel, (b) CCIR Good channel, and (c) CCIR Poor channel.

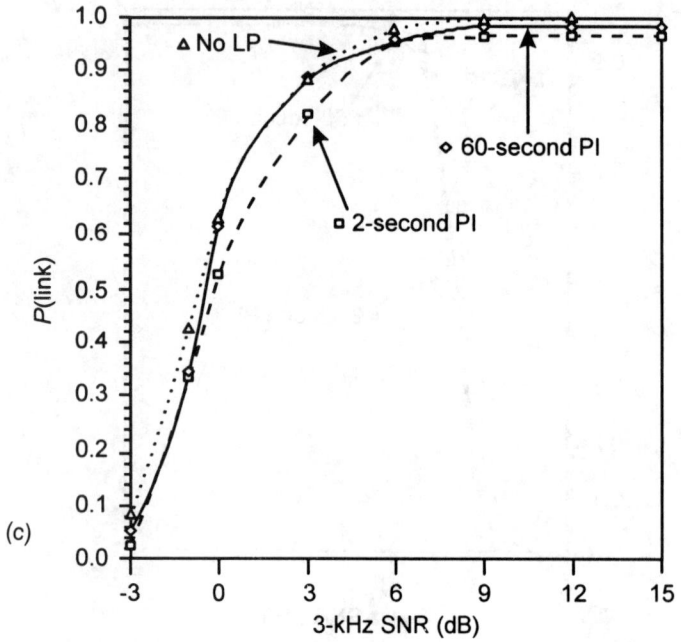

Figure 7.9 (continued)

tions were (1) individual calls with no embedded messages, (2) unanimous vote threshold in FEC module set to 0, (3) two channel per second scanning rate (500-ms dwell time), and (4) channel fading gains recomputed at intervals of 8 ms. From these results, it is apparent that the initial implementation of LP used in these simulations produced a small but measurable degradation in linking probability. In nearly error-free channels, the linking probability achieved when using LP (especially AL-2) falls short of 100%. This result occurs because of the increased sensitivity of the LPCM to a false-word sync problem (described next) that is also present in the nonprotected ALE implementation.

7.4.2.2 False Word Sync Problem

When the word sync algorithm is continuously reading symbols from the modem, there is a non-negligible probability that both Golay decodes will succeed at an erroneous word phase. In the FEC mode in which six errors are detected and one error is corrected, this probability is approximately 0.03 for each candidate word checked.

When the Golay decoder accepts a misaligned word, the preamble and ASCII subset checks will usually reject the word. In particular, fewer than 1% of all 24-bit words contain three ASCII-38 characters and one of the three preambles accepted by the word sync algorithm in the simulator. However, if the Golay decoder determines that a misaligned word is correctable, and the resulting ALE word passes the preamble and ASCII checks, then the word sync algorithm will accept the erroneous word phase and the linking attempt will probably fail.

Denote the probability of Golay decoder success on random inputs as p_G and the probability that a random word passes the preamble and ASCII checks as p_{pac}. As noted earlier, $P_G \approx 0.03$ and $P_{pac} \approx 0.01$. The probability of a word sync error in nonprotected mode, p_{wse}, can then be estimated by

$$\begin{aligned}
p_{wse} &= (1 - P_{\text{no false Golay decode before correct word phase}}) p_{pac} \\
&= \left[1 - \sum_{i=0}^{48} (P_{i \text{ symbols before correct word phase}} \right. \\
&\quad \left. \cdot P_{\text{no false Golay decode in first } i \text{ symbols}}) \right] p_{pac} \\
&= \left[1 - \frac{1}{49} \sum_{i=0}^{48} (1 - p_G)^i \right] p_{pac} \\
&= \left\{ 1 - \frac{1}{49} \left[\frac{1 - (1 - p_G)^{49}}{p_G} \right] \right\} p_{pac} \\
&\approx 0.005
\end{aligned} \qquad (7.1)$$

This result assumes (1) a uniform relative word phase distribution and (2) that the FEC process is invoked for each arriving symbol. Thus, in the long run, a nonprotected ALE station should experience a linking failure on ideal channels about once in 200 attempts due to false word sync, unless the word sync algorithm is improved over that described earlier. Use of a higher unanimous-vote threshold should reduce p_{wse} at the possible expense of reduced linking probability over marginal channels. Another possible improvement is "soft" word sync detection.

When LP is added to the receiver, each candidate ALE word is decrypted under several PI/word number combinations (either six or eight). As a result, *several* words are presented to the preamble and ASCII checks for each candidate word that passes the FEC sublayer checks. The value of p_{wse} is therefore increased by the number of seeds examined, resulting in a decrease in linking probability by a few percent. In the simulation results shown in Figures 7.9(a), (b), and (c), P(link) reached a plateau of 0.96 to 0.97. When diagnostic simulations were made at 30-dB SNR, the linking failures in every case were due to false word sync acquisition.

7.4.2.3 Full-Performance Linking Protection

Subsequent simulations have shown that adjustment of the unanimous vote threshold to a value of 25 provides 99% to 100% linking performance in high SNR channels, with minimal performance degradation at low values of SNR.

7.4.3 Measurement-Simulation Comparison

Figure 7.10 shows the results of a comparison of measurements performed by NTIA at the SNR values for which performance standards are specified in FED-STD-1045, compared to the results of the simulations with the unanimous vote threshold set to a value of 25. The measurements are qualitatively similar to the simulation results, with somewhat lower performance in general. This behavior was anticipated, however, because the simulator supports only a subset of the protocols supported by the tested equipment. Recall that the simulator

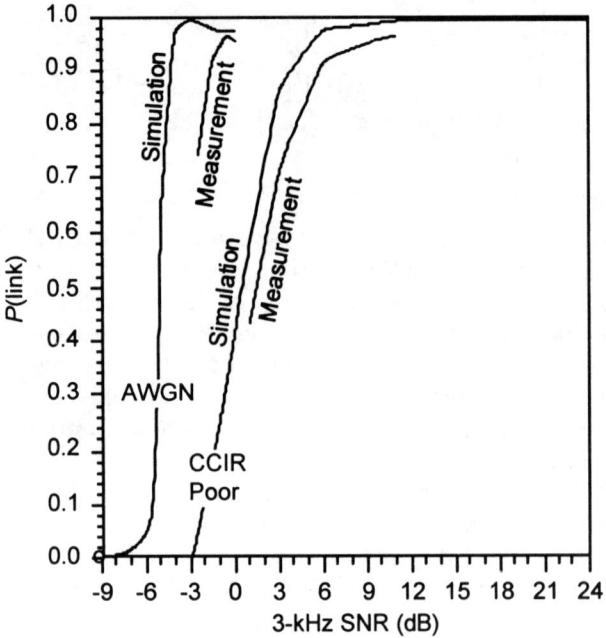

Figure 7.10 Simulation versus measurement for LP level AL-2.

does not support group calls or multiple-word addresses. As a result, the number of possible protocol branches that it must examine during word sync acquisition and during T_{sc} is reduced to exactly one: Every ALE word must be identical to the last ALE word received until T_{lc} is reached. Furthermore, only one format of T_{lc} must be accommodated by the simulator. Examination of the simulator output suggests that the elimination of these possibilities for going astray in following the protocol may be at least partially responsible for the better results produced by the simulator.

7.4.4 Summary of Linking Protection Performance Evaluation

Simulation of HF LP technology has shown that LP can minimize the disruption caused by imitative ALE calls with minimal degradation in linking performance. A unanimous-vote threshold near a value of 25 provides a good balance of protection and ALE linking performance over a broad range of channel conditions.

References

[1] Johnson, E. E, "Analysis of High Frequency Radio Linking Protection," *MILCOM'92 Conf. Proc.*, San Diego, CA: IEEE, Oct. 1992.
[2] CCIR, "HF Ionospheric Channel Simulators," Report 549-2, *Reports to the CCIR, 1986,* Vol. III, Part A, Geneva: International Radio Consultative Committee, 1986, pp. 59–67.
[3] CCIR, "Use of High Frequency Ionospheric Channel Simulators," Recommendation 520-1, *Recommendations of the CCIR, 1986,* Vol. III, Part B, Geneva: International Radio Consultative Committee, 1986, pp. 57–58.
[4] Watterson, C., et al. "Experimental Confirmation of an HF Channel Model," *IEEE Trans. Comm. Technol.*, Vol. COM-18, No. 6, 1970, pp. 792–803.

The High-Frequency Data Link Protocol   8

8.1 INTRODUCTION

8.1.1 Background

The approval cycle for U.S. military and federal standards includes a period of review and comment by engineers in interested government agencies, with participation often sought from knowledgeable engineers in industry as well (see Sec. 1.1.2). As MIL-STD-188-141A [1] was undergoing this process, some participants questioned the choice of an FSK modem for ALE. They believed that the introduction of adaptive equalization, decision-directed decoding, and other techniques had recently improved PSK modems to yield higher performance in HF channels than could be achieved by FSK.

For example, the *serial tone* PSK modem specified in MIL-STD-188-110A [2] operates at an uncoded data rate of 4800 bps versus the 375-bps rate employed by the 8-ary FSK modem used by ALE. In addition, the PSK modem can dynamically vary the coded data rate to match channel conditions. Under ideal channel conditions, this PSK modem can sustain a 2400-bps data rate. In its robust 75-bps mode, the PSK modem can "punch" data through such poor channels that the ALE FSK modem cannot establish a link.

The key reasons why a robust FSK modem was chosen over the modern MIL-STD-188-110A PSK modem for the ALE standard follow from the very technology that boosts PSK modem performance on HF channels. The PSK modem requires a preamble of at least 600 ms at the beginning of each transmission to train the demodulator. For particularly challenging channels, the interleaver length and the preamble length are both extended to 4.8 sec. This preamble not only lengthens the frame, but requires that potential receivers be already "on channel" when the preamble begins. Thus, two substantial problems arise when using an advanced PSK modem for ALE:

1. During the link establishment handshake, data rate is less important than link *turnaround time*, the time needed to complete two-way communication and establish an HF link. During the time required for the PSK modem to send a 600-ms preamble, the ALE modem can send over half of a frame containing one-word addresses. During the time required for the 4.8-sec preamble, the ALE modem could complete more than half of its entire handshake.
2. Because receivers must be present on channel at the beginning of the PSK modem preamble, the asynchronous design of the current ALE system could not be supported. To use PSK modems for link establishment, stations would need to be synchronized to some extent. The durations of time slots used for link establishment calls could be no shorter than the allowed network timing uncertainty. Thus, a PSK-based ALE system would require tighter synchronization than does the current linking protection standard to achieve the same "time to link" as the current ALE system.

Thus, the superior data rates and better SNR performance of the PSK modem are of reduced advantage for link establishment. However, once the link is established, data users who value data throughput over turnaround time will prefer to switch to these modern data modems.

The preceding chapters have discussed the data link layer protocols for ALE as well as station coordination functions that can be carried by the ALE waveform. We now turn our attention to a high-performance data link protocol that uses advanced HF data modems. Figure 8.1 shows the position of this protocol function in the hierarchical HF system model.

8.1.2 History

In 1992, the National Telecommunications and Information Administration (NTIA) requested that the HF industry develop and submit for standardization a data link protocol suitable for use with modern HF data modems. A number of groups had been working on such protocols for some time, and those groups used this opportunity to meet and discuss their work in the neutral setting provided by the technical advisory committee formed by NTIA.

By the end of 1993, the groups had come to general agreement on the protocol, with only a few minor points of difference between the two leading contenders for standardization. The two competing protocols were implemented and tested, with the test results brought to a meeting at New Mexico State University in February 1994 at which the remaining differences were addressed. The result of this meeting was a single HF data link protocol

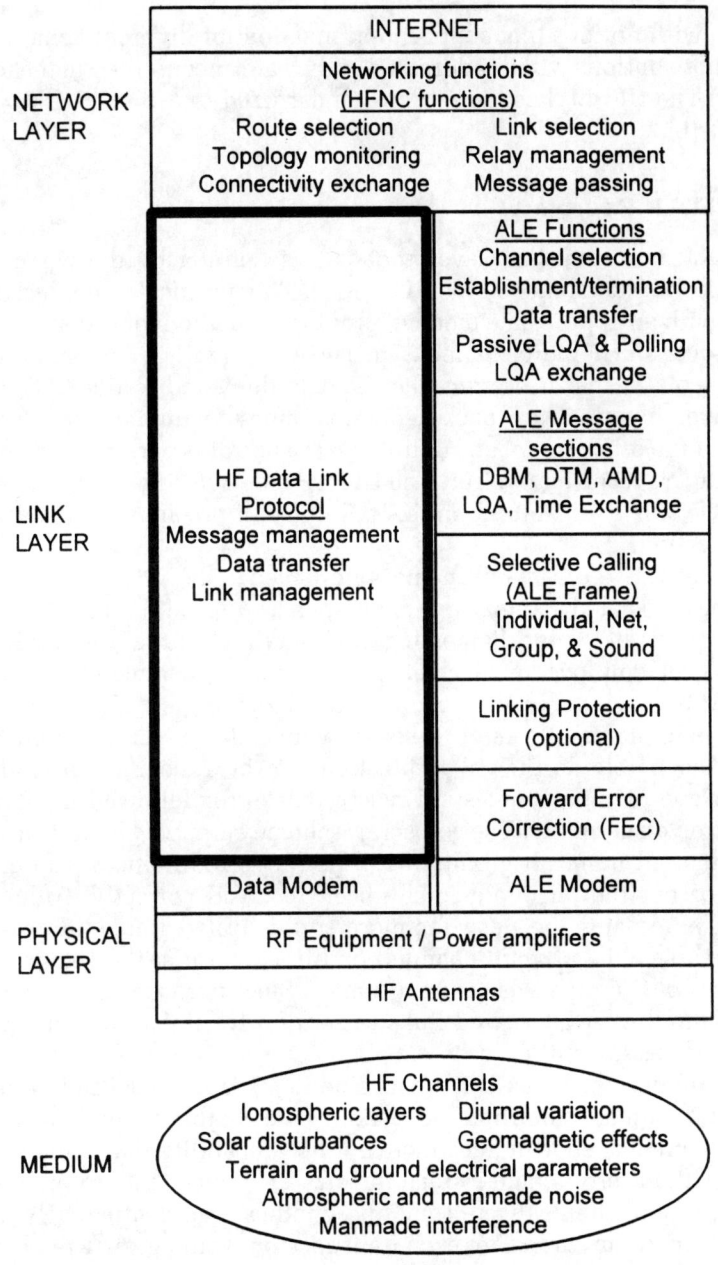

Figure 8.1 The HF data link protocol in the hierarchical-layered view of HF communications.

(HFDLP), with options that support optimization for different communities of users. (These options will be addressed as they are encountered in the following sections.) The HFDLP has since been standardized in MIL-STD-187-721C and FED-STD-1052.

8.1.3 HFDLP Overview

The HFDLP is a data link layer protocol as defined by the Open Systems Interconnection Reference Model (OSI-RM). This protocol, when used in conjunction with an appropriate modem, provides a method for transmitting error-free data over an HF radio channel. The HFDLP is similar to other modern data link protocols in that *user data*, that is, data delivered to the HFDLP from a higher layer function, are packaged into frames for delivery to the HFDLP terminal at the other end of an HF link. Error control is performed by repeating frames that are received in error. The HFDLP uses *selective repeats* rather than a *go-back-N* mechanism to minimize the usage of precious HF bandwidth for error correction.

By contrast, a go-back-N scheme is employed in the X.25 link layer protocol [3] and the associated amateur version known as AX.25 [4]. The latter protocol, which employs fields for digipeating not contained in the original X.25 standard, has been employed extensively by government, commercial, and civilian users (HAMs) as a radio link layer protocol. AX.25 is highly effective on wireline or continuous nonfading radio links for which the number of retransmitted frames is negligible. In the HF application, however, channel dropouts occur in which one or more frames are received in error followed by subsequent frames received correctly. The go-back-N scheme employed by X.25 and AX.25 requires that all frames beginning with the first erred frame be retransmitted, whether or not they had previously been received correctly. Although this approach eliminates the need for more sophisticated frame addressing and message reassembly software, channel conditions can arise that require multiple retransmissions of otherwise correct frames. Such retransmissions can reduce the throughput of AX.25-based links to zero on HF links over which HFDLP successfully passes traffic.

The number of bytes per data frame is variable in HFDLP and can be adapted to channel conditions to maximize data throughput. In high error rate environments, short frames increase the probability that some frames get through without errors, and reduce the size of frames that are retransmitted. However, if short frames were sent singly rather than contiguously, then the required turnaround time after each transmission would greatly reduce overall link throughput. Consequently, groups of frames are sent together in what are termed *data series*. The number of frames per data series is also variable, giving an additional degree of freedom for optimizing data link performance.

Although the HFDLP can be employed in the usual data link layer *bit pipe* mode, its most powerful capabilities are message oriented. In other words, HFDLP treats messages as units, and it can negotiate use of the data link based on message priority as well as reserve buffer space at the receiving terminal using the known sizes of messages. The HFDLP also supports multiplexing multiple network layer connections over a single data link through the use of a *connection ID* that is appended to each message sent.

These control capabilities are implemented through the use of *control frames*, which are sent over the same data link as the data frames. Control frames consist of *header* fields which support three subprotocols within the HFDLP:

1. *Message management protocol*, which works with complete messages as units and negotiates the transfer of messages over the data link, including priority-based preemption and resumption of messages.
2. *Data transfer protocol*, which manages the transfer of data and control frames over the link, including acknowledgments and flow control, as well as negotiation of modem parameters (data rate and interleaver depth), frame sizes, and the number of frames sent in each transmission. Of the three protocols, the data transfer protocol corresponds most closely to the usual notion of a data link protocol.
3. *Link management protocol*, which establishes and monitors continued activity on the physical layer links that support the HFDLP. Except for ARQ circuit mode operations, the data link is dropped when neither terminal has traffic to send. Link timeouts cause link termination when receptions from the distant terminal have ceased, even though a message may be in progress. The message management protocol (or a higher layer entity) may then invoke reestablishment of the physical link followed by reestablishment of the DLP link and resumption of the message.

The HFDLP can operate with either fixed-length control frames, which always contain all header fields, or with variable-length control frames, which contain only those fields necessary in each instance (actually one of four predefined sets of fields) as shown in Table 8.1.

The HFDLP includes several operating modes. The primary mode of operation is the *ARQ* mode, which provides for error-free point-to-point data transfer. The two secondary modes of operation are the *broadcast* mode and the *ARQ circuit* mode. The broadcast mode allows unidirectional data transfer to one or more receivers. The ARQ circuit mode allows a link to be established and maintained in the absence of traffic. Only the ARQ mode is mandatory.

Each of the three subprotocols is discussed in turn later in the chapter. In the discussion of the HFDLP subprotocols, the terms *transmitting terminal*

Table 8.1
HFDLP Modes of Operation

Mode	ARQ (?)	Control Frame Sizes
ARQ	Y	Fixed or variable
ARQ circuit	Y	Variable
Broadcast	N	Fixed

and *receiving terminal* will be used. It is important to understand that the transmitting terminal is not necessarily the terminal that is instantaneously sending a frame; instead, it is the terminal that has negotiated the right to send data frames. (This negotiation uses the message management protocol.) Thus, the transmitting terminal sends data frames as well as control frames that "herald" the characteristics of additional data frames to be sent. However, the transmitting terminal *receives* control frames that acknowledge data frames and Heralds.

8.1.4 HFDLP Operation

A message transfer using the HFDLP commences when a message to be delivered arrives at one of the HF station terminals. The sequence of steps in conveying the message to its destination is illustrated in the following example:

1. If the HFDLP terminal at the station that needs to deliver the message (the *source terminal*) has no HF link to the destination station, an ALE controller at the source station is requested to establish a suitable link.
2. When an HF link is available, the source terminal begins HFDLP link establishment. This protocol establishes unambiguously which of the two terminals has the higher priority message to deliver, and that terminal becomes the transmitting terminal: (a) The source terminal sends a *call control* frame that contains a request to establish an HFDLP link, a *message management* header that describes the message to be sent (including its priority), and a *herald* that describes the data rate, frame size, and number of frames per data series that the sending terminal proposes to use in sending the message. (b) If the destination terminal (1) responds with a *call acknowledgment* (ACK) control frame and (2) has a higher priority message that it (coincidentally) needs to deliver to the source station, it will describe that message in its response in the message management header fields; this constitutes *reverse preemption*. Usually, however, it has no traffic for the other station or it would have called itself! Thus,

the destination terminal will normally respond with a control frame that accepts the incoming message, possibly with a request to use a different data rate or frame size.

3. On completion of this negotiation of message priorities and data transfer parameters, the terminal that will send the first data series is understood to be the transmitting terminal, and the other terminal is considered to be the receiving terminal. The transmitting terminal sends the first series of data frames and awaits acknowledgment (a *Data-ACK* frame).
4. The receiving terminal checks the received data frames for errors, and returns a *Data-ACK control* frame that carries individual acknowledgment bits for each frame received.
5. If the transmitting terminal does not receive the Data-ACK frame within a reasonable time, it sends a Data-ACK-Request control frame to prompt retransmission of the Data-ACK. When the Data-ACK does arrive, the transmitting terminal forms a new data series, which starts with any frames that the receiving terminal has asked to be retransmitted, and sends it. Unless negotiation is needed during the message transfer, the channel simply carries alternating data series and Data-ACKs.
6. If a higher priority message arrives at the transmitting terminal before the current message is completely sent, the transmitting terminal can initiate forward preemption of its own message by sending a *herald control* frame describing the new message instead of a data series. If the receiving terminal accepts the preemption, the previous message is suspended, and transfer of the new message begins.

The diagrams shown in Figure 8.2 illustrate several of the possible cases of frame exchanges that occur during HFDLP message transfers. The remainder of this chapter discusses the operation of this protocol.

8.2 DATA TRANSFER PROTOCOL

The data transfer protocol provides the usual data link layer services of data delivery over a single link with flow control, error detection, and error correction via automatic retransmission (ARQ).

8.2.1 Data Transfer Protocol Operations

The data transfer protocol is implemented through the exchange of control frames and data frames over the data link. This exchange is governed by the ARQ mode discussed in the following paragraphs. The two optional modes

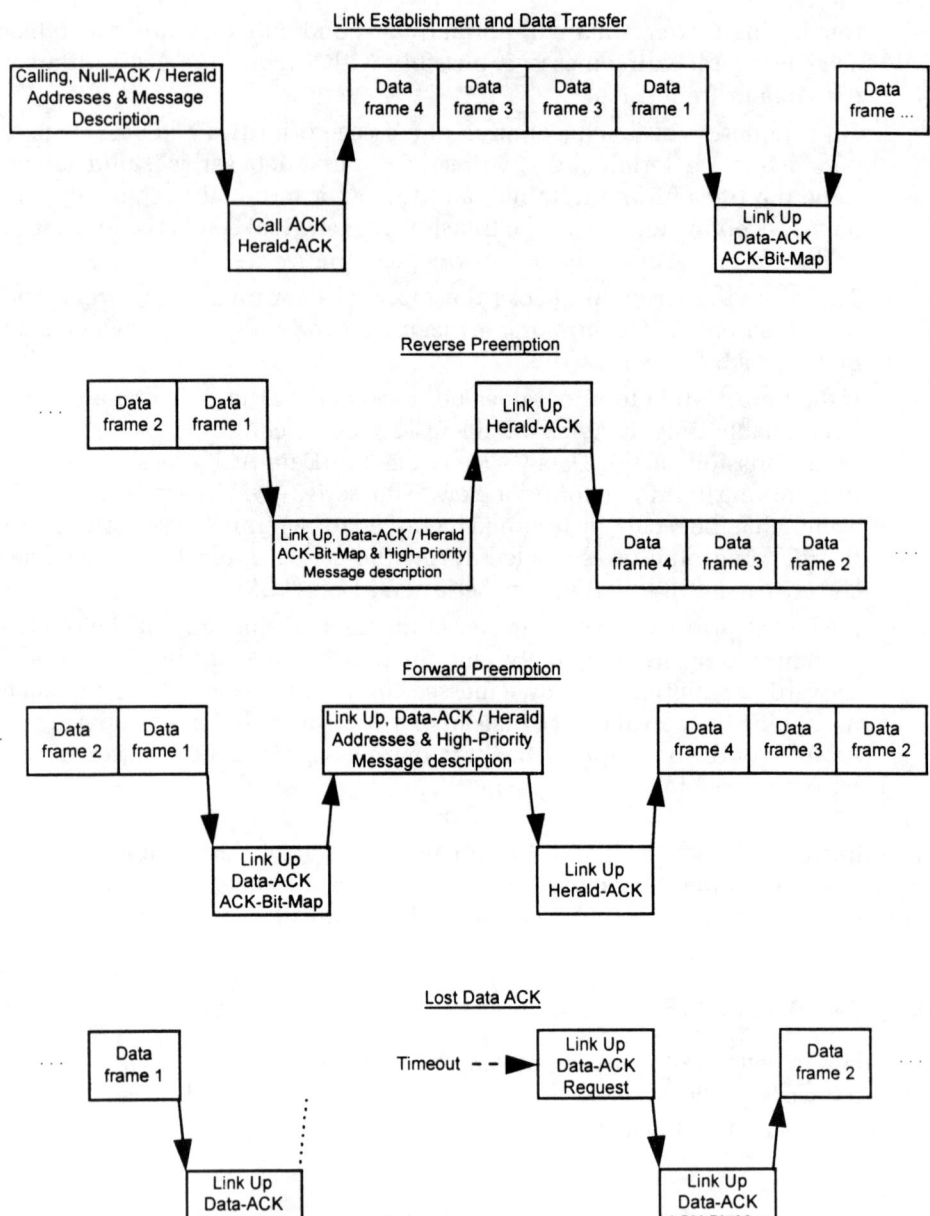

Figure 8.2 Examples of HFDLP frame exchanges.

(broadcast mode and ARQ circuit mode) represent only minor variations on this mode and are discussed later in the chapter.

The ARQ mode cycles among three phases: the *negotiation* phase, the *data transfer* phase, and the *data acknowledgment* phase. Negotiation can be combined with data acknowledgment (e.g., to reverse the link for high-priority reverse-channel traffic).

8.2.1.1 Negotiation Phase

HFDLP terminals use the negotiation phase of the data transfer protocol to resolve flow control and transfer characteristics of the data link connection prior to the transfer of data frames over the link. The negotiation phase starts with the transmission of a control frame containing a herald, which announces a data series to be transferred. This phase ends with the transmission of an acknowledgment of the herald, which signifies acceptance of the announced data series.

The negotiation phase is required in the following circumstances:

- Before the first data frames are sent over a link (except immediate mode message transfers, described later).
- Before each data series if the most recent control frame from the terminal that will receive the data frames invoked in continuous negotiation mode.
- When flow control prevents the transfer of any data frames.
- Before any change in previously negotiated values in the herald fields, except that no negotiation is necessary to change only the data rate and frames-in-next-series if the following rules are followed: (1) If the data rate increases by 2^n, the frames-in-next-series value must increase by the same factor, keeping the ratio of frames-in-next-series to data rate constant. If this would result in a frames-in-next-series value greater than 255, frames-in-next-series is set to 255, but the ratio of frames-in-next-series to data rate without this limiting operation is stored for future reference. (2) If the data rate decreases by 2^n, the frames-in-next-series value must decrease by the same factor (then rounded down to the nearest integer), keeping the ratio of frames-in-next-series to data rate constant, except that if the previous frames-in-next-series value was 255, the new frames-in-next-series value is computed from the new data rate using the previously stored ratio of frames-in-next-series to data rate, constrained as before to be no greater than 255. (3) If the series carrying the last unacknowledged frame of a message does not require the full complement of frames. Note that this can happen repeatedly if the final frames are retransmitted.

Initiation

A terminal initiates the negotiation phase of the protocol by transmitting a control frame containing a herald that announces the data rate, interleaver setting, number of frames in the data series, and the number of data bytes per data frame. The receiving terminal will answer the herald with one of two responses: (1) a Herald-ACK signifying acceptance of the offered data series or (2) a Null-ACK indicating that the receiving terminal cannot accept the offered data series. The transmitting terminal will retransmit the herald if it does not receive a valid response within the response timeout period (see Sec. 8.6.3).

Acceptance

A receiving terminal signifies its acceptance of the data series offered in a herald frame by sending a Herald-ACK frame in response. The receiving terminal may negotiate (override) any of the herald fields, except that bytes per frame may not be changed if any data frames from a previous data series are to be resent in the current data series.

On receipt of the Herald-ACK frame, the transmitting terminal will normally transition to the data transfer phase and begin transferring the data frames of the accepted series. Note, however, that a transmitting terminal may respond to a Herald-ACK by immediately transmitting a new and different herald (e.g., for forward preemption). The receiving terminal must, therefore, treat each herald received as if it contained new information and not merely as an identical repetition of a previously received herald.

Data Refusal

A HFDLP terminal refuses an offered data series by sending a Null-ACK frame in response to a herald frame announcing a new data series. A transmitting terminal will normally abort a message transfer when a data series is rejected: If a change in herald parameters would have resulted in an acceptable data series, the receiving terminal would have so indicated.

If the cause for refusal is a temporary lack of local buffer space, however, the receiving terminal may impose flow control by setting the flow control flag in the Null-ACK. In this case, the transmitting terminal will repeatedly retransmit the herald frame in order to determine when the flow control restriction is lifted (see description of flow control later).

8.2.1.2 Data Transfer Phase

On transition to the data transfer phase, the transmitting terminal will transmit the data frames announced in the acknowledged herald. The transmitting terminal is expected to send the total number of data frames announced in the herald without delay or interruption. All data frames are of the same size as was announced in the herald. This implies that the last data frame of a message may need to be padded with *fill* bits. Receiving terminals use message size information from the message management header to determine where the message is to be truncated in order to remove the fill bits from its output data stream.

8.2.1.3 Data Acknowledgment Phase

The data acknowledgment phase begins after the last data frame of the data series has been transmitted. The transmitting terminal will stop transmission and wait for a Data-ACK frame from the receiving terminal. A Data-ACK frame contains an ACK-Bit-Map indicating which data frames were received error free and which frames require retransmission. After receipt of the Data-ACK, the transmitting terminal will prepare a new data series containing all prior frames requiring retransmission and enough new data frames to fill the data series. Unless negotiation is required, the terminals will return to the data transfer phase on completion of the data acknowledgment phase.

If the transmitting terminal fails to receive a Data-ACK frame from the receiving terminal within the response timeout period, it will transmit a Data-ACK-Request frame to inform the receiving terminal that a Data-ACK was not received. Receipt of an Data-ACK-Request by a receiving terminal causes it to resend the last Data-ACK frame sent. If the ACK sequence number received by the transmitting station in a Data-ACK is incorrect, the receiving terminal must have missed the entire preceding data series, which must therefore be resent by the transmitting terminal.

8.2.1.4 Flow Control

A receiving terminal may impose flow control by returning a control frame (Null-ACK, Data-ACK, or Herald-ACK) with the flow control flag set to 1. (The flow control flag is present only when the extended addressing flag is set to 0.) If the ACK-Bit-Field in a Data-ACK with a flow control flag set to 1 indicates that some data frames should be retransmitted, the transmitting station will resend those data frames. This situation arises when the receiving station's reassembly buffers are full, pending receipt of the requested data frames. In no case will a transmitting station send new data frames in response to a control frame containing a flow control flag set to 1.

When a transmitting terminal is under flow control and has no unacknowledged data frames to resend, it will poll the receiving terminal by sending heralds describing the data remaining to be sent. The receiving terminal will maintain flow control as necessary by responding to heralds with Null-ACKs. The Null-ACKs may be minimum-length control frames if permitted by the control mode; it is not necessary to continue to send the flow control flag to continue previously imposed flow control. However, if longer control frames are sent, the flow control flag must continually be set to continue flow control.

A receiving terminal lifts flow control by responding to a frame from the transmitting terminal with a control frame containing a flow control flag set to 0. The type of control frame sent to lift flow control must be an appropriate response to the frame from the transmitting terminal (e.g., a Herald-ACK for a herald, or a Data-ACK for re-sent data frames or a Data-ACK-Request).

8.2.2 Data Transfer State Machine

The *data transfer state machine* (DTSM) manages the frame-by-frame, series-by-series transfer of data from the transmitting terminal to the receiving terminal. It negotiates the herald parameters as necessary and uses response timeouts to prompt retransmissions from a transmitting terminal. The state diagram for the DTSM is shown in Figure 8.3. Transitions in Figure 8.3 are labeled as follows: The event causing a transition is identified above the line, with the type of frame(s) sent as a result listed below the line. Internal DTSM flags that can prompt a transition are listed between angle brackets (<flag>).

8.3 MESSAGE MANAGEMENT PROTOCOL

The *message management protocol* negotiates the order of transmission of messages over the data link based on message priority and provides for the multiplexing of multiple connections over a single data link. It provides mechanisms for preemption of a lower priority message to transfer a higher priority message (in either the forward or reverse direction), and the resumption of the transfer of the lower priority message after the complete transfer of the higher priority message. The protocol provides a means for the terminal receiving a message to specify the resumption point within the preempted message, or to request complete retransmission if the preempted message was discarded.

8.3.1 Message Management Protocol Operations

HFDLP terminals uses the message management protocol to coordinate the transfer of data messages, the priority preemption of messages, and the resumption of preempted messages. The message management protocol is implemented through an exchange of control frames that contain the *message management*

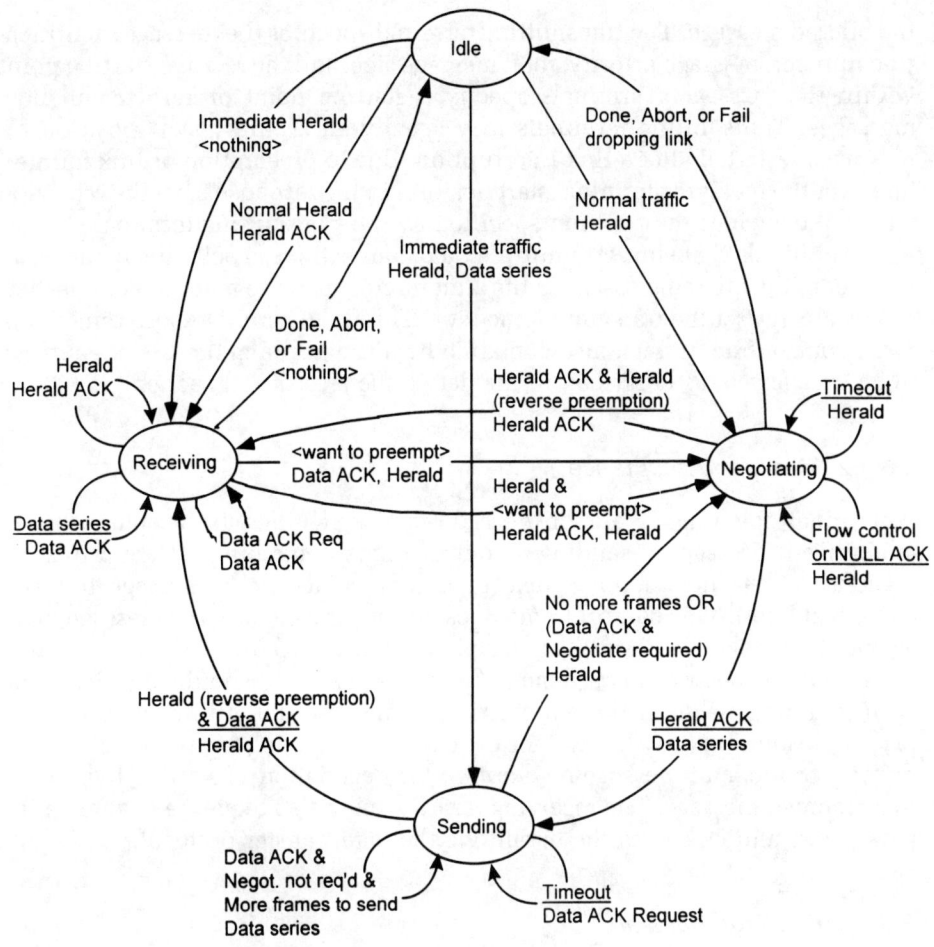

Figure 8.3 Data Transfer state machine.

header fields. The message management header fields contain a complete description of the message to be transferred or resumed. Terminals are able to selectively accept or reject messages over the data link based on the message source, priority, length, or connection number.

8.3.1.1 Message Announcement

A terminal announces a message to be transferred over the HF data link by transmitting a control frame with a message management header that describes the offered message. The transmitting terminal specifies the message identification number, message priority, total message size, and the relative starting point within the message. Terminals specify a starting point of zero for all new messages. Transmitting terminals may specify an arbitrary start position for messages restarted after a link interruption (due to preemption or link failure); however the receiving terminal start position in the response from the receiving terminal overrides the position specified by the transmitting terminal.

Terminals that support multiple data connections specify the connection number of the offered message in the announcing message management header; others always set the transmit connection ID field to zero. *Message announcement control* frames must also contain a herald describing the first data series to be transferred, as described in the data transfer protocol section.

8.3.1.2 Message Acceptance

A terminal operating in one of the ARQ modes signifies its acceptance of an announced message by sending a control frame containing a Herald-ACK in response to the herald frame (except in immediate mode message transfer, described later). This control frame does not need to contain the message management header fields: Message acceptance is the default. If the Herald-ACK does contain message management header fields, they should duplicate the corresponding fields in the announcing frame. See the following section on priority resolution (Sec. 8.3.1.4) for the case of a receiving terminal with higher priority traffic, and *message resumption* for negotiation of starting byte offset in a resumed message. The receiving terminal may also request a change in the parameters announced in the herald via the data transfer protocol.

8.3.1.3 Message Refusal

A terminal operating in one of the ARQ modes may refuse an announced message by returning a control frame with the "receive message next byte offset" field set to a value greater than or equal to the announced transmit message size divided by eight. A refused message may be discarded by the transmitting terminal. Immediate mode message transfers do not pause for a Herald-ACK, however, and therefore cannot be refused until after the first data series has been sent.

8.3.1.4 Priority Resolution

Terminals resolve data link contention by exchanging message management headers announcing the highest priority traffic available for transfer. The terminal with the highest priority traffic is allowed to be the first to use the data link. Thus, a terminal receiving announcement of a lower priority message will preempt that incoming message as described in the later section on reverse preemption, except that the message management header is sent with a Null-ACK to refuse the first data series (rather than the Data-ACK needed when preempting a message in progress).

8.3.1.5 Link Sharing

If both terminals have traffic of equal priority, the terminals may optionally share the data link on an equal basis. In such sharing, the two terminals alternate the transmission of data series. Each Data-ACK includes a herald. The terminal receiving the Data-ACK/herald responds with a Herald-ACK. The terminal receiving this Herald-ACK responds by transmitting its data series. The terminal receiving the data series responds with a Data-ACK/herald, and the link direction reverses again.

8.3.1.6 Message Preemption

The message management protocol allows terminals to preempt lower priority traffic with a new message of higher priority. Preemption is supported in both the forward direction (higher priority traffic in the same direction as the preempted traffic) and the reverse direction (higher priority traffic in the direction opposite to the preempted traffic).

Terminals may only preempt ongoing message transfers with messages of higher priority than the preempted message. Nested preemption is allowed provided the preemption meets this relative priority requirement. If a terminal resumes preempted messages, it must resume them in the reverse order of preemption, that is, the transmission of the most recently preempted message is resumed first. This ensures that the relative priority requirement is maintained throughout the preemption and resumption process. Terminals supporting preemption and resumption will usually implement a preempted messages stack.

Forward Preemption

A transmitting terminal may preempt the ongoing transfer of a message with a higher priority message by sending a new message management header announcing the new message. The message management header announcing the new message should be sent at the first logical opportunity for a control frame transfer. The receiving terminal can refuse the offered message as described earlier. Upon acceptance of the preempting message, the transmitting terminal will suspend transfer of the preempted message until the higher priority traffic is transferred.

Reverse Preemption

A receiving terminal may preempt the transfer of an incoming message by sending a message management header announcing a higher priority message at the first available control frame opportunity. The transmitting terminal can refuse the offered message as described earlier. Upon acceptance, the terminals will reverse the link direction and transfer the higher priority message until completion; transfer of the preempted message may then be resumed.

8.3.1.7 Message Resumption

A terminal may resume the transfer of a preempted message after the completion of the higher priority traffic (or of a message interrupted by link failure after link recovery). In both cases, a message management header is sent that announces the resumption of the message. The message management header will contain the same message description data as the original announcement of the message with the exception of the "transmit message next byte offset" field.

The transmitting terminal should specify the starting offset within the message upon resumption to be the first byte of the message not acknowledged by the receiving terminal prior to the preemption (or link failure). The receiving terminal may override this starting position by sending a message management header specifying the desired start position in the "receive message next byte offset" (may be zero to force restart instead of resumption).

All bytes of the message following the negotiated starting position must be sent, rather than just the bytes in frames not acknowledged prior to preemption or interruption. The transmitting terminal should not assume that the receiving terminal has retained any frames beyond the negotiated starting position.

8.3.1.8 Null Message Management Headers

When none of the preceding situations applies, no message management header need be sent. In this case, unused message management header fields are set to all 0s by the transmitting terminal; the receiving terminal ignores message management headers having transmit message size set to zero.

8.3.2 Message Management State Machine

The message management protocol is conceptually implemented as a *message management state machine* (MMSM). The MMSM provides the top-level interface to the HFDLP, which is called on by the local network layer entity. The MMSM at each terminal reflects the current status of that terminal: idle, receiving terminal, transmitting terminal, or prospective transmitting terminal.

The transmitting terminal and receiving terminal states correspond to the definitions given previously: A terminal is the transmitting terminal on a link when it has negotiated the right to send data frames, and it is the receiving terminal when it has agreed to receive data frames. A terminal only enters the prospective transmitting terminal state when it has previously agreed to receive data frames, but now needs to become the transmitting terminal. This occurs when it is either reversing the link after each data series (link sharing with equal priority messages) or when it wishes to preempt the message in progress to deliver a higher priority message to the current transmitting terminal (reverse preemption).

In either case, the receiving terminal sends a control frame that indicates its desire to reverse the direction of message flow, but it cannot then go directly to the transmitting terminal state until the current transmitting terminal acknowledges the link turnaround. Use of the prospective transmitting terminal state ensures that the two terminals on a link are never simultaneously in the transmitting terminal state, even in the presence of lost control frames. The state diagram for this finite state machine is shown in Figure 8.4.

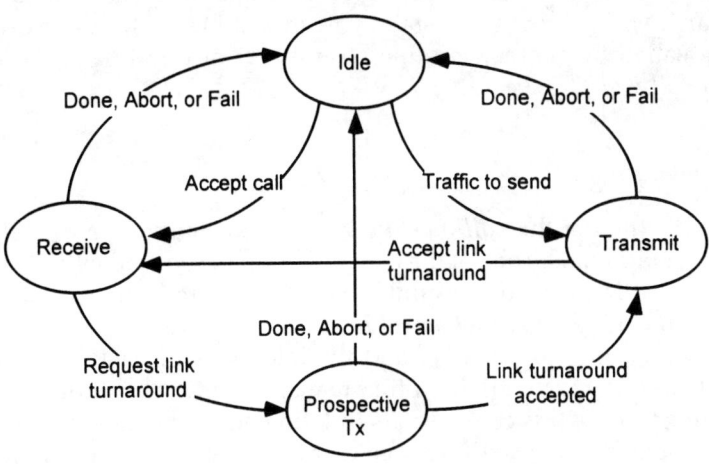

Figure 8.4 Message Management state machine.

8.4 LINK MANAGEMENT PROTOCOL

HFDLP terminals employ an optional link management protocol to coordinate terminal state and resolve access contention prior to beginning the transfer of data over the link. The link management protocol is implemented through an exchange of control frames containing the link management header fields.

During the link establishment phase, the two terminals attempting to link will exchange terminal addresses and link state information so that both terminals are fully aware of the link state of the other terminal and therefore do not attempt to transfer data before the other terminal is ready to accept the data.

The link establishment phase is optional and can be bypassed. When the link establishment phase is bypassed, the terminal sets the link state field in all control frames to the *link-up state*. Immediate message transfer refers to this bypass mode of operation.

8.4.1 Link Management Protocol Operations

The link management protocol requires the two terminals attempting to link to sequence through a set of link states, as defined next, to ensure that both terminals reach the link-up state before either attempts to transfer data traffic.

8.4.1.1 Idle State

The *idle state* of the link management protocol is the resting state of the protocol. The terminal will reside in this state whenever it has no traffic to send and is not being called by another terminal. Terminals also return to the idle state upon link failure.

8.4.1.2 Calling State

A terminal will enter the *calling state* when it is requested to establish a link with another terminal by one of the data link users associated with the terminal. When in the calling state, the terminal sends full-length (520-bit) control frames containing the following: (1) a control frame header indicating the control mode of the link it wishes to establish and the addresses of itself and the distant terminal (short or extended); (2) a link management header indicating its link state (calling state in this case) and its link timeout; (3) a message management header announcing the (first) message that it wishes to transfer to the destination; and (4) an extended function header containing the user ID of the calling terminal. After sending the control frame, the terminal will wait for a response from the called terminal. Upon reception of a valid control frame from the called terminal containing a link management header acknowledging the link request, the terminal will advance to the link-up state. If the terminal does not receive a timely call acknowledgment frame, it will retry the call.

8.4.1.3 Call Acknowledge State

A terminal will enter the *call acknowledge state* upon reception of a valid control frame containing a link management header indicating that a terminal is attempting to establish a link with it. When in the call acknowledge state, the terminal sends control frames in response to the calling terminal's control frame that indicate the local link state and link timeout and the accepted link modes and characteristics. If the calling terminal accepts variable-length control frames, a Type 1 (minimum-length) control frame may be returned to accept the announced message, data series, and user ID (overrides to control mode, and negotiation mode may be included in a control frame of any size). If a longer control frame is sent, refer to the descriptions of the other protocols for contents of the message management and data transfer headers.

The *extended function* header, if present, contains a user ID set to one of the following:

- Zero, indicating the interoperable mode (which overrides any other user ID), or
- Identical to the User ID of the calling terminal to accept the corresponding specialized mode.

After sending the control frame, the terminal will wait for a response from the calling terminal.

If the acknowledging terminal does not receive a timely valid reply to its response, and it has higher priority traffic (i.e., it initiated reverse preemption), it will retransmit the call acknowledge announcing this traffic after the response timeout. If it does not have higher priority traffic, it will await a repetition of the call. If it has not received a repeated call within the link timeout specified in the call frame, it will return to the idle state.

8.4.1.4 Link-up State

The *link-up state* is the fully operational state of the terminal. The terminal may begin the data transfer protocol when it reaches the link-up state. The terminal will enter the link-up state upon receipt of a valid control frame indicating that the other terminal is in either the call acknowledge or link-up state, or upon receipt of a local request for immediate message transfer when it is in the idle state. When in the link-up state, the terminal will include a link management header in all control frames sent indicating that its link state is "link-up." The terminal may implement any operational feature of the data transfer and message management protocols when in this state. The terminal will remain in the link-up state until the link is dropped (or fails) whereupon it will return to the idle state.

8.4.1.5 Link Failure

If the link fails while in progress (detected by link timeout), the protocol will act as if the link was aborted and the terminal will return to the idle state. (As an option, the terminal may first attempt to reestablish the link to continue the message transfer.) The terminal should be capable of signaling this sudden change in status to a higher layer controller and to the operator. Terminals may retain the successfully received portion of the uncompleted message that was in progress when the link failed in the same manner as a preempted message. The length of time to hold the uncompleted message should be an operator-selectable parameter.

8.4.1.6 Link Termination

The link may be dropped due to a local request, a link timeout, or the absence at both terminals of messages to be sent (when not in circuit mode). Terminals should send a control frame with link state set to *dropping link/idle* before dropping the link.

8.4.1.7 Immediate Mode

An immediate mode message transfer bypasses link establishment and the initial data transfer negotiation. The first transmission begins with a 520-bit control frame carrying a link state of link-up, a message management header announcing a message, and a herald describing the characteristics of the first data series, with the user ID field normally set to 0. This control frame is immediately followed by two sync bytes, followed immediately by the first data series of the announced message, with no change in data rate. Following this first data series, the data transfer protocol uses either ARQ mode or ARQ circuit mode, as specified in the initial control frame.

8.4.2 Link Management State Machine

The *link management state machine* (LMSM) executes the state transitions specified in Table 8.2 and employs a link timeout mechanism to detect a failed

Table 8.2
HFDLP Link Management State Transitions

State	Action	Transition Criteria	Next State
Idle	None	Normal link request from local operator or connection	Calling
		Immediate link request from local operator or connection	Link-up
		Receive control frame with Link State = Calling	Call acknowledge
		Receive control frame with Link State = Link-up	Link-up (immediate mode; send no Herald-ACK)
Calling	Emit link establishment header with Link State = Calling	Response timeout expired	Same state
		Neighbor Link State = Calling	Call acknowledge
		Neighbor Link State = Call	Link-up
		Neighbor Link State = Link-up	Same state
		Link timeout expired	Idle
Call acknowledge	Emit link establishment header with Link State = Call acknowledge	Response timeout expired	Same state
		Neighbor Link State = Calling	Same state
		Neighbor Link State = Call	Link-up
		Neighbor Link State = Link-up	Link-up
		Arrival of data series	Link-up
		Link timeout expired	Idle
Link-up	Emit link establishment header with Link State = Link-up	Neighbor Link State = Link-up	Link-up
		Transmit terminal timeout, internal repeat countdown > 0	Link-up (retransmit)
		Link timeout expired	Idle or calling
		Neighbor Link State = Calling	Call acknowledge
		No more traffic to send	Idle
		Neighbor Link State = Dropping	Idle

link. When transmitting terminals drop a link, they should send a courtesy *dropping link* frame to release the other terminal.

8.5 TIMEOUTS

Timeouts serve two functions in the HFDLP: They ensure that terminals do not wait indefinitely for responses, and that transmissions from terminals that simultaneously attempt to link with each other have a low probability of colliding at every retransmission and therefore failing to link. A *response timeout* is used by the transmitting terminal to prompt retransmissions, and by the receiving terminal to detect link failure. A *link timeout* is used by the transmitting terminal to detect link failure.

8.5.1 Turnaround Times

After reception of a valid transmission a terminal must initiate the responding RF transmission no sooner than 1 sec and no longer than 10 sec after cessation of the received RF signal.

8.5.2 Response Timeout

After each valid reception the terminal will establish a time at which it will take action if further valid receptions are not forthcoming. These times differ for transmitting terminals and receiving terminals (defined in Sec. 8.1.3). A further distinction must be made between acknowledged transmitting terminals and prospective transmitting terminals. When a receiving terminal determines that it has traffic to send that is of equal or higher priority than the current transmitting terminal (e.g., when such higher priority traffic is delivered to it by the local user), it heralds this event following the next legitimate reception from the transmitting terminal. Until this receiving terminal receives an acknowledgment of its new status, this terminal is a prospective transmitting terminal. A terminal that has received acknowledgment of its transmitting terminal status is an acknowledged transmitting terminal. (Note that under certain conditions there may be no acknowledged transmitting terminal, but there will never be more than one acknowledged transmitting terminal.)

During link establishment, terminals in the call acknowledge state act as either receiving terminals (if not attempting reverse preemption) or as prospective transmitting terminals (if preempting).

Timeout values are set as follows for the interoperable mode (user ID = 0):

- After each legitimate reception, an acknowledged transmitting terminal sets a response timeout equal to the ending time of its responding transmission plus 25.4 sec.
- Following each retransmission (or the initial transmission of a Data-ACK-Request), an acknowledged transmitting terminal sets its response timeout to the starting time of that transmission plus 48.8 sec.
- A prospective transmitting terminal sets its response timeout to 48.8 sec after the starting time of its transmission, whether this transmission follows a legitimate reception or was a retransmission following a missed response.
- A receiving terminal sets its response timeout following each legitimate reception from the transmitting terminal to a value equal to the ending time of its transmission in response to the reception plus the estimated ending time of its next reception plus the most recent link timeout value received from the transmitting terminal during the current link. If expecting a control frame, 25.4 sec are allowed for the next transmission; otherwise (data series expected), the time allowed for the next transmission is the sum of a 10-sec turnaround time plus $N + 1$ interleaver times, where N is the number of interleavers of the announced size required to hold the announced data series.
- After a call (or immediate transmission) a terminal sets its timeout to be 25.4 sec after the end of the first transmission. If no valid reception has occurred prior to this timeout, the terminal will act in accordance with Section 8.5.4. After retransmissions the timeout is set to the starting time of the retransmission plus either 48.8, 63.8, or 78.8 sec. The choice among these three values is randomly generated in such a manner that each has an approximately equal probability of being selected, with little correlation among the choices either at a single terminal or between pairs of terminals.

Terminals may use timings different from those just given only if they have negotiated a nonzero user ID field, which is associated with different timing rules. (Manufacturers are encouraged to publish generally useful sets of timing rules for industry-wide implementation.)

8.5.3 Link Timeout

The *link timeout* value sent in control frames is computed as the time from the end of the first transmission of a frame until all retransmissions that would result from no response to that frame have been completed, including the response timeout after the last retransmission, and accounting for any automatic data rate or interleaver changes that would be performed by that station.

8.5.4 Action Following Timeout

The action taken by a terminal when no legitimate response is received is as follows:

1. If a transmitting terminal (acknowledged or prospective) has received no signal by the time specified by the response timeout, it will decrement its internal retry countdown and transmit a frame that requests a retransmission from the other terminal. If the missing response is a Data-ACK frame, the transmitting terminal sends a frame with the ACK type set to Data-ACK-Request. If the missing response is a Herald-ACK, it will repeat the herald frame. If the missing response is that associated with a call (or an immediate transmission), the repeated frame is the call (or the herald that was transmitted at the start of the immediate transmission).
2. If a transmitting terminal (acknowledged or prospective) receives a transmission with an invalid CRC (i.e., its modem synchronized to a received preamble, but the CRC in the received frame was invalid), including a transmission that begins before but concludes after the response timeout, it will take the actions specified in the preceding paragraph, except that the transmission will take place as soon as possible within the constraints imposed on turnaround times.
3. Upon expiration of the link timeout, a transmitting terminal will drop the link and either return to the idle state, discarding any unfinished message with notification to the (local) source of that message, or request reestablishment of the physical layer link (possibly through an ALE controller) with the intent to resume the message if relinking is successful.
4. A receiving terminal that has not received a valid transmission prior to the expiration of its response timeout will drop the link. A receiving terminal may respond immediately (within the constraints imposed by Sec. 8.6.1) to transmissions with invalid CRC. If the receiving terminal is in the process of reception when the timeout expires, it will complete the reception but will drop the link if no valid CRC is obtained.

In general, whenever a terminal receives a valid transmission (i.e., contents that are reasonable in the context of the protocol and carries a valid CRC), it will reset its retry countdown to its original value and set the response timeout as discussed above.

8.6 BROADCAST MODE

The broadcast mode is used for one-way transfer of data from a single transmitting terminal to one or more receiving terminals. Since multiple terminals can receive the data, acknowledgments of the data frames are not allowed. A terminal initiates a broadcast transmission by transmitting a 520-bit control frame containing appropriate data transfer and message management header fields. The link state is set to "link-up" and the control mode is set to "broadcast."

The terminal will immediately follow the control frame with data frames. This process is repeated until the entire message is transmitted. Herald frames between each data series are optional. (Broadcast mode is essentially ARQ mode with fixed-length control frames and no acknowledgments from receiving terminals.)

8.7 ARQ CIRCUIT MODE

The ARQ circuit mode operates identically to the variable-length control frame ARQ mode, except that terminals maintain the link in the absence of user data until directed to drop the link by the user. Terminals maintain the data link connection in the absence of data by sending herald and Null-ACK control frames that announce no message. Terminals respond to this null herald by sending a Null-ACK control frame along with a herald of their own. Terminals should use the error statistics of the null herald/Null-ACK exchange to maintain a valid estimate of the supportable data rate of the data link. When available, new user messages are announced according to the requirements of the ARQ data transfer mode. Note that the ARQ circuit mode does not include a fixed-length control frame option.

8.8 FULL-DUPLEX OPERATION

The formats and protocols for full-duplex operation are identical to those specified for simplex operations except that timeouts and retransmissions for one traffic direction must work around the data series transferred in the other direction. Simultaneous traffic should be supported whenever traffic of any priority exists at both terminals. Note that reverse link preemption is not needed in full-duplex operation.

8.9 ADAPTING DATA RATE

A simple but effective algorithm for adapting the data rate of the modem during data link operation is as follows: (1) If all of the frames are transferred without error, the data rate should be raised; or (2) if fewer than 50% of the frames are transferred without error, the data rate should be lowered.

References

[1] Department of the Army, Information Systems Engineering Command, *MIL-STD-188-141A: Interoperability and Performance Standards for Medium and High Frequency Radio Equipment*, Philadelphia, PA: Naval Publications and Forms Center, Attn. NPODS, Notice 2, Sep. 1993. Available on the Internet at URL ftp://tracebase.nmsu.edu/pub/hf/pubs/mil_std_188_141a.

[2] Department of the Army, Information Systems Engineering Command, *MIL-STD-188-110A: Interoperability and Performance Standards for Data Modems*, Philadelphia, PA: Naval Publications and Forms Center, Attn. NPODS, Sep. 1991. Available on the Internet at URL ftp://tracebase.nmsu.edu/pub/hf/pubs/mil_std_188_110a.

[3] CCITT, *Interface Between Data Terminal Equipment (DTE) and Data-Circuit Terminating Equipment (DCE) for Terminals Operating in the Packet Mode on Public Data Networks*, CCITT Recommendation X.25.

[4] Fox, T. L., AX.25 *Amateur Packet-Radio Link Layer Protocol*, Version 2.0, Newington, CT: American Radio Relay League, Oct. 1984.

High-Frequency Radio Networking   9

9.1 INTRODUCTION

The integration of collections of HF stations into networks produces synergistic effects in improved efficiency in use of the spectrum and improved ability to route traffic in spite of fluctuating connectivity. Improved spectral efficiency accrues from the ability to use broadcasts to send messages to many stations with a single transmission. Furthermore, when several stations are located in close proximity, the quality of their links to other stations may be sufficiently well correlated that only one of these stations needs to consume channel time by sounding to monitor link quality.

On the other hand, if the stations in a network are sufficiently dispersed, then their link outages are essentially independent. In this case, the ability to route traffic through indirect paths, that is, the ability to use other network members to relay messages, reduces latency in message delivery. Such path diversity supports the routing of traffic around link outages due to poor propagation. This diversity also permits routing traffic around resources that are congested or have failed, been destroyed, or otherwise compromised. Figure 9.1 shows the portion of the HF system model addressed in this chapter.

A simple example demonstrates the benefit of network communications to overcome link outages. Assume that all links in the network shown in Figure 9.2 have independent outages[1] with probability of outage at any instant equal to α. An outage occurs whenever no frequency will support communications between a pair of stations, whether due to propagation, jamming, interference, or use by another member station. We will calculate the outage probability between stations X and Y when either (1) only a direct path may be used

---

[1]Groups of stations whose link outages are correlated should be considered a single station with increased link capacity in this analysis.

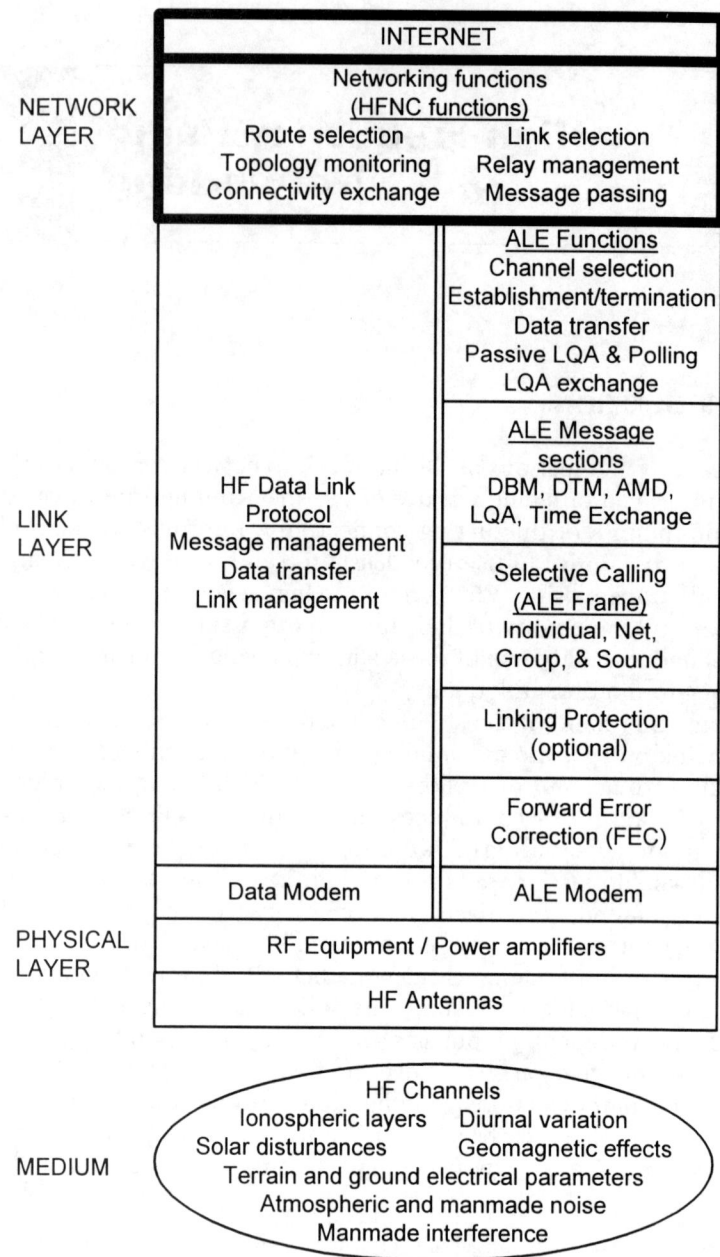

Figure 9.1 HF Network-layer protocol in the hierarchical-layered view of HF communications.

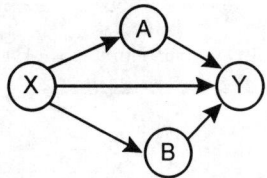

Figure 9.2 Sample HF network.

between the stations, or (2) paths that use at most one relay station may be used. For the case limited to the direct path, the probability of outage is simply α. Using the notation $P_o(XY|n)$ to represent the outage probability between X and Y when n potential relay stations are available, the probability of outage considering only the direct path is $P_o(XY|0) = \alpha$.

To compute the probability that no direct or indirect paths are available, we first compute the probability that the path through any particular relay station, say A, is experiencing an outage. This probability is the likelihood that either of the links XA and AY is "out," that is, undergoing an outage, less the probability that both links are out. This subtraction is necessary to avoid double-counting the probability that both links are out. Denoting the outage probability for the path from X to Y as P(XAY), we have $P(XAY) = \alpha + \alpha - \alpha^2 = 2\alpha - \alpha^2$. Since all link outage probabilities are assumed independent, $P_o(XY|n) = \alpha \prod_{i=1}^{n} P(XiY) = \alpha(2\alpha - \alpha^2)^n$, where the index i represents each of the potential n relay stations in turn.

Values of $P_o(XY|n)$ for several values of n and α are shown in Table 9.1. Clearly, even a small number of alternate routes greatly reduces the probability that a message cannot be delivered to its destination, even when individual links are unreliable. Note, however, that the assumption of link outage independence becomes increasingly suspect for increasing values of n within a fixed geographic area due to increased correlation between individual link environments.

Path diversity is less effective, of course, in overcoming widespread ionospheric disturbances unless it is combined with media diversity (e.g., satellite, wireline, fiber, etc.). The latter notion is discussed further in Sections 9.4 and 9.5, while the remainder of this chapter addresses the tools needed to find, monitor, and use indirect paths through networks of HF radio stations. Section 9.2 describes nuances of traditional networking techniques which are also fundamental to the success of automated networks. Section 9.3 introduces the emerging technologies now being standardized and implemented in worldwide automated HF networks.

Table 9.1
Outage Probability Versus Number of Available Relays

Number of relays available	α (link outage probability)					
	50%	20%	10%	5%	2%	1%
0	0.5000	0.2000	0.1000	0.0500	0.0200	0.0100
1	0.3750	0.0720	0.0190	0.0049	0.0008	0.0002
2	0.2812	0.0259	0.0036	0.0005	0.0000	0.0000
3	0.2109	0.0093	0.0007	0.0000		
4	0.1582	0.0034	0.0001			
5	0.1187	0.0012	0.0000			
6	0.0890	0.0004				
7	0.0667	0.0002				
8	0.0501	0.0001				
9	0.0375	0.0000				
10	0.0282					

9.2 AUTOMATED NETWORKING TECHNOLOGIES

The availability of inexpensive microprocessor technology has not only enabled the development of low-cost ALE technology, but it is also bringing automation to HF network operations. As of this writing, a number of architectures for automated HF networking have been developed, including the MIL-STD-187-721C [1] and MIL-STD-188-141A [2] standards (used in the U.S. Air Force *SCOPE COMMAND* project, among others), as well as techniques developed by GEC Marconi for the Swedish KV-90 project. MIL-STD-187-721C serves as a planning standard for MIL-STD-188-141A, the mandatory U.S. military standard for automated HF systems.

Each such architecture must address the need to find, establish, monitor, and employ indirect paths for HF traffic. The remainder of this section outlines the general techniques available to address each of these functions using terminology adapted from MIL-STD-187-721C. A detailed exposition of the MIL-STD-187-721C architecture follows in Section 9.4.

9.2.1 Finding Indirect Paths

An "indirect" path from station A to station B is formed by a sequence of links through one or more relay stations. One of the key functions of networking schemes is finding such indirect paths for use when the direct path becomes unusable. Techniques for finding indirect paths include (1) queries from stations seeking indirect routes, (2) unsolicited exchange of routing and connectivity information for use in constructing routing plans, and (3) discovery of connectivity by observing the paths followed by traffic passing through a station. Each of these functions is further defined and described below.

9.2.1.1 Routing Queries

Consider a station SAM that needs to find a path (i.e., a sequence of links) to a second station, JOE, for which it has no routing information. Station SAM may query other network member stations with, "Who can reach JOE?" The responses to such queries may simply indicate connectivity or no connectivity, or they may provide more detailed information about the channel quality available on their link to JOE. As described, such a technique is sufficient for finding paths that require only one relay station. Finding longer paths formed from more than one intervening relay station can be achieved through the use of recursive queries (e.g., "Who can reach someone who can reach JOE?"). Alternatively, other mechanisms can be employed by at least some Net members so that they are always aware of longer paths and can respond with them when queried.

9.2.1.2 Connectivity Exchange

When the delays implicit in the routing-query approach cannot be tolerated, stations can periodically exchange lists of the stations that they can reach. Again, such reports can contain either detailed path qualities or simply lists of stations that can be reached by the reporting station. Because the volume of traffic for such exchanges grows as the square of the number of network stations, this technique should generally be used only within small networks. In larger networks, it may be used for tracking connectivity among a small number of key stations (e.g., routing "hubs" and "gateways").

9.2.1.3 Recording Back Routes

Stations that receive a message may record the implicit connectivity to all stations along the path that the message has traversed in reaching them. This technique is facilitated when relay stations append their addresses to each message that passes through them.

9.2.1.4 Example

As an example, consider the imaginary network known as RMTNET (Rocky Mountain Network) shown in Figure 9.3. The address of the net control station is

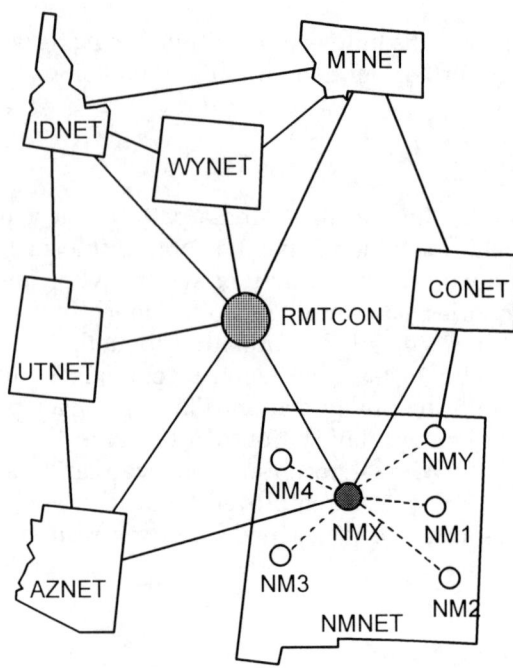

Figure 9.3 The Rocky Mountain Net.

RMTCON. In the subnetwork RMXNET, RMTCON communicates with stations NMX, COX, WYX, IDX, UTX, and AZX. Each of the latter stations communicates with a distinct group of field stations. For example, NMX shares the NM set of channels with stations NMY, NM1, NM2, NM3, and NM4, forming NMNET. NMY is the alternate net control station for NMNET and is also the alternate for NMX in the RMXNET.

RMXNET uses a high-bandwidth terrestrial backbone as its principal physical-layer medium, with HF radio providing a narrowband backup capability. The state networks, such as New Mexico Net (NMNET), however, are entirely dependent on HF links. The relatively high bandwidth of the terrestrial backbone in RMXNET is used to support periodic connectivity exchanges among the members of RMXNET. For example, NMX reports to RMXNET every 15 minutes with a list of NMNET stations with which it can communicate by HF.

The HF subnetworks, however, are programmed to use query routing, rather than connectivity exchange, for finding indirect paths. For example, whenever station NM1 cannot communicate with NM3, station NM1 broadcasts a query to the NMNET asking, "Who can reach NM3?" Station NM1 then uses the responding station with the best link to NM3 as a relay until direct connectivity with station NM3 is restored. However, if station NM1 has messages destined for station ID1, it is programmed to route such "cross-net" traffic to station NMX. Station NMX then selects a path to ID1 using its dynamically updated routing table.

9.2.2 Monitoring Connectivity

The ability to adapt to changing connectivity and path qualities requires that stations monitor the quality of paths in use, as well as other mission-critical paths even when not in use. The monitoring function requires several additional functions as described in the following paragraphs.

The principal source of *path quality* data in a network is the data link controllers. These link controllers automatically monitor noise levels on the channels in use, or the frequency of packet retransmissions, to gauge the quality of active data links. Such link-level measurements are reported by the data link controllers to their local networking entities, which then update their tables of path qualities. The collection of this data introduces no overhead traffic, but it provides no information about the distant data links in a path. Stations track the connectivity of inactive links using sounding or polling as described in Chapter 5.

The periodic reporting technique for finding indirect paths described above will also suffice for path quality monitoring. However, the resulting overhead traffic volume continues to be a concern when limited bandwidth HF links are used to carry this traffic.

Rather than periodically exchange connectivity data, stations can request *asynchronous notification* when path quality changes significantly, and can further restrict overhead by placing such requests only for critical links or paths.

Similarly, stations can notify other net members when they are about to reduce power or shut down for any reason, and when they resume normal operation. This approach provides timely and reliable data of imminent topology changes so that routing algorithms can anticipate such changes, rather than react as other stations sporadically discover and report them.

Continuing with our example from the previous section, after station NM1 has selected a relay for traffic to station NM3, it may request that the selected relay station asynchronously report loss of connectivity to NM3. Station NM1 monitors its connectivity *to that relay station* using a timeout that is reset every time it communicates with or receives a sound from the relay station. Thus, it can track its connectivity to NM3 without burdening the channels with overhead traffic.

Since station NMX is the default router for cross-net traffic, all net members need to be alerted when station NMX has to shut down for preventive maintenance. Station NMX may achieve a smooth transition of control to NMY by first sending a full connectivity report to station NMY, and then broadcasting a station status report indicating that station NMX is "going down." Station NMY would then broadcast a station status report to announce that it was temporarily assuming control of the NMNET, including the function of a default router.

9.2.3 Using Indirect Paths

HF links can be connected in tandem at many levels as shown in Figure 9.4. For example, an audio repeater connects links at the *physical* layer, a frame repeater connects links at the *data link* layer, and a router operates at the *network* layer. In this discussion, we distinguish only between functions related to making routing decisions (routers) and those that are not (repeaters).

9.2.3.1 Routing

In addition to techniques for discovering and monitoring indirect paths, a router must include an algorithm for selecting paths for traffic. This algorithm must include a means for choosing among HF links as well as informing the chosen "next station" along the path whether traffic is intended for that station or for another distant station.

9.2.3.2 Repeater Control

A repeater is a relay that does not make routing decisions. To use repeaters, a mechanism must be employed to identify those HF links to be "connected" at that repeater and to establish such a connection.

9.3 AN EXAMPLE IMPLEMENTATION

The U.S. Department of Defense recently published MIL-STD-187-721C, a planning standard that identifies a suite of technologies to automate the operation and management of HF radio networks. At the time of this writing, similar work is underway for the development of an analogous range of U.S. federal standards. This section describes the MIL-STD-187-721C technology suite as an example of a modern HF networking architecture.

9.3.1 HF Node Controllers

For the purposes of discussion, the network layer functionality of an automated HF station is considered to reside in an HF *node controller* (HFNC).

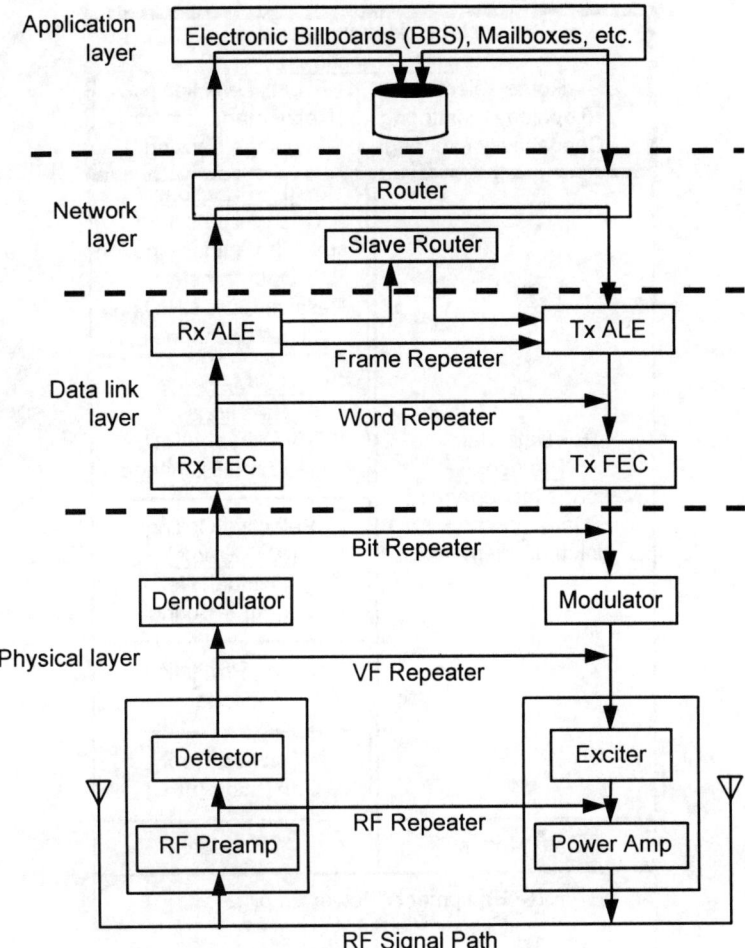

Figure 9.4 Relaying alternatives.

Figure 9.5 shows the conceptual organization of the physical, data link, and network layers that provide the functionality underlying HF automation. Figure 9.6 shows a block diagram of the HFNC. The "S&F" in Figure 9.6 refers to the store-and-forward function, which is responsible for finding a route through a portion of the HF network, or subnetwork, using relays and other media as necessary. In the figure, "AME" refers to the automatic message exchange function, which is responsible for conveying messages over each data link in the path through a subnetwork. The HFNC maintains two important data bases describing HF network connectivity as follows:

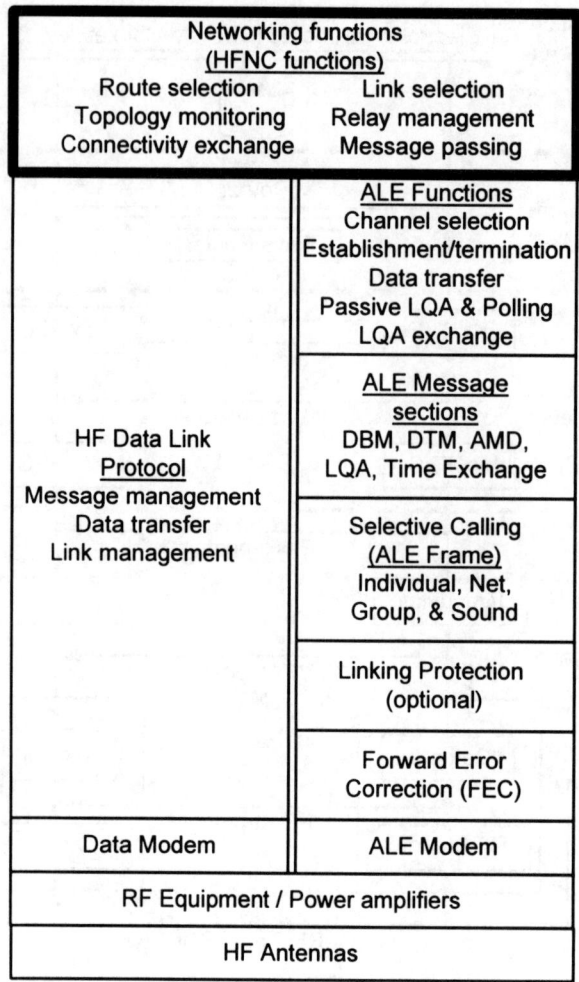

Figure 9.5 Overview of HF automation showing HFNC functions.

- *Routing Table.* A routing table is a listing of previously computed Destination — Next Station pairs for use in routing incoming messages. For each message destination, a routing table contains the address of one or more recommended relay stations to use when direct transmission to that destination is not possible. Table 9.2 shows a sample routing table for station NM1 in the RMTNET example discussed in the previous section.
- *Path Quality Matrix.* A path quality matrix is a dynamically-updated table of aggregate (single or multi-link) voice and data path qualities to various

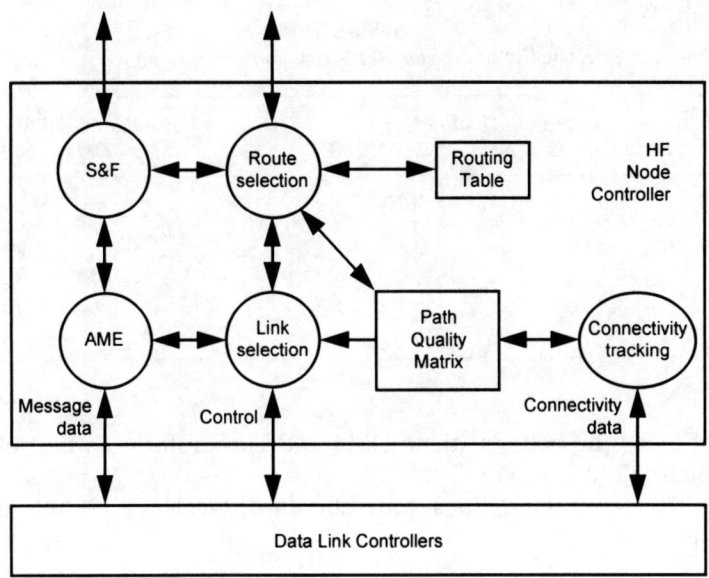

Figure 9.6 HF Node Controller functions.

Table 9.2
Example Routing Table for NM1

Destination	Next Station (Data)	(Voice)
NM1	<Self>	<Self>
NM2	NM2	NM2
NM3	NM4	NMX
NM4	NM4	NM4
NMX	NMX	NMX
NMY	NMY	NMY
All others	NMX	NMX

destinations via the best known paths to those stations. Table 9.3 shows a sample path quality matrix for station NM1 in the RMTNET example discussed in the previous section.

In the RMTNET example, station NM4 has the best-performing link to NM3 and is therefore the best choice for routing data traffic to NM3. However, only NMX and NMY are equipped for voice relaying. Of these two stations,

Table 9.3
Example Path Quality Matrix for NMn (Age and Relay Counts not shown)

Destination	Data Path Quality via:					Voice Path Quality via:				
	NM2	NM3	NM4	NMX	NMY	NM2	NM3	NM4	NMX	NMY
NM2	6	0	5	5	5	10	–	–	8	7
NM3	2	1	6	5	4	–	0	–	7	4
NM4	3	0	7	6	5	–	–	8	7	6
NMX	5	0	6	7	6	–	–	–	12	9
NMY	4	0	5	6	7	–	–	–	1–	11
ID1	–	–	–	5	–	–	–	–	2	–

station NMX has the better path to NM3 and is therefore the best choice for voice traffic to NM3.

MIL-STD-187-721C defines four standard levels of HFNC functional capability [3]:

- A Level 1 HFNC has no routing table, path quality matrix, nor store-and-forward functionality. It can route messages only to directly reachable stations. Thus, indirect routing must be generated by an external router, which will explicitly name the next relay station when passing a message to a Level 1 HFNC for delivery.
- A Level 2 HFNC includes a routing table and store-and-forward ability, but no path quality matrix. Indirect routing is performed automatically, but the routing table will usually be generated externally. In a typical application, a Level 2 HFNC would receive its routing table from a central network control site, with updates provided either manually by local operators, over the air from the network control station, or from its own connectivity tracking function. store-and-forward functionality in Level 2 (and higher) HFNCs is supported by connectivity monitoring and routing queries. When connectivity is lost to a station listed in the routing table, messages destined for that station are queued until a connection is re-established. The HFNC can also actively seek new relay stations through the routing query protocol, and post the results to its routing table.
- A Level 3 HFNC adds the path quality matrix and the connectivity exchange protocol to the Level 2 capabilities. Level 3 controllers discriminate among possible paths for messages using path quality formulas that consider link degradation due to congestion as well as natural phenomena. Routing decisions thus adapt more quickly in Level 3 HFNCs than in Level 2, because the latter respond only to link loss, while the former can detect a deteriorating link and switch before the link becomes unusable.

- A Level 4 HFNC adds an Internet router to the capabilities of a Level 3 HFNC, and can therefore act as a gateway between HF and other subnetworks.

Unlike the mesh topologies commonly found in wired wide-area networks (WANs), HF networks will often be hierarchies of star topologies. One possibility shown in Figure 9.7, in which Level 3 or 4 stations serve as hubs of stars of less-capable stations, and are themselves linked into a "backbone" star. This configuration is a generalization of the example network from Section 9.3.

HF subnetworks in Internet applications will typically employ Level 4 HFNC gateways at interface points between HF and other media, with Level 2 or 3 HFNCs at other stations in the network. Note that Figure 9.7 depicts only operational connectivity and does not represent the geographical locations of stations. In many cases, the hubs of stars achieve maximum connectivity with their Net members when they are located well away from those net members, rather than centered among them.

9.3.2 Network-Layer Protocol Overview

The military standard suite of protocols at the network layer supports automatic message exchange, connectivity exchange, relay management, and station status reporting. All messages sent between two HFNCs are preceded by a single ASCII

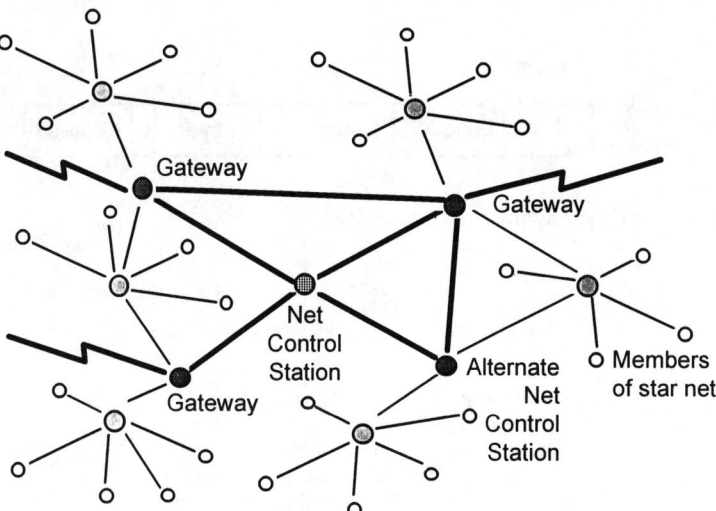

Figure 9.7 Example HF subnetwork topology.

character denoting the type of message to follow. When an HFNC receives a message, it examines this one-character header to determine the format and protocol to use in interpreting the remainder of the message. The defined network layer header characters are listed in Table 9.4. Network-layer addresses are defined as follows:

- *Network-Layer Address Records.* Station addresses in most network-layer protocols are carried in network-layer address records. The format of these records is shown in Figure 9.8, with a breakdown of the address record type codes shown in Table 9.5.
- *Network Broadcast Address.* Where permitted by network-layer protocols, the broadcast address "@?@" (identical to the ALE Allcall address) is used to refer collectively to all reachable stations.

Table 9.4
Network-Layer Header

Header	Message Type
C	Connectivity exchange (CONEX)
M	User message (with AME header)
R	Relay management
S	Station status message

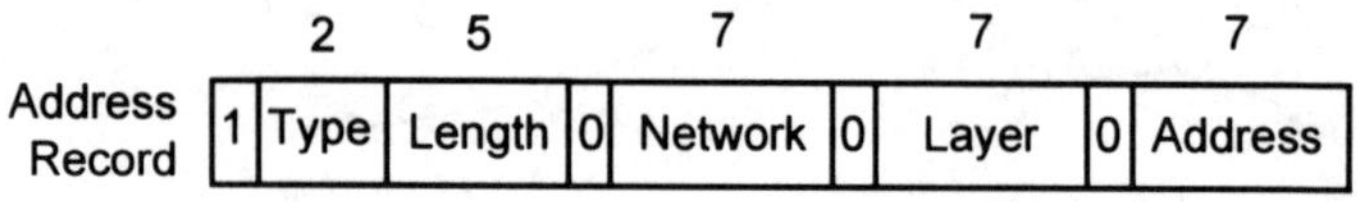

```
                        2     5        7        7        7
         Address    ┌─┬────┬──────┬─┬───────┬─┬──────┬─┬────────┐
                    │1│Type│Length│0│Network│0│Layer │0│Address │
         Record     └─┴────┴──────┴─┴───────┴─┴──────┴─┴────────┘
```

Figure 9.8 Network-layer address record format.

Table 9.5
Network-Layer Address Record Type Codes

Type Code	Message Type
0	Source address
1	Relay station (mandatory)
2	Relay station (suggested)
3	Destination address

9.3.3 Automatic Message Exchange

9.3.3.1 Overview

The automatic message exchange protocol [4] provides a simple, connectionless, datagram service. A port number is included in the AME header (see Figure 9.9 and Table 9.6) to support internal routing of AME datagrams to higher-layer entities such as the Internet protocol, the HF network management protocol,

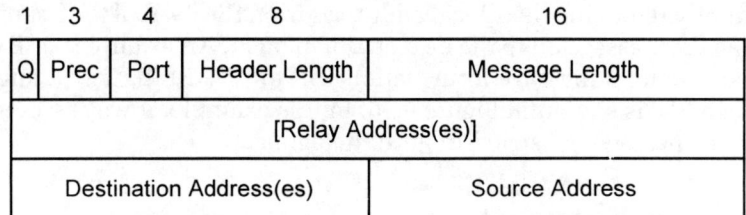

Figure 9.9 Automatic message exchange header.

Table 9.6
AME Header Fields

Field	Description
QOS	A single bit indicating whether to emphasize speed of delivery (QOS = 0) or minimum probability of loss (QOS = 1) in handling message.
Prec	Three-bit code with '0' indicating the lowest precedence. It is used for queuing at station HFNCs, thus determining order of link establishment, order of delivery, etc.
Port	Four-bit code designating destination port within network controller, analogous to Network Service Access Point (NSAP) in the OSI model. It is assigned port numbers listed in Table 9.7.
Header length	An 8-bit count of the bytes in the AME header, starting with the precedence/port byte and ending with the last character of the source address record.
Message length	A 16-bit count of the bytes in the transport message following the AME header (does not include the network layer or AME header bytes).
Relay(s)	Zero or more address records (see address record formatting in Figure 9.8 and Table 9.5). When relays are specified, they must be either suggested relays or mandatory relays. When mandatory relays are specified, the message must be routed through those relays in the specified order.
Destination(s)	Composed of one or more address records. The message body will be delivered to all destination addresses.
Source	One address record used to specify the address of the station that originated the message.

or the operator display for orderwire messages. A complete listing of the assigned port numbers is shown in Table 9.7. Each AME message header contains the network-layer address of the message source and the address of at least one destination. The header may also list mandatory or optional intermediate stations through which the message must, or may, be routed. This header provides a flexible mechanism for appending source routing to messages.

For example, in a network consisting only of Level 1 HFNCs, source routing could be performed by operators or by host computers that maintain routing tables. The Level 1 HFNCs then implement the specified routing scheme by sequentially removing their local addresses from the head of the list of relays and passing the message on to the next station named. An example AME header of an electronic mail message from station JOE to CHARLIE, with a suggested path through ED, is shown in Figure 9.10. In this example, it was assumed that the transport message is carried in an IP datagram.

Table 9.7
AME Port Number Designations

Port Number	Transport Message Destination
0	Automatic message exchange control channel
1	Operator terminal
2	Operator storage (e.g., local disk)
3	HF transport protocol (HFTP)
4	"Connectionless" network protocol (CLNP)
5	Internet protocol (IP)
6	HF network management protocol (HFNMP)
All others	Reserved until standardization

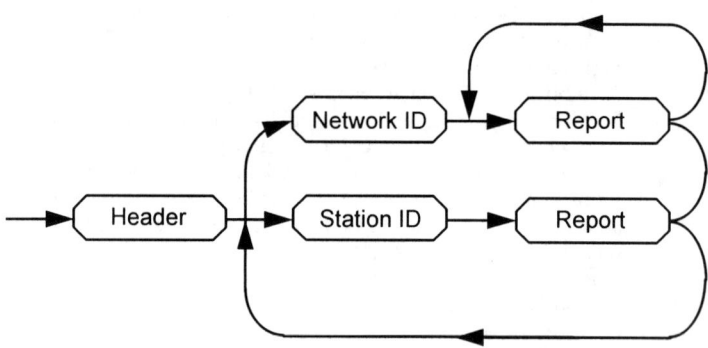

Figure 9.10 Connectivity exchange message syntax.

9.3.3.2 Outgoing Message Processing

Each network message passed to the AME process from the S&F process (see next section) contains an AME header and a transport message. The AME process interprets the first address record in the AME header as the desired destination for that message. This first address record will either name a relay station or the ultimate destination.

Outgoing messages from the S&F process are queued by the AME process for transmission in order of precedence. The AME process requests a link to the destination of each message from the link-selection function. When the link-selection function indicates that a link to that destination is available, the AME process (1) "prepends" a network-layer header (the character "M") to the message, (2) translates the network-layer address to a data-link-layer address appropriate for the selected data link controller, and (3) provides the network message and the translated address to that controller for transmission. The network message now consists of the network-layer header followed by the AME header and the transport message.

If a direct link cannot be established to the destination, the AME process may take one of two actions: (1) Return the message to the S&F process as undeliverable, which is appropriate if the S&F process can attempt alternate routing, or (2) store the message for later delivery when a link can be established. In the latter case, messages should be stored in separate queues for each destination so that only one series of linking retries is made for each destination, rather than one series for each queued message. The first retry should be made after a time sufficient for a busy station to have resumed listening for linking attempts. To minimize use of the spectrum for futile linking attempts, subsequent retries should occur only at intervals sufficient for channel conditions to have measurably improved. This retry interval may be shortened when a queue contains high-priority messages. When contact is eventually made with the desired destination, whether through a successful linking retry or through connectivity discovered by reception of a message from that station, messages queued for that destination are sent in decreasing priority order.

9.3.3.3 Incoming Message Processing

Incoming messages are delivered to the AME process when their network-layer header is "M" (user messages). The AME process simply strips this single-character network-layer header, and then passes each received message to the S&F process for processing.

9.3.4 Store-and-Forward Operation

9.3.4.1 Outgoing Message Processing

The message S&F process accepts messages from, and delivers messages to, users or *transport-level* processes. Each transport message to be sent is accompanied by the network-layer addresses of its destinations. For each transport message to be sent, the S&F function groups the network-layer destination addresses according to the first station along the path to each network destination. These addresses are obtained from the route-selection function, and may be the final destination, if a direct path is the best path. For each such group, an AME header is formed containing the destinations that share an initial relay station. The transport message is then appended to this header to compose a network message. These network messages are passed to the local AME process for delivery.

9.3.4.2 Incoming Message Processing

As network messages are received from other stations and are delivered to the S&F function by the local AME process, the AME header of each message is removed and processed as follows: (1) If any of the destination addresses in the header are self addresses, then the embedded transport message is delivered locally as specified in other fields (i.e., precedence and port) of the header; (2) all self addresses are removed from the header; and (3) if any destinations remain, those addresses and the transport message are handled as discussed above for a new outgoing message.

9.3.5 Null S&F Functions

A *null* S&F function may be used in place of the message store-and-forward process described above when automatic message routing is not supported (e.g., in Level 1 HFNCs). The null S&F function simply forms AME headers for outgoing messages and processes the AME headers from incoming messages as described below.

9.3.5.1 Outgoing Messages

For each outgoing message, the null S&F function creates an AME header with preprogrammed values in all fields, except that the header length and message length fields are computed for the actual message and AME header. The user (or transport-layer process) should be able to override default values. Normally, the user will override only the default destination address and the precedence and the port fields. A user may insert relay addresses for manual source routing. If the user is able to override the source address, this capability should normally be restricted to selecting one of a set of preprogrammed addresses to preclude impersonation of other stations.

9.3.5.2 Incoming Messages

The destination and relay address records in the AME header of each incoming message are examined. If a self address is found in any of these address records, the message and the AME header are delivered to the user. This permits users to manually relay messages.

9.3.6 Connectivity Exchange

The connectivity exchange (CONEX) protocol [5] is used by Level 3 and Level 4 HFNCs to share path quality data. For example, if station A receives a report from station B about the path BC, station A can combine its measurement of the link quality AB to compute the end-to-end quality of the path ABC. Path qualities are then used in selecting routes for traffic. For the details of this protocol, see MIL-STD-187-721C.

9.3.7 HF Relay Management Protocol

9.3.7.1 Overview

The HF relay management protocol (HRMP) [6] is used to remotely control repeaters and to query directly reachable stations for connectivity data regarding a locally unreachable station. Precedence and preemption are supported for optimum use of network resources. HRMP is a connectionless protocol, although it can be used to set up analog tandem circuits or data virtual circuits.

Every HRMP message refers to three stations: (1) the relay station involved (actually or potentially), (2) the control station managing the relay, and (3) the distant station to which access is provided by the Relay. HRMP messages are always exchanged between the control station and the relay station. The HRMP message format includes the addresses of three stations (see Figure 9.11[a]): (1) the station sending the message (Msg Source), (2) the station to which an indirect path is sought or established (Desired Dest), and (3) the station receiving the message (Msg Dest). The addresses of all three stations are encoded as network-layer address records described in Section 9.3.2. Figures 9.11(b) and 9.11(c)

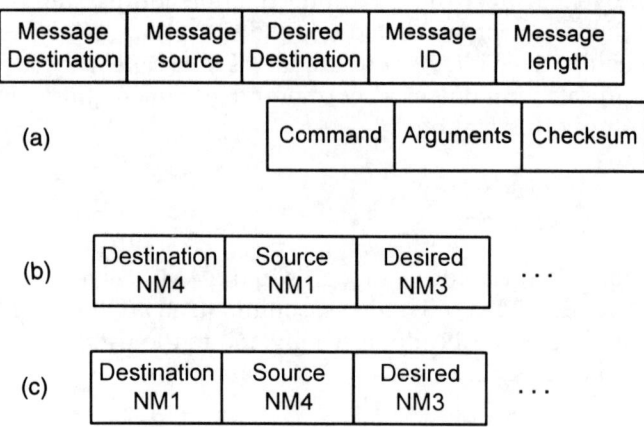

Figure 9.11 Network management messages: (a) general relay management message, (b) request for relay from NM1 to NM4, and (c) response from NM4 to NM1.

illustrate the addresses that would be used in the RMTNET example as station NM1 requests NM4 to relay traffic to station NM3.

9.3.7.2 HRMP Operation

The following paragraphs list the allowed exchanges for each class of relay control actions supported by the HRMP. HRMP exchanges are one of two types:

- *Request-Response:* A control station sends a request to a relay station and starts a timer. If a response is received before the timer expires, the protocol completes successfully. Otherwise, the control station will abort the exchange. The lack of a response indicates loss of connectivity to the relay station. For this reason, a retransmission will not be initiated until sufficient time has elapsed for connectivity to be restored either from improved propagation or resumption of operations at the relay station. (When a request is sent via an HF link, the timeout will allow sufficient time for the ALE controller to try many possible channels. This process may require one to two minutes.)
- *Notification:* A relay station sends an unsolicited notification message to a control station to announce an event asynchronously. Such events include preemption of a repeater in use by the control station or loss of connectivity to a distant station monitored at the request of the control station.

9.3.7.3 Routing Queries

A station seeking to find an indirect path to a distant station sends a routing query to prospective relay or relays. If a station has available facilities adequate to perform the requested service, and if it has connectivity to the requested distant station, then it returns a query response that describes the type and quality of service it can provide. Note that routing queries may be made for either message store-and-forward service or repeater service.

9.3.7.4 Repeater Control

Both analog and digital repeaters may be remotely controlled through the use of the repeater control commands. When a repeater is engaged using HRMP, the relay station assigns it a repeater number (analogous to a virtual circuit number) for unambiguous reference to the circuit established. When a repeater is "seized" in this fashion, it is unavailable for other HF paths until released or preempted.

9.3.7.5 Connectivity Monitoring

HRMP also includes a connectivity monitoring mechanism that can be used to track indirect connectivity without the overhead of the full CONEX protocol. This feature is important because the CONEX protocol can consume sizable fractions of HF channel time. For the RMTNET example, station NM1 routes messages to station NM3 through station NM4. Station NM1 may request that NM4 asynchronously report loss of connectivity to NM3, so that station NM1 could then find (or activate) an alternate route to NM3.

Connectivity monitoring is a notification-based alternative to connectivity exchange (see Sec. 9.3.5). A monitor connectivity message requests that the named relay station monitor, or cease monitoring, its connectivity to the named distant station. The relay station will then broadcast, report (to the named control station), or cease reporting changes in its ability to reach that distant station. If the distant station field contains the network broadcast address (see Sec. 9.3.2), the relay station is requested perform the indicated command for all stations.

9.3.8 HF Station Status Protocol

The HF station status protocol (HSSP) [7] was designed to support a notification-based mechanism for tracking the status of network member stations with less overhead traffic than a polling-based approach. Status reports are sent when a station changes scan set, begins radio silence, goes out of service, assumes network management duties, returns to normal operation, and so on. Some examples of changes of status that may be reported are listed in Table 9.8.

9.4 AUTOMATED HF NETWORK MANAGEMENT

The evolution of HF networking will result in increasingly complex networks with equipment often placed at remote sites. Moreover, an objective of ongoing programs is to consolidate high-power HF assets among the military services and install them in unmanned "lights out" facilities. Clearly, these HF system configurations will emphasize the need for a standardized protocol for the remote control of HF stations and for remotely diagnosing problems in HF networks.

Similar needs in the existing Internet have led to the development of the simple network management protocol (SNMP) [8]. However, the hostility of the HF medium presents clear challenges to the development of a mechanism for reliably monitoring and controlling distant radio stations. MIL-STD-187-721C describes a protocol that addresses these challenges, while maintaining compatibility with the standard Internet SNMP network management architecture.

9.4.1 Background

Automation of HF radio networks to date has simplified the task of establishing links using HF radios. However, ALE and other HF automation technologies have brought a new problem to managing radio networks. The automatic controllers use a number of intricate data structures that must be kept consistent

Table 9.8
Station Status Codes

Code	New Status
0	Normal operations
1	Assuming Net Control
2	Relinquishing Net Control (from non-NCS)
3	Radio silence
4	Reduced power
5	Alternate Scan Set 1
6	Alternate Scan Set 2
7	Alternate Scan Set 3
127	Out of service

throughout a network if operations are to proceed smoothly. Another aspect of network management, that has not been addressed by the ALE standards, is the need to observe network connectivity and equipment status from network control sites (see Figure 9.12). This remote observation capability is necessary so that corrective action can be initiated promptly when malfunctions or other disruptions occur.

Managers of packet networks have been at work on these problems for some time. The most mature and widespread example of the existing network management architectures is the Internet-standard Network Management Framework, which was developed in the late 1980s. This technology is more often referred to by the protocol that it employs for managing network nodes, the simple network management protocol (SNMP).

SNMP was designed so that it "explicitly minimizes the number and complexity of management functions realized by the management agent itself" [8]. That is, the development costs of including SNMP in managed equipment are minimized at the expense of (perhaps) increasing the complexity of the software at the distant management stations. Fortunately, the ratio of managed nodes to management stations is large, so the benefit of widespread implementation has greatly outweighed the cost of implementing the management software.

The following briefly summarizes the salient points of the SNMP approach:

- Network management stations monitor and control network elements by communicating with "agents" in those elements. This interaction uses SNMP to *get* and *set* the values of defined data objects. Agents may also

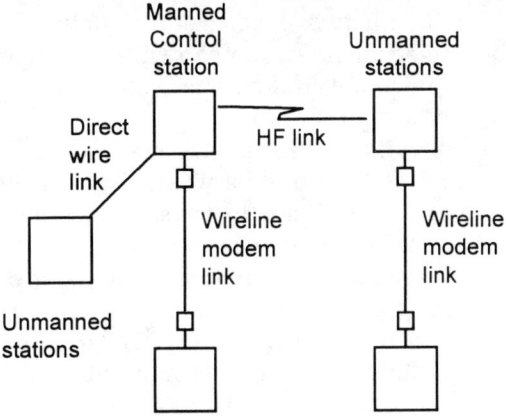

Figure 9.12 Network management example.

send "trap" messages to management stations to announce important events asynchronously.
- The defined data objects are described in the *management information base* (MIB). The original MIB is strongly oriented to the TCP/IP protocol suite, although it is easily extensible. Object definitions are expressed formally in Abstract Syntax Notation 1 (ASN.1) [9]. Object names and values are encoded for transmission in accordance with a set of ASN.1 Basic Encoding Rules [10].
- When elements do not implement SNMP, they may still be managed by using "proxy agents" that translate the standard SNMP messages into messages understood by these elements.
- Authentication is included in the standard, although current practice uses only trivial authentication. The mechanism is extensible using ideas similar to HF linking protection (see Chapter 7).
- SNMP requires only a connectionless datagram transport service (for example, the user datagram protocol (UDP) [11] employed by the Internet).

These and other SNMP features make it well suited for HF network management functions.

9.4.2 HF Network Management Requirements

The assets to be managed in an automated HF network include media-specific equipment such as transceivers, modems, ALE controllers, and HF node controllers. Figure 9.13 depicts a network management station and a controlled HF network node. The management station shown in the figure uses HF links to control this station (node), but it could just as well employ a wide area network, wireline modems, or other types of links.

An automated network management system must support the efficient control of automated HF stations and networks, including the following functions: (1) monitoring and reporting network status (connectivity, capabilities, congestion, faults, etc.); (2) updating network routing tables; (3) manipulating the operating data of automated communications controllers (such as ALE controllers); (4) identifying software versions and updating the software, in ALE and other communications controllers; (5) rekeying linking protection scramblers; and (6) remotely operating all communications equipment, which includes adjusting transmitter power of linked stations, reading meters, and rotating antennas. Because of the mission-critical nature of the networks to be controlled, authentication is integral to the network management protocol.

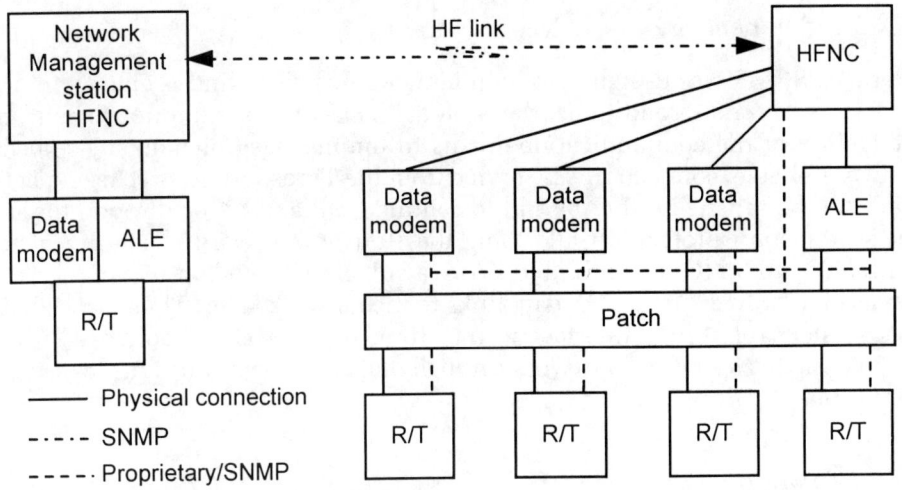

Figure 9.13 Management of HF network nodes.

9.4.3 Applicability of SNMP for HF Network Management

Several questions were addressed in assessing SNMP for use in HF network management [14]: (1) Can it support all of the functions required (functionality)? (2) Can it work over the HF medium (compatibility)? and (3) Can it perform acceptably without imposing unacceptable overhead (performance)? Each of these issues is discussed below.

9.4.3.1 Functionality

Over-the-air data manipulation functions clearly depend upon the interrogation and revision of data at remote sites, which is precisely the model supported by SNMP. Over-the-air rekeying can also be cast as an authenticated transfer of new values to defined objects, such as the storage locations of keys in this case. Remote control is less clearly supported by the SNMP approach. As noted in the SNMP standard, however, all control actions may be implemented as "side effects" of writing to appropriate variables in an automated controller. For example, rotating an antenna occurs as a side effect of updating a variable that specifies the azimuth of that antenna. Thus, all of the network management functions can be supported by SNMP.

9.4.3.2 Compatibility

Because SNMP was designed to help network managers find problems in networks under crisis conditions, the protocol makes few assumptions about the reliability of the communications paths to the managed elements. It expects only unreliable, connectionless service from the Transport layer. This expectation can be satisfied either by implementing UDP and IP, or through the use of no transport protocol at all, using the store-and-forward capability of the MIL-STD-187-721C network layer. Any available HF modem technology can be used to provide usable HF data links to the network layer. Thus, SNMP will be isolated from direct interface to the HF medium. More importantly, it has been designed specifically to work through the unreliable conditions that sometimes plague HF links.

9.4.3.3 Performance

Network management stations monitor and control network elements by communicating with agents in those elements. This communication is carried in SNMP messages, so both management stations and agents must execute the SNMP protocol. The protocol is deliberately "lightweight;" that is, it is intended to keep the costs of implementing and executing SNMP sufficiently low so that all elements in a network can be directly managed (i.e., that all equipment of interest will implement SNMP).

The result of minimizing the complexity of the software that implements SNMP is that most of the complexity of network management is transferred to the management station software. However, due to the relatively low rate of messages expected in managing typical HF radio networks, even an inexpensive notebook computer should posses adequate processing power to serve as a network management station.

A key requirement for any management protocol to be used over HF channels is that it minimize the number of bits communicated in performing its functions. The minimization of traffic was a goal of the SNMP developers, but the networks for which it was designed place far lower costs on each bit sent. Thus, while SNMP is generally regarded to be a lightweight protocol in the Ethernet environment (10 Mbps), it is not clear that it is sufficiently lightweight to be used over HF channels where all overhead traffic is viewed much more suspiciously.

Consider the number of bits in an SNMP message. Each message in the current version of SNMP (version 2) includes: (1) Privacy and Authentication Header fields, (2) an SNMP command (e.g., a "Get" to read an object, a "Set" to update an object, or a "Response" to return a value), (3) a "Data" field that holds a request ID, (4) "Error" status and "Index" fields, and (5) a list of variable "bindings." Each variable binding contains an "object identifier" as well as a value in the case of Set requests or a Get response. It is these variable bindings that carry the management information.

In each variable binding, an object identifier specifies the name of a managed object by describing where it is defined in a tree of standards. This formal system of naming objects typically consumes 10 or more octets for each object. "Values" of objects can be encoded in as few as three octets (for integers less than 128); however, strings of N characters will require N+2 octets for their encodings. No provision is made in SNMP for getting or setting entire tables; however, each entry must be individually named in requests and responses.

If the headers for UDP and IP are included, an SNMP message that responds to a request for a three-character address from an ALE controller address table will require on the order of 70 octets, plus the HF automatic message exchange header and data link layer header. Eliminating UDP and IP would remove 28 octets from this total. Reduction of this overhead is one of the key goals of the HF variant of SNMP.

9.4.4 HNMP: The HF variant of SNMP

Because the initial version of SNMP did not provide sufficient authentication capability for HF network management, MIL-STD-187-721C is based on version 2 of SNMP [13–24], usually denoted SNMPv2. This section describes the differences between SNMPv2 and the HF variant of SNMP (termed HNMP). In addition, it addresses questions of managing existing assets that do not implement SNMP, controlling access to managed assets, and integrating the management protocol with the existing HF protocol suite.

HNMP is identical with SNMPv2, with the following variations: (1) Object Identifiers for objects defined in the HF MIB are encoded for transmission using a truncated encoding scheme that reduces overhead; (2) a "GetRows" variant of the *GetBulk* message is introduced; (3) a PIN authentication scheme is mandatory, while the SNMPv2 MD5 authentication scheme is optional; and (4) retransmission timeouts in network management programs are adjusted to allow time for link establishment, as well as for the transmission of requests and responses over modems that achieve throughputs of 100 bps or less.

The relationship of the network management protocol to the other protocols in use within an HF station is shown in Figure 9.14, where the HNMP uses UDP for a transport-layer protocol, IP for an Internet-layer protocol, and the HF automatic message exchange protocol as the network-layer protocol. Figure 9.14 also shows integration of IEEE 802 protocols as an illustration of the use of HNMP over an Ethernet LAN. Other LAN and WAN protocols may be integrated similarly. When interoperation with management stations outside

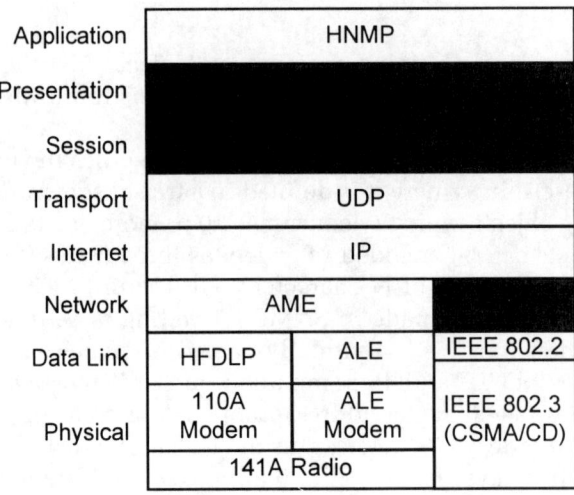

Figure 9.14 Interrelationship of protocols.

the local HF sub-network is not required, UDP and IP may be eliminated to reduce the overhead of network management messages.

9.4.4.1 Objects Used in HF Network Management

SNMP functions by reading and writing Data Objects defined for each functional element (e.g., HF node controller, ALE controller, modem, or radio). These data structures are defined using an abstract syntax so that the details of how the data are stored by individual network components are hidden. When data are exchanged over the air or some other medium, however, it is necessary that all parties to the exchange use the same encodings for the data. Object names and values sent in HNMP messages are encoded using the Basic Encoding Rules for ASN.1 [9], with a truncated encoding used for object identifiers of objects from the HF management information base [25].

RFC-1450 defines the objects commonly used to manage TCP/IP internets. The standard objects for HF network management are defined in the HF MIB. This MIB module contains groups of objects for radios (and related RF equipment), ALE controllers, linking protection, HF data modems (and associated data link controllers), and networking controllers. Objects specific to each manufacturer's equipment are specified in MIB provided by that manufacturer. A management station integrates MIB modules from the elements it manages, resulting in access to a wide-ranging and dynamic set of management data. The structure of MIBs is defined in RFC-1442 [14].

9.4.4.2 GetRows Mechanism

In addition to the protocol data units (PDUs) defined in SNMPv2, HNMP includes GetRows Request and GetRows Response PDUs. The GetRows operation is similar to the SNMPv2 GetBulk operation, except that the response to a GetRows operation is a new compact PDU. A GetRows response includes the Object ID only of the first object in each row, followed by the values of all objects requested in that row. This elimination of the largely redundant transmission of Object IDs can dramatically reduce the number of bits sent when reading tables. This capability is an important consideration for managing ALE controllers and radios over the relatively low bandwidth of HF channels. A similar idea could be employed for efficiently setting rows of tables, but it is not part of the standard.

9.4.4.3 Access Control

Access to the management information of network elements is controlled in HNMP at two levels. The first level is an administrative model that restricts the objects at each element that are accessible to other parties and the operations that may be performed by those parties. The second level of access control is authentication of messages; that is, determination that a message actually comes from the party named in the message. The following three mechanisms are available to authenticate HNMP messages:

- *Trivial Authentication:* This checks the transport-layer address of the originator of the message to verify that the named originator actually sent the message.
- *Personal Identification Number Authentication:* This requires operator entry of a PIN, which is appended to every message and checked by agents.
- *Cryptographic Authentication:* This attaches a Digest of each authenticated message at the beginning of the message (*authInfo* in the *SnmpAuthMsg*). This Digest is computed from the message contents and a secret initialization vector in such a way that it is considered computationally not feasible to spoof the authentication system [26]. A time-of-day mechanism is included as well to limit the effects of replay attacks (see Chapter 7).

An extra level of access control is provided for HF access to managed stations when linking protection is used to authenticate ALE calls. In this case, anonymous, distant HF stations can be denied the ability to even establish links to the managed network.

9.4.4.4 Proxy Management

When elements do not implement HNMP, they may still be managed by using Proxy Agents that translate the standard HNMP messages into messages understood by the non-HNMP (foreign) elements. Initially, few network elements will implement HNMP. Proxy Agents will be needed to extend the management capability to current-generation equipment. As a general rule, the Proxy Agent for any foreign network element should reside in the lowest-level controller that has a control path to that element, often an HFNC.

The provision for Proxy Agents in HNMP will greatly ease its use in HF networks. A phased approach to integrating HNMP into automated HF networks is arguably the most prudent. This approach would initially limit the penetration of HNMP to no level lower than HFNCs. Proxy Agent software would execute within each HFNC to translate HNMP messages into the particular command sequences used by the other equipment at each site. This phased approach has the clear advantage of limiting the initial round of new software development only to equipment that is software-based (HFNCs), rather than requiring upgrades to firmware-based equipment such as fielded ALE controllers.

9.4.4.5 Performance

The performance of HNMP may be gauged by how many bits are transferred to perform common operations. A fairly complex station, such as that shown schematically in Figure 9.15, may be used for computing some example bit counts.

A similar station containing one ALE controller, seven radios, ten antennas, six HF data link protocol (HFDLP) controllers, one antenna matrix, and one automated BLACK patch panel was analyzed [27]. The number of octets needed for each type of equipment is shown in Table 9.9. It was found that at a data rate of 1200 bps, and assuming 50% overhead for ARQ, a complete download of the management information for this station would consume approximately 200 seconds.

Of course, over a LAN, a WAN, or even a high-speed modem link, the time for this download would be on the order of one second. For high-speed links, this latency is primarily determined by the overhead of the lower-level protocols rather than the HNMP overhead.

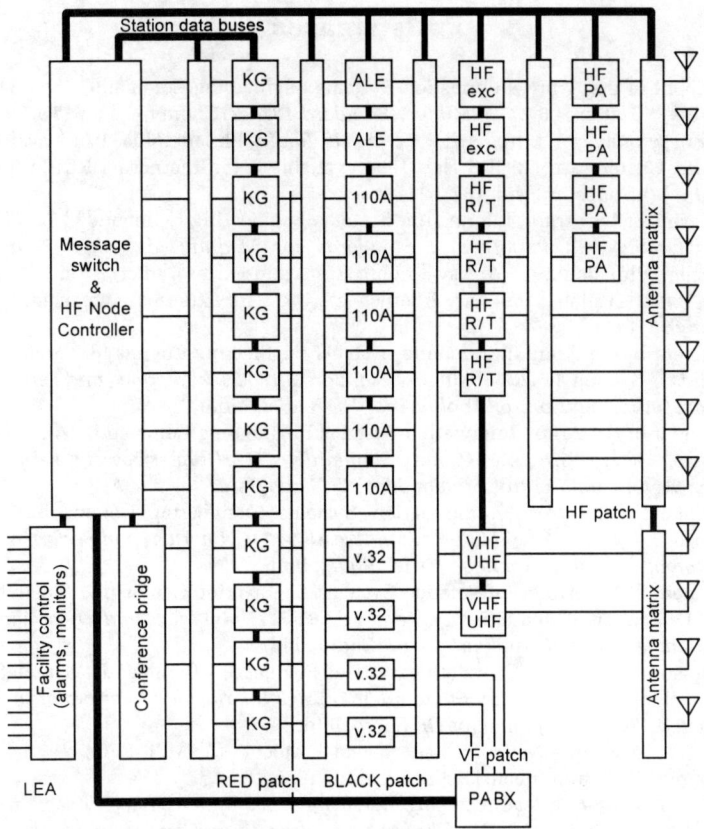

Figure 9.15 Large-scale HF Internet gateway.

Table 9.9
Sample Traffic Volume for Network Management

Element	Octets
ALE Controller	3,000
Radios	15,000
Antennas	500
HFDLP (identical programming)	200
Antenna matrix	500
BLACK patch	500
Total (rough estimate)	20,000

References

[1] Department of the Army, Information Systems Engineering Command, "Standard Levels of Capability," Section 4.6.2.7 of *Military Standard 187-721C: Interface and Performance Standard for Automated Control Appliqué for HF Radio,* Philadelphia, PA: Naval Publications and Forms Center, Attn. NPODS, Nov. 1994, available on the Internet at URL ftp://tracebase.nmsu.edu/pub/hf/pubs/mil_std_187_721c.

[2] Department of the Army, Information Systems Engineering Command, *MIL-STD-188-141A: Interoperability and Performance Standards for Medium and High Frequency Radio Equipment,* Philadelphia, PA: Naval Publications and Forms Center, Attn. NPODS, Notice 2, Sept., 1993, available on the Internet at URL ftp://tracebase.nmsu.edu/pub/hf/pubs/mil_std_188_141a.

[3] Department of the Army, Information Systems Engineering Command, "Standard Levels of Capability," Section 4.6.2.7 of *Military Standard 187-721C: Interface and Performance Standard for Automated Control Appliqué for HF Radio,* Ibid.

[4] Department of the Army, Information Systems Engineering Command, "Automatic Message Exchange," Section 5.7.5.4 of *Military Standard 187-721C: Interface and Performance Standard for Automated Control Appliqué for HF Radio,* Ibid.

[5] Department of the Army, Information Systems Engineering Command, "Connectivity Exchange," Section 5.7.4 of *Military Standard 187-721C: Interface and Performance Standard for Automated Control Appliqué for HF Radio,* Ibid.

[6] Department of the Army, Information Systems Engineering Command, "Relay Management Protocol," Section 5.7.6 of *Military Standard 187-721C: Interface and Performance Standard for Automated Control Appliqué for HF Radio,* Ibid.

[7] Department of the Army, Information Systems Engineering Command, "Station Status Protocol," Section 5.7.7 of *Military Standard 187-721C: Interface and Performance Standard for Automated Control Appliqué for HF Radio,* Ibid.

[8] RFC-1157, "A Simple Network Management Protocol" (SNMP). RFCs may be obtained by anonymous "ftp" from nis.nsf.net or nic.ddn.mil.

[9] ISO/IEC 8825, *Information Processing Systems — Open Systems Interconnection — Specification of Abstract Syntax Notation One (ASN.1),* International Organization for Standardization, 1990.

[10] ISO/IEC 8824, *Information Processing Systems — Open Systems Interconnection — Specification of Basic Encoding Rules for Abstract Syntax Notation One (ASN.1),* International Organization for Standardization, 1990.

[11] RFC-768, "User Datagram Protocol."

[12] Johnson, E. E., "SNMP for HF Radio Network Management," New Mexico State University, 1993.

[13] RFC-1441, "Introduction to Version 2 of the Internet-standard Network Management Framework."

[14] RFC-1442, "Structure of Management Information for Version 2 of the Simple Network Management Protocol (SNMPv2)."

[15] RFC-1443, "Textual Conventions for SNMPv2."

[16] RFC-1444, "Conformance Statements for SNMPv2."

[17] RFC-1445, "Administrative Model for SNMPv2."

[18] RFC-1446, "Security Protocols for SNMPv2."

[19] RFC-1447, "Party MIB for SNMPv2."

[20] RFC-1448, "Protocol Operations for SNMPv2."

[21] RFC-1449, "Transport Mappings for SNMPv2."

[22] RFC-1450, "Management Information Base for SNMPv2."

[23] RFC-1451, "Manager-to-Manager Management Information Base."
[24] RFC-1452, "Coexistence between Version 1 and Version 2 of the Internet-standard Network Management Framework."
[25] Johnson, E. E., "Management Information Base for HF Radio Networks," USAISEC Technical Report ASQB 94089, Nov., 1994, (appended to MIL-STD-187-721C).
[26] RFC-1442, *Ibid*.
  RFC-1321, "The MD5 Message-Digest Algorithm."
[27] Johnson, E. E., "HNMP: Remote Control and Management Protocol for HF Radio Networks," *MILCOM '95 Conf. Proc.*, IEEE, 1995.

About the Authors

Eric E. Johnson is president of Johnson Research, a consulting firm specializing in the design, modeling, and analysis of networking techniques for automated radios, including protected automatic link establishment, link quality analysis, connectivity exchange, sounding, message relay and store-and-forward, and mitigation of interference to and among adaptive radios. He is associate professor in the Department of Electrical and Computer Engineering at New Mexico State University at Las Cruces, New Mexico, and director of the Parallel Architecture Research Laboratory. The laboratory provides a research environment for the development of mathematical models of concurrency, a laboratory facility for developing and teaching concurrent programming techniques, and a hardware test bed for implementing, evaluating, and demonstrating hardware and software concurrency techniques. Dr. Johnson has served as a chief scientist and subject matter specialist for Science Applications International Corporation (SAIC). He has been involved in the modeling and analysis of adaptive techniques for automated communications and the analysis and design of cryptographic architectures. In this capacity, Dr. Johnson was a key contributor to the development of federal and military standard technology for HF ALE. He analyzed two competing designs for the ALE standard, identified performance advantages of the design ultimately chosen (the MITRE scheme created by Gene Harrison), and contributed enhancements to that design. Dr. Johnson developed, analyzed, and evaluated a cryptographic algorithm used in standard LP (the Johnson algorithm). He worked with Robert I. Desourdis, Jr., to develop techniques for improved analysis of HF channel conditions using only the ALE modem. Dr. Johnson was the technical lead for a nationwide effort to identify requirements and develop technology for automating HF radio networks and the automated management and remote control of HF networks. Dr. Johnson's educational background includes a Ph.D. in electrical and computer engineering, 1987, New Mexico State University; an M.Sc. in electrical engineering, 1980, Washington University; a B.S.E.E., 1979, Washington University; and a B.S. in physics, 1979, Washington University. Dr. Johnson contributed to numerous government standards, including FED-STD-1049, FED-STD-1045, MIL-STD-188-141A, and MIL-HDBK-232A. He has over 100 publications in journals, conferences, and technical reports. Dr. Johnson is a member of Institute of Electrical and Electronics Engineers (IEEE), IEEE Computer Society, Association for Computing Machinery (ACM), ACM Special Interest Group on Computer Architecture (SIGARCH), and ACM Special Interest Group on Computer System Performance (SIGMETRICS).

Dr. Johnson is the principal author of this book. He authored Chapters 5 through 9 and Sections 1.1.2, 1.1.3, and 1.2.

Robert I. Desourdis, Jr. received a B.S. in mathematics from the Worcester Polytechnic Institute (WPI) in 1977 and an M.S. in electrical engineering from WPI in 1979. He received an M.S. in technology and policy from the Massachusetts Institute of Technology in 1980 while serving as a research assistant for the Center for Policy Alternatives and the Center for International Studies. Since that time, Mr. Desourdis has performed extensive modeling, simulation, analysis, systems integration, test, and evaluation of land-mobile and over-the-horizon RF communication systems. From 1980 until 1982, Mr. Desourdis was employed by Signatron, Inc., in Lexington, Massachusetts, and Science Applications International Corporation (SAIC) in Stow, Massachusetts, from 1982 to 1984. Mr. Desourdis worked for the GTE Strategic Systems Division in Westborough, Massachusetts from 1984 to 1986 where he developed a VHF/UHF land-mobile radio network simulation used to optimize RF system parameters for the Hard Mobile Launcher deployment of the Small ICBM. Mr. Desourdis returned to SAIC from 1986 to 1996 and performed a series of large-scale VHF meteor-scatter modeling and experimental RF communication system integration and evaluation programs. He has become a recognized expert in meteor-burst communications and has published extensively on the subject, including MILCOM publications in each year from 1988 to 1994 and a major portion of *Meteor Burst Communications: Theory and Practice* (Wiley). In HF communications, Mr. Desourdis developed draft materials for MIL-STD-188-721C "Advanced Link Quality Analysis" and performed modeling and analysis of sky-wave propagation in Antarctica for the Naval Undersea Warfare Center and the National Science Foundation Division of Polar Programs. More recently, Mr. Desourdis has been actively involved in the development and evaluation of the Intelligent Transportation System (ITS) and has served as the secretary/treasurer of the Massachusetts State Chapter of ITS America. Mr. Desourdis now works in the Telecommunications Division of CACI in Fairfax, Virginia, as director of systems engineering for wireless applications development. In this role, he is performing radio system analysis, modeling, simulation, and case studies of operational systems to determine optimal spectrum-use strategies for public safety wireless communications.

Mr. Desourdis developed the concept for this book, solicited key authors and contributors, acquired the book contract, managed and integrated author contributions, performed detailed editing of each chapter, finalized all figures and tables, and assembled the book for publication. In addition, Mr. Desourdis authored several sections, including Section 1.1.1, Section 2.2.4.2, part of Section 2.2.5.2, portions of Section 2.2.5.3, and the PCA-induced HF-outage results reported in Section 2.3.2.4.

Gregory D. Earle received his B.S. degree from the honors program in electrical engineering at Purdue University in 1981. He subsequently completed the M.S.E. in engineering physics at Purdue in 1983 and the Ph.D. in electrical engineering at Cornell University in 1988. Currently, Dr. Earle is assistant professor in the W.B. Hanson Center for Space Sciences and adjunct professor of electrical engineering at the University of Texas at Dallas. Prior to accepting this position he was a research scientist for eight years in the Experimental Research and Technology Development Division of Science Applications International Corporation. Dr. Earle's ongoing research and development projects include development and testing of diagnostic systems for low density plasmas; microprocessor-based implementation of neural network algorithms for instrumentation, prediction, and controls; and applications and design of novel microelectromechanical (MEM) sensors. Dr. Earle is a member of the physics and engineering undergraduate honor societies. He was a national merit scholar and was the recipient of a NASA research fellowship from 1985 to 1987. He is the author of over 15 refereed publications and has one patent pending. Dr. Earle is a member of IEEE, the American Geophysical Union, and the American Physical Society.

Dr. Earle authored much of Chapter 2, particularly Section 2.2. He also coauthored Section 2.3 with significant contributions from Dr. Leonard Wagner and Mr. Jens Ostergaard.

Stephen C. Cook amassed ten years' experience in industry in the fields of electronic design, spacecraft system studies, and project management in both Australia and the UK, before joining the Defense Science and Technology Organization (DSTO) in the Australian Department of Defense in 1988 as head of Radio Networks. In that position, he led a group that sought to improve military communications through research in coded modulation and adaptive protocols. This work, which has been recognized by several awards, proved highly successful and has been applied to HF, troposcatter, and meteor burst communications systems. The commercialization of the technology has had significant impact on the architecture and dimensioning of several defense communication projects.

Since June 1994, Dr. Cook has been research leader, Military Information Networks. In this position, he is responsible for the management and scientific leadership of a branch of around 65 research staff and the relevance of its work to the Australian Defense Organisation. Dr. Cook has been active on the TTCP technical panel STP-8 on terrestrial communications since 1989, serving six years as Australian National Leader. He is also adjunct associate professor at the University of South Australia where he supervises postgraduate students in communications engineering and measurement science.

Dr. Cook authored Chapter 4, which he had originally published as an HF '95 paper entitled "Advances in High-Speed HF Radio Modem Design." Material from this HF '95 article was also used in Section 2.4. Dr. Cook also authored Sections 1.3 and 1.4.

Jens C. Ostergaard is currently employed as director of research and development for Greenland Telecom. He was previously employed as a research associate at the Research Foundation at the Center for Atmospheric Research (ULCAR) of the University of Massachusetts at Lowell. In this position, Mr. Ostergaard served in a lead technical role for meteor scatter propagation and communication research programs conducted by the Phillips Laboratory's Geophysical Directorate of the U.S. Air Force (USAF) under a cooperative research contract between the USAF and the university. He was designated principal investigator for the meteor scatter research by the university and also participated in the HAARP research program in Alaska, which was designed to investigate ionospheric modification by RF heating. Previously Mr. Ostergaard was employed by ELECTONIKCENTERALEN, Danish Academy of Technical Sciences, the Danish Center for Applied Electronics Research and Development, where he served as a consultant in the field of ionospheric and transionospheric radiowave propagation. He was employed by the Danish Meteorological Institute as the leader of the Geophysics Observatory in Thule, Greenland. Mr. Ostergaard holds a M.Sc.E.E. from the Danish Technical University (TUD) in Copenhagen, Denmark.

Mr. Ostergaard authored Sections 2.4 and 2.5, with contributions from Dr. Cook's HF'95 paper. He coauthored Section 2.3, and established Dr. Leonard Wagner as a key contributor to that section.

Malcolm J. Packer has over 12 years of experience in theoretical and applied electromagnetics, analysis, design, and development of antennas from MF to SHF. He received a B.S.E.E. from New Jersey Institute of Technology in 1984 and an M.S.E.E. in electromagnetic fields, waves, and optics from Northeastern University in 1988. Mr. Packer is a member of Eta Kappa Nu, IEEE Antennas and Propagation, IEEE Biomedical, Applied Computation Electromagnetic Society, and the American Radio Relay League, Inc. From 1984 to 1988, Mr. Packer was with GTE Government Systems in the antenna development group located in Westborough, Massachusetts. There he designed many HF, VHF, and UHF antennas and performed an extensive theoretical analysis of electromagnetic propagation through layered materials. From 1988 to 1995 he was with Science Applications International Corporation (SAIC) located in Marlborough, Massachusetts, as a scientist in applied electromagnetics. Mr. Packer led a program to analyze special antennas for the Foreign Technology Division of the U.S. Air Force, with the goal of accurately developing numerical models to predict antenna performance in both the near and far field. During this time, he also managed the design, development, and installation of wideband HF sensors for Rome Laboratory. He was also the technical lead for the development of the SAIC Electromagnetic Antenna Modeling (EAM) software, an MS Windows-based graphical user interface for the Numerical Electromagnetics Code and the Basic Scattering Code. He has written numerous algorithms to analyze the effects of antennas embedded within matter. Mr. Packer is presently employed as a principal systems engineer with Harris RF Communications in Rochester, New York. Mr. Packer is currently focused on computational electromagnetics, analysis of antennas and their interaction with propagation, biological effects of electromagnetic radiation, and analysis of transients events and their effects on HF antennas and radio equipment.

Mr. Packer authored Chapter 3, employing many qualitative and quantitative results from his recent work in numerical electromagnetic modeling of HF antenna systems.

Leonard S. Wagner received a B.S.E.E from the Polytechnic Institute of Brooklyn in 1951, and M.S.E.E. and Ph.D. degrees from Cornell University in 1958 and 1962, respectively. From 1951 to 1958 he was employed by the General Electric Company as an engineer working on ground-based early warning radar systems. In 1962, he accepted a position as assistant professor of electrical engineering at Cornell University and was involved in research in ionospheric physics. From 1966 to 1968 he was on the staff of the Arecibo Ionospheric Observatory and was engaged in research on ionospheric effects of solar flares leading to the first direct measurements of flare-induced E and F layer electron density enhancements. In 1969, Dr. Wagner accepted an appointment as research physicist at the Naval Research Laboratory in Washington, DC where he remained until 1995. While at the Naval Research Laboratory, Dr. Wagner was associated with programs involving high-frequency radio surface wave propagation in marine environments, channel probe investigations of the HF sky-wave channel, with emphasis on difficult channels such as the auroral channel, and ionospheric modification effects of high-power HF radio-wave heating of the ionosphere. Currently, Dr. Wagner is employed as a consultant for KAMAN Sciences Corporation and engaged in research at the Naval Research

Laboratory on ionospheric plasma responses to high power HF radio-wave heating.

Dr. Wagner contributed significantly to Section 2.3 and provided important material for Section 2.3.2.

Index

52-tone modems, 30
Absorption, 47–49
   auroral, 100–101
   deviative, 47
   electron Landau damping and, 48–49
   nondeviative, 47, 48
   polar cap, 109
Acknowledgment
   ALE frame, 235
   one-to-one calling, 235
   responders receiving, 246
Additive noise, 168
Additive white Gaussian noise
      (AWGN), 29, 167, 175
   channels, 176, 192
   pure channel, 291
Addresses, 206–12
   basic word, 207–8
   character sets, 206–7
   extended, 208
   individual station, 207–8
   multiple station, 209
   null, 212
   self, 211
   wildcard, 246–47
   See also ALE
Addressing modes, 209–12
   Allcalls, 210
   Anycalls, 210–11
   null, 212
   self-address, 211
   stuffing, 209–10
   wildcards, 211

Advanced link quality analysis
      (ALQA), 255–57
   channel quality measures, 256
   defined, 255
   link performance measures, 256–57
   See also Link quality analysis (LQA)
Advanced Stand-Alone Prediction System
      (ASAPS), 76–77
   comparison, 77–79
   defined, 76
   input requirements, 76
   output integration, 77
   predicted MUF values, 77
   See also Virtual geometry models
ALE, 4, 191–215
   acknowledgment frame, 235
   addresses, 206–12
   coding, 198–203
   controllers, 6, 194
   data link layer, 279
   data structures, 247–48
   defined, 191
   error-correction suite, 200–202
   frames, 191
   frame structure, 9, 212–14
   group calls, 9–10, 242
   individual calls, 233
   linking performance, 192–94
   net calls, 9, 239
   nonlinking transmissions, 236–37
   operation, 9
   orderwire functions, 248–74
   protected system block diagram, 279, 283
   protocols, 217–74
   response frame, 234

ALE (continued)
  signal structure, 192, 193
  slotted responses, 239–41
  sound call structure, 222
  sounds, 10, 221–23
  state diagram, 228
  state machine, 205
  station addresses, 191
  top functional level, 229
  unprotected station vulnerability, 280–82
  word format, 194–98
  words, 191–98
  word size, 192
  word synchronization, 203–6
  See also High–frequency (HF)
ALE–capable stations, 6
ALE operations, 217–19
  illustrated, 218
  listen before transmit, 219
  parameters, 248
  rules, 219
Allcalls, 210
  frames, 245
  protocol, 244–45
  See also Anycalls
AMBCOM program, 79
Amplitude probability distribution (APD), 116
Antenna loads, 151
Antennas, 129–62
  "angular" designs, 152
  broadband, 135–45
  cavity–backed, 150
  cone–type, 137
  crossed dipole, 133
  dielectric/ferrite coatings, 150
  electrically small, 146
  electrical parameters, 160–61
  half–wavelength dipole, 133
  in hierarchical HF systems structure, 130
  horizontal log–periodic, 141–44
  inverted–cone, 136
  log–periodic (LPAs), 136
  long–wire, 140
  monopole pattern, 132
  monopole pattern over various surfaces, 154
  mounting effects, 157–60
  multimode, 144–45
  narrowband, 132–35
  noise factors, 115
  performance, 131, 151
  physically constrained, 146

  physically small, 146
  physical size of, 161
  receive–only, 146–49
  rhombic, 141
  selection of, 131
  short–helix vs. short dipole, 151
  "small" HF, 145–51
  small transmit design, 149–51
  spiral cone, 147–48
  systems, 131
  traveling–wave, 136, 138–41
  V, 140–41
  vertical log–periodic, 138, 139
  V–whip, 133
  whip, 132
  See also Antennas, environmental effects
Antenna tuners, 133–35
  function of, 150
  use of, 151
  See also Antennas
Antennas, environmental effects, 151–60
  ground effects, 152–57
  mounting effects, 157–60
  overview, 151–52
Anycalls, 210–11
  Double Selective, 211
  Global, 210–11
  procedure, 245–46
  Selective, 211
  See also Allcalls
"Any media" networks, linking to, 14
ARQ circuit mode, 303, 323
  operation of, 323
  See also HF data link protocol (HFDLP)
ARQ mode, 303
ARQ modems, 185–86
Asynchronous notification, 331
Atmospheric noise, 57, 115–17
  characterization, 116
  See also Noise
Auroral region, 95–107
  absorption, 100–101
  channel performance map, 106, 107
  communication characteristics, 101–7
  defined, 95
  diffuse effects, 96–97
  discrete effects, 96
  electron precipitation effects, 98
  geometry of, 95
  ionization, 98
  irregularities and blobs, 99

nighttime F–layer trough, 101
polar patches, 99
sporadic E, 100
See also Latitude–dependent phenomena
Australian DSTO, 167, 181, 182
Automated HF, 6–10
   hierarchical layers, 7
   key components, 6
   networks, 8
   overview, 6–8
   scanning, 8–9
   selective calling, 9–10
   station block diagram, 6
   See also High–frequency (HF)
Automated HF network management, 346–55
   background, 346–48
   example, 347
   HNMP, 351–55
   requirements, 348
   SNMP, 347–48, 349–51
   See also HF networks
Automatic link establishment. *See* ALE
Automatic message display (AMD), 265–66
Automatic message exchange (AME), 339–41
   connectivity exchange message syntax, 340
   defined, 339
   header, 339
   header fields, 339
   incoming message processing, 341
   outgoing message processing, 341
   port number designations, 340
AX.25 link layer protocol, 302

Binary phase shift keying (BPSK), 169
Bit error rate (BER), 30, 169
   LQA, 249
   LQA report, 252
Broadband antennas, 135–45
   defined, 135
   horizontal polarization, 138–44
   multimode antennas, 144–45
   vertical polarization, 136–38
   See also Antennas
Broadcast mode, 303, 322–23
   use of, 322
   See also HF data link protocol (HFDLP)

Calling cycle, 212–13
CCIR channel
   characteristics, 114
   performance, 172–74
Channel performance map
   defined, 106
   illustrated, 107
Channel quality measures (CQMs), 256
Channels, 35–123
   distortion, 167–68
   dropout, 237
   evaluation, 220–25
   ionospheric measurements, 60–71
   link performance prediction, 71–84
   LQA–scored, 226
   noise, 114–20
   path geometry, 43
   performance, 35
   radio interference, 121–23
   selection, 225–26
   sky–wave propagation, 41–60
   table, 247
Character sets, 206–7
   basic–38 subset, 206–7
   expanded–64 subset, 207
   full–128 set and binary data, 207
CMD formats
   analog port selection, 261
   channel select, 258
   CRC, 272
   crypto selection, 261
   data port selection, 262
   digital LINCOMPEX zeroization, 262
   digital squelch, 263
   frequency select, 257
   generic scheduling, 264
   LQA, 251
   LQA report, 252
   LQA report request, 254
   mode control, 259
   modem selection, 260
   power control, 259
   user–unique, 273
   See also Words
Cochannel interference, 168
   carrier–wave, 176–78
   performance with, 176–79
   pulse, 178–79
Coding, 198–203
   Golay FEC, 202–3
   overview, 199–202
   See also ALE
Cold plasma dispersion relation, 83
Comparative approach, 27, 28–29
   defined, 27
   example, 28–29

Comparative approach (continued)
    See also HF systems design
Connectivity exchange (CONEX)
        protocol, 343, 345
Connectivity monitoring, 331–32
Control frames, 303
    Data–ACK, 305
    defined, 303
Controllable single sideband (SSB) radios, 6
Coronal mass ejection (CME), 89
Corrected geomagnetic local time (CGLT), 95
Customer–supplier relationship, 22–25
    choice of, 25
    partnership approaches, 24–25
    system design by customer and, 23
    system design by customer and contractor and, 24
    system design by prime contractor and, 23–24
Cyclic redundancy check (CRC), 271–73
    analysis, 272
    ARQ, 273
    CMD format, 272
    defined, 271
    See also Orderwire functions

Data block message (DBM), 269–71
    asymptotic throughput, 271
    basic-size data block, 270
    characteristics, 269
    extended-size data block, 270–71
    fragmentation, 271
    structure, 269
    See also Orderwire functions
Datagram service, 14
Data link protocol. *See* HF data link protocol (HFDLP)
Data text message (DTM), 266–69
    asymptotic throughput, 268–69
    characteristics, 267
    corrupted frames, 268
    modes, 267
    structure, 267
    transfer, 268
    See also Orderwire functions
Data transfer protocol, 303, 305–10
    data acknowledgment phase, 309
    data transfer phase, 308–9
    DTSM, 310
    flow control, 309–10
    negotiation phase, 307–8
    operations, 305–10

    See also HF data link protocol (HFDLP)
Data transfer state machine (DTSM), 310
    defined, 310
    illustrated, 311
Day/night terminator, 86–91
    CME, 89
    defined, 86
    effects, 87
    magnetic storms/substorms, 89–91
    solar flares, 88–89
    sporadic E, 87
    TIDs, 87
Deviative absorption, 47
Digital HF modems, 165–88
    52-tone, 185–86
    additive noise, 168
    ALE, 193, 194
    ARQ, 185–86
    background, 165–67
    channel distortion, 167–68
    cochannel interference, 168
    contemporary designs, 172–82
    enhanced, 182–87
    equipment specifications, 180–81
    FSK, 170
    in hierarchical HF systems structure, 166, 278
    MIL–STD–188–110A, 172, 175
    on-the-air trials, 179–80
    parallel-tone, 170–72
    peak-to-average ratio, 180
    performance in impulsive noise, 174–76
    performance with cochannel interference, 175–79
    single-tone, 172
    space diversity reception, 182
    synchronization, 181
    TCM–16, 183–84
    traditional designs, 169–72
Discrete Fourier transform (DFT), 123
D region, 38
    path loss mechanism, 47
    plasma free electrons in, 47
    See also Ionospheric layers
Dwell time, 8

Effective isotropic radiated power (EIRP), 31
Effective radiated power (ERP), 27, 30–31
Eigenmodes, plasma, 48
Electromagnetic Antenna Modeling (EAM), 129
Electron

density profiles, 40
    Landau damping, 48–49
    precipitation effects, 98
Emergency connections, 11
Enhanced HF modems, 182–87
    52-tone, 185
    design issues, 182–84
    parallel-tone, 184–86
    single-tone, 186–87
    See also Digital HF modems
Equatorial spread F (ESF), 91–92
E region, 41
    auroral, 99
    enhanced electric currents, 98
    See also Ionospheric layers
Es clouds, 94
Evolutionary acquisition, 25
Extreme ultraviolet (EUV), 35

Faraday rotation effect, 57
Fast Fourier transform (FFT), 171
File compression, 15
    examples, 16
    LZW algorithm and, 15, 17
First principles models, 79–84
    AMBCOM, 79
    IONORAY, 79–84
    SKYCOM, 84
    See also Link performance prediction
Forward error correction
        (FEC), 30, 171, 277–78
Frames, 191
    acknowledgment, 235
    Allcall, 245
    calling cycle and, 212–13
    conclusion section, 214
    control, 303
    corrupted, 268
    illustrated, 212
    message section, 213
    net-call, 239
    response, 234
    sounds, 214
    structure, 9, 212–14
    See also ALE
Free Euclidean distance, 182
F region, 38–39
    blobs, 99
    high-latitude boundary, 99
    nighttime trough, 101
    peak plasma density in, 58
    reflection, 52

See also Ionospheric layers
FSK modems, 170
Functional decomposition, 22

Galactic noise, 119–20
Global information infrastructure, 10–19
Global phenomena, 85–91
    day/night terminator, 86–91
    overview, 85–86
Global positioning system (GPS), 3
Go-back-N scheme, 302
Golay decoders, 295
Golay FEC coding, 202–3
    decoding, 203
    encoding, 202
    examples, 203
    See also Coding
"Gray-line propagation," 37, 87
Ground effects, 152–57
    gain improvement and, 155
    ground electrical parameters and, 156
    ground radial design and, 156
    ground screens and, 155, 156
    monopole pattern over various
        surfaces, 154
    on horizontal dipole, 157
    See also Antennas
Group calls, 9–10, 242
Gyrofrequency, 55

Handshaking, 225
    incoming, 231
    outgoing, 230
    ping-pong, 235
HF data link protocol (HFDLP), 299–323
    adapting data rate, 323
    ARQ circuit mode, 303, 323
    ARQ mode, 303
    broadcast mode, 303, 322–23
    connection ID, 303
    controllers, 354
    data transfer protocol, 303, 305–10
    defined, 300–302
    frame exchange examples, 306
    full-duplex operation, 323
    in HF hierarchical structure, 301
    history, 300–302
    link management protocol, 303, 316–20
    LMSM transitions, 319
    message management
        protocol, 303, 310–15
    messages and, 303

HF data link protocol (HFDLP) (continued)
  operating modes, 303–4
  operation, 304–5
  overview, 302–4
  receiving terminal, 304
  selective repeats, 302
  subprotocol support, 303
  timeouts, 320–22
  transmitting terminal, 303–4
HF Industries Association (HFIA), 214
HF networks, 325–55
  automated management, 346–55
  automated technologies, 328–32
  connectivity monitoring, 331–32
  example implementation, 332–46
  hierarchies of star topologies, 337
  illustrated sample, 327
  indirect paths, finding, 328–31
  indirect paths, using, 332
  sample network volume, 355
  subnetwork topology, 337
HF node controllers (HFNC), 332–37
  block diagram, 335
  data bases, 333
  functions overview, 334
  path quality matrix and, 334–35
  routing table and, 334
  standard levels, 336–37
  See also HF networks
HF relay management protocol
       (HRMP), 343–45
  connectivity monitoring, 345
  exchange types, 344
  message stations, 343
  operation, 344
  overview, 343
  repeater control, 345
  routing queries, 345
HF sky–wave channels, 35–123
  distortion, 167–68
  dropout, 237
  evaluation, 220–25
  ionospheric measurements, 60–71
  link performance prediction, 71–84
  noise, 114–20
  path geometry, 43
  performance, 35
  radio interference, 121–23
  selection, 225–26
  sky–wave propagation, 41–60
  table, 247

HF station status protocol (HSSP), 345–46
HF systems
  for mobile networks, 11, 12
  parameters, 27
  for partitioned networks, 12
HF systems design, 25–32
  by customer, 23
  by link budgets, 27, 29–32
  by prime contractor, 23–24
  comparative approach, 27, 28–29
  formation of candidates, 26–28
  requirements definition, 25–26
  shared by customer and contractor, 24
  See also HF systems
HF systems engineering, 19–25
  customer–supplier relationship
       and, 22–25
  defined, 20
  as front–end process, 20
  functional decomposition, 22
  process, 20–22
  purpose, 19–20
  requirements analysis, 20–21
  system acquisition process integration
       and, 22
  system requirements and, 20
  See also HF systems
High–frequency (HF)
  antennas, 129–62
  automated communications, 6–10
  blackout, 89
  data link bandwidth, 10, 15
  data link protocol. See HF data link
       protocol (HFDLP)
  in global information infrastructure, 10–19
  interface to Internet, 17–19
  Internet gateway, 18, 355
  military standards, 4–5
  radio development, 3–6
  standards evolution, 5–6
  See also ALE; HF systems; HF systems
       design; HF systems engineering
High visible sun, 67
HNMP, 351–55
  access control, 353
  defined, 351
  GetRows mechanism, 353
  performance, 354
  Proxy Agents, 354
  proxy management, 354

See also Automated HF network management
Homosphere, 38
Horizontal log–period antennas, 141–44
    double–curtain, 145
    fixed types, 144
    forward tilt of, 143
    illustrated, 142–43
    rotatable, 142–43
    single–curtain, 144
    uses, 142
    See also Antennas
Horizontal polarization, 42, 43

ICEPAC, 76
Impedance loading, 150
Impedance transformers, 135
Incremental acquisition, 25
Indirect paths, 328–31, 332
    connectivity exchange, 329
    example, 329–31
    finding, 328–31
    recording back routes, 329
    repeater control, 332
    routing, 332
    routing queries, 329
    using, 332
    See also HF networks
Integrated Services Digital Network (ISDN), 208
Interference
    carrier–wave, 176–78
    cochannel, 168
    migration techniques, 122
    pulse, 178–79
    radio, 121–23
    tone–burst, 179
Interleaving, 30, 169
    ALE word, 201
    maximum, 173–74
    minimum, 172–73
International Geophysical Reference Field (IGRF) magnetic field model, 81
International Reference Ionosphere (IRI), 81
Internet, 12–19
    compatibility, 14
    HF gateway, 18, 355
    HF interface to, 17–19
    performance limitations, 14–17
    to rapid–deployment forces, 13
Internet protocol (IP), 14
IONCAP, 74–76
    algorithm, 75

    approach, 74–75
    comparison, 77–79
    defined, 74
    development, 75
    ICEPAC, 76
    IONWIN, 76
    predicted MUF values, 77
    PROPVUE, 76
    VOACAP, 75–76
    See also Virtual geometry models
Ionograms, 60–66
    critical frequency and, 60
    critical frequency/height parameters, 64
    defined, 60
    heights associated with, 63
    interpretation of, 63
    measurements, 62
IONOLINK, 66, 77
IONORAY, 60, 79–84
    algorithm, 80
    Auroral Oval module, 81
    code outputs, 83–84
    defined, 80
    flowchart, 82
    IGRF magnetic field model, 81
    IRI, 81
    layers module, 81
    MSIS atmospheric model, 81
    tradeoffs, 84–85
    uses, 84
    See also Link performance prediction
Ionospheric Communications Analysis and Predictions model. *See* IONCAP
Ionospheric disturbances summary, 113
Ionospheric layers, 38–41
    D region, 38
    E region, 41
    F region, 38–39
    illustrated, 39
    "sporadic," 41
Ionospheric measurements, 60–71
    ionograms, 60–66
    MUF calculations, 66–71
Ionospheric path loss, 57–60
Ionospheric plasma, 35–38
Ionospheric propagation, 45–46
Ionospheric tilts, 79
IONOSTATS, 66
IONWIN, 76

Latitude–dependent phenomena, 91–112
    auroral region, 95–107

Latitude–dependent phenomena (continued)
   low–latitude (equatorial) region, 91–92
   mid–latitude region, 92–95
   polar region, 107–12
Lempel–Ziv–Welch (LZW) algorithm, 15, 17
Link budget approach, 27, 29–32
   calculating with, 29
   defined, 27
   ERP calculation, 30–31
   See also HF systems design
Link establishment, 226–48
   available state, 227
   linked state, 227
   linking state, 227
   one–to–many calling, 237–47
   one–to–one calling, 227–37
   overview, 226–27
Linking performance, 192–94
   measured and simulated results, 194
   minimum performance, 192–94
   probability requirements, 194
   simulations, 292–97
   simulator stack, 292
   See also ALE
Linking protection (LP), 277–97
   ambiguity resolution, 286–87
   defined, 277
   effect of, 293–94
   full performance, 296
   performance analysis, 291–97
   performance evaluation summary, 297
   PI analysis, 290–91
   procedure overview, 282–83
   processing example, 287–90
   protection interval (PI), 282
   receive processing, 284–86
   receiving state diagram, 285
   technique, 280–90
   TOD, 282, 283–86
   transmit processing, 283–84
   transmitting state diagram, 285
Linking protection control module
      (LPCM), 279–80
   defined, 279
   receiving, 283
   transmitting, 283, 287
Link management protocol, 303, 316–20
   call acknowledge state, 317
   calling state, 316–17
   defined, 316
   idle state, 316

   immediate mode, 318
   link failure, 318
   link termination, 318
   link–up state, 317–18
   LMSM, 318–20
   operations, 316–18
   See also HF data link protocol (HFDLP)
Link management state machine
      (LMSM), 318–20
Link performance measures (LPMs), 256–57
Link performance prediction, 71–85
   applications, 84–85
   background, 71–74
   first principle models, 79–84
   model trade–offs, 84–85
   virtual geometry models, 74–79
Link quality analysis (LQA), 224–25
   advanced, 255–57
   bit error ratio, 249
   CMD function summary, 250
   CMD word format, 251
   multipath (MP) measurements, 225
   overview, 249
   PBER, 224–25
   scores, 226
   SINAD, 225, 249–51
   table, 247
Link termination
   automatic, 236
   manual, 236
   one–to–one calling, 235–36
Link timeouts, 321
L mode, 53, 55, 57
Log–periodic antennas (LPAs), 136
   horizontal, 141–44
   vertical, 138, 139
   See also Antennas
Low–latitude (equatorial) region, 91–92
Low visible sun, 67
LQA report, 251–54
   BER field, 252
   CMD format, 251–52
   control bit assignments, 253
   formats, 252
   request CMD format, 253–54
   See also Link quality analysis (LQA)
      Orde wire functions

Magnetic field effects, 52–57
Magnetic storms/substorms
   causes for, 89
   day/night terminator, 89–91

illustrated, 90
mid–latitude region, 94
Mass Spectrometer and Incoherent Scatter (MSIS) atmospheric model, 81
Maximum observed frequency (MOF), 66
Maximum useable frequency (MUF), 50–51
   calculations, 66–71
   IONCAP/ASAPS–predicted values, 77
   numerical distribution, 78
   standard values, 68, 69
Message management protocol, 303, 310–15
   defined, 310
   link sharing, 313
   message acceptance, 312
   message announcement, 311–12
   message preemption, 313–14
   message refusal, 312
   message resumption, 314
   MMSM, 315
   null message management headers, 314
   operations, 310–14
   priority resolution, 312–13
   See also HF data link protocol (HFDLP)
Message management state machine (MMSM), 315
Message preemption, 313–14
   forward preemption, 313–14
   reverse preemption, 314
Mid–latitude region, 92–95
   magnetic storms, 94
   sporadic E, 93–94
   spread F, 94–95
   See also Latitude–dependent phenomena
MIL–STD–188–110A modem, 172, 175
   error–control code, 186
   on–the–air trails, 179–82
   performance, 175
   performance in presence of sine–wave interference, 177
   waveform, 175
"Mitre scheme," 4
Mode control, 259–63
   analog port selection, 261–62
   crypto negotiation and handoff, 260–61
   data port selection, 262
   digital LINCOMPEX zeroization, 262
   digital squelch, 263
   modern negotiation and handoff, 260
   overview, 259–60
   See also Orderwire functions
Modems. *See* Digital HF modems

Mounting effects, 157–60
   fixed installations, 158
   mobile platforms, 158–60
   See also Antennas
Multimode antennas, 144–45
Multipath (MP)
   measurements, 225
   propagation, 114, 167
Multiple station addresses, 209
Multipoint topology, 238

Narrowband antennas, 132–35
   horizontal polarization, 133–34
   tuners, 133–35
   vertical polarization, 132
   See also Antennas
Near–vertical–incidence sky–wave (NVIS), 67, 132
Net calls, 9, 239
Network–layer protocol
   address record format, 338
   header, 338
   in HF hierarchy, 328
   overview, 337–39
Noise, 114–20
   additive, 168
   atmospheric, 115–19
   background, 114–15
   cdf, 118
   effective antenna, 115
   galactic, 119–20
   impulsive, modem performance in, 174–76
   man–made, 115–19
Noise report, 254–55
   broadcast technique, 254–55
   format, 255
   noise power measurements, 255
   See also Orderwire functions
Nondeviative absorption, 47, 48
Null addresses, 212
Null S&F functions, 342–43

O mode, 53, 54, 55, 56
   linearly polarized, 55
   ray paths, 74
   vector diagram, 54
One–to–many calling, 237–47
   Allcall protocol, 244–45
   Anycall protocol, 245–46
   multipoint topology, 238
   slotted responses, 239–41

One–to–many calling (continued)
    star group calling, 241–44
    star net calling, 238–39
    wildcard calling protocol, 246–47
    See also Link establishment
One–to–one calling, 227–37
    acknowledgment, 235
    collision detection, 237
    link termination, 235–36
    nonlinking ALE transmissions, 236–37
    receiving individual call, 231–34
    response, 234–35
    sending individual call, 228–31
    See also Link establishment
Open Systems Interconnection Reference
        Model (OSI–RM), 6, 302
Operational support, 25
Orderwire functions, 248–74
    ALQA, 255–57
    automatic message display (AMD), 265–66
    channel and frequency
        commands, 257–58
    cyclic redundancy check (CRC), 271–73
    data block message (DBM), 269–71
    data text message (DTM), 266–69
    link quality analysis (LQA), 249–51
    LOA report, 251–54
    mode control, 259–63
    noise report, 254–55
    power control, 258–59
    scheduling commands, 263–65
    time exchange, 274
    user unique functions (UUFs), 273–74
Other address table, 248

Parallel polarization, 42
Parallel–tone modems, 170–72
    16–tone/39–tone, 171, 172
    enhanced, 184–86
    signaling waveform, 171
    See also Digital HF modems
Path loss, ionospheric, 57–60
Peak–to–average ratio, 180
Perpendicular polarization, 42
Photodissociation, 37
Planning standard, 5
Plasma
    defined, 35
    density variability, 39
    D region and, 38
    eigenmodes, 48
    electromagnetically active, 57

    formation of, 35–36
    frequency, 52
    ionospheric, 35–38
    peak density, 58
    solar wind, 90
Polar arcs, 107–8
Polarization
    horizontal, 42, 43
    parallel, 42
    perpendicular, 42
    vertical, 42, 43
Polar patches, 99, 108–9
Polar region, 107–12
    arcs and patches, 107–9
    cap absorption, 109
    communication characteristics, 109–12
    polar cap absorption, 109
    See also Latitude–dependent phenomena
Polling, 225
Power control, 258–59
    CMD format, 259
    command bits, 259
    defined, 258
    See also Orderwire functions
Prediction programs, 31–32
Principle of similitude, 135–36
PROPVUE, 76
Protection interval (PI)
    analysis, 290–91
    defined, 282
    length choice, 290
Pseudo bit error rate (PBER), 224–25

Radio interference, 121–23
Rapid–deployment networks, 11
Ray–path geometry, 42–46
Refraction, 49–52
    index, 50–51
    Snell's law geometry, 50
Requirements specification, 25–26
Response timeouts, 320–21
R mode, 53, 55, 56, 57
RMXNET, 330
Round–the–world (RTW) propagation
        modes, 45

Scalar variables, 248
Scanning, 8–9, 221
    calling cycle, 226
    calls, 280
    call station, 226–27
    period, 231

Scheduling commands, 263–65
  groups, 264
  time codes, 264
  See also Orderwire functions
Scott Base ionosonde, 74, 77
Selective repeats, 302
Self–addresses, 211
  multiple, 244
  table, 247–48
  See also Addressing modes
Serial tone PSK, 299
Short–wave fadeout (SWF), 89
Side–scattering, 101
Signal–to–interference ratio (SIR), 178
Signal–to–noise ratio (SNR)
  establishing minimum, 30
  minimum required, 60
  PSK modem, 300
  for standard channels, 291
  throughput vs., 30
Simple network management protocol (SNMP), 347–48
  applicability, 349–51
  compatibility, 350
  functionality, 349
  message bits, 250
  minimization of traffic and, 350
  performance, 350–51
  summary, 347–48
  See also Automated HF network anagement
Simulations, 292–97
  false word sync problem, 294–96
  full–performance linking protection, 296
  measurement vs., 296–97
  overview, 292–94
  simulator, 292
  See also Linking protection (LP)
SINAD measurements, 225, 249–51
Single–tone modems, 172
  enhanced, 186–87
  experimental vs. standard, 187
  TCM and, 186
SKYCOM, 84
Sky–wave propagation, 41–60
  absorption, 47–49
  atmospheric effects, 47
  in benign ionosphere, 25–85
  in disturbed ionosphere, 85–114
  ionospheric path loss, 57–60
  magnetic field effects, 52–57

refraction, 49–52
representative paths, 46
spreading loss, 41–42
wave propagation, 41–42
See also HF sky–wave channels
Sky–wave ray–path geometry, 42–46
  illustrated, 43
  modeling of, 44
  Rayleigh–fading, 114
  virtual height geometry, 44
Slotted responses, 239–41
Slotted–response timing, 243–44
Small antennas, 145–51
  categories of, 145–46
  receive–only, 146–49
  transmit design, 149–51
  vertical/horizontal, 149
  See also Antennas
Snell's law refraction geometry, 50
Solar flares, 88–89
Solar proton event (SPE), 109
Sounding, 10, 221–23
  ALE call structure, 222
  example interval calculation, 224
  rate, 223
  signal, 221
Space diversity reception, 182
Spiral cone antennas, 147–48
Sporadic E, 87
  auroral, 100
  mid–latitude region, 93–94
Stairway to Heaven, 5
Star group calling, 241–44
  derived slots and, 242–43
  long responses, 244
  multiple self–addresses, 244
  naming group and, 241–42
  slotted–response timing, 243–44
Star net calling, 238–39
Star topology, 238
Store–and–forward (S&F) operation, 342
  incoming message processing, 342
  Null, 342–43
  outgoing message processing, 342
Sudden phase anomalies (SPAs), 89
Sunspot number (SSN), 36–37
  cycle, 37
  determination, 36
  maximum, high sun, 72, 73
  maximum, low sun, 69, 70, 71
  minimum, low sun, 68

Sunspot number (SSN) (continued)
  minimum/maximum daily average, 67
Synchronization
  modem, 181
  at receiver, 205–6
  TOD, 285
  word, 203–6
Tables
  channel, 247
  LOA, 247
  other address, 248
  self-address, 247–48
Thule riometer, 111–12
Time-division multiple-access (TDMA), 239
Time exchange, 274
Time of day (TOD), 282, 283–86
  receive processing, 284–86
  synchronization, 285
  transmit processing, 283–84
Timeouts, 320–22
  action following, 322
  link, 321
  response, 320–21
  turnaround times and, 320
Titanic description, 1–3
Traveling ionospheric disturbances (TIDs), 87
Traveling-wave antennas, 136, 138–41
  long-wire antennas, 140
  rhombic antennas, 141
  V antennas, 140–41
  See also Antennas
Trellis-coded modulation (TCM),
  scheme construction, 183
  TCM-16 modem, 183–84

User unique functions (UUFs), 273–74
  addressing, 273–74
  CMD format, 273
  defined, 273
  See also Orderwire functions

Vertical-incidence reflection, 52
Vertical log-periodic antennas, 138, 139
Vertical polarization, 42, 43
Virtual geometry models, 74–79
  ASAPS, 76–77
  comparison with measurement, 77–79
  IONCAP, 74–76
  See also Link performance prediction
Virtual height of reflection, 44

Voice/data service
  to mobile platforms, 11
  to remote locations, 11
Voice of America Coverage Analysis Program
    (VOACAP), 75–76
V-whip antenna, 133

Wave modes
  L mode, 53, 55, 57
  O mode, 53, 54, 55, 56
  R mode, 53, 55, 56, 57
  X mode, 53, 54, 55, 56
Wildcard calling protocol, 246–47
Wildcards, 211
Wireless
  legacy, 1–3
  lessons learned, 3–6
Words, 191
  24-bit, 199
  basic address, 207–8
  bit coding and interleaving, 201
  CMD, 197–98
  coded 49-bit, 204
  DATA, 197
  false sync problem, 294–96
  format, 194–98
  FROM, 198
  leading three bits, 195
  LSB, 194
  MSB, 194
  REP, 197
  size of, 192
  structure of, 196
  stuffing, 209–10
  synchronization, 203–6
  THRU, 198
  TIS, 196–97
  TO, 196
  transmitter phrase, 203–5
  TWAS, 197
  types of, 195–98
  unprotected sync acquisition, 286–87
  valid sequences, 198
  See also ALE

X.25 link layer protocol, 302
X mode, 53, 54, 55, 56
  elliptically polarized, 55
  vector diagram, 54

The Artech House Telecommunications Library

Vinton G. Cerf, Series Editor

Advanced High-Frequency Radio Communications, Eric E. Johnson, Robert I. Desourdis, Jr., et al.

Advanced Technology for Road Transport: IVHS and ATT, Ian Catling, editor

Advances in Computer Systems Security, vol. 3, Rein Turn, editor

Advances in Telecommunications Networks, William S. Lee and Derrick C. Brown

Advances in Transport Network Technologies: Photonics Networks, ATM, and SDH, Ken-ichi Sato

An Introduction to International Telecommunications Law, Charles H. Kennedy and M. Veronica Pastor

Asynchronous Transfer Mode Networks: Performance Issues, Second Edition, Raif O. Onvural

ATM Switching Systems, Thomas M. Chen and Stephen S. Liu

Broadband: Business Services, Technologies, and Strategic Impact, David Wright

Broadband Network Analysis and Design, Daniel Minoli

Broadband Telecommunications Technology, Byeong Lee, Minho Kang, and Jonghee Lee

Cellular Mobile Systems Engineering, Saleh Faruque

Cellular Radio: Analog and Digital Systems, Asha Mehrotra

Cellular Radio: Performance Engineering, Asha Mehrotra

Cellular Radio Systems, D. M. Balston and R. C. V. Macario, editors

CDMA for Wireless Personal Communications, Ramjee Prasad

Client/Server Computing: Architecture, Applications, and Distributed Systems Management, Bruce Elbert and Bobby Martyna

Communication and Computing for Distributed Multimedia Systems, Guojun Lu

Communication and Computing for Distributed Multimedia Systems, Guojun Lu

Computer Networks: Architecture, Protocols, and Software, John Y. Hsu

Computer Mediated Communications: Multimedia Applications, Rob Walters

Computer Telephone Integration, Rob Walters

Corporate Networks: The Strategic Use of Telecommunications, Thomas Valovic

Digital Beamforming in Wireless Communications, John Litva, Titus Kwok-Yeung Lo

Digital Cellular Radio, George Calhoun

Digital Hardware Testing: Transistor-Level Fault Modeling and Testing, Rochit Rajsuman, editor

Digital Switching Control Architectures, Giuseppe Fantauzzi

Digital Video Communications, Martyn J. Riley and Iain E. G. Richardson

Distributed Multimedia Through Broadband Communications Services, Daniel Minoli and Robert Keinath

Distance Learning Technology and Applications, Daniel Minoli

EDI Security, Control, and Audit, Albert J. Marcella and Sally Chen

Electronic Mail, Jacob Palme

Enterprise Networking: Fractional T1 to SONET, Frame Relay to BISDN, Daniel Minoli

Expert Systems Applications in Integrated Network Management, E. C. Ericson, L. T. Ericson, and D. Minoli, editors

FAX: Digital Facsimile Technology and Applications, Second Edition, Dennis Bodson, Kenneth McConnell, and Richard Schaphorst

FDDI and FDDI-II: Architecture, Protocols, and Performance, Bernhard Albert and Anura P. Jayasumana

Fiber Network Service Survivability, Tsong-Ho Wu

A Guide to the TCP/IP Protocol Suite, Floyd Wilder

Implementing EDI, Mike Hendry

Implementing X.400 and X.500: The PP and QUIPU Systems, Steve Kille

Inbound Call Centers: Design, Implementation, and Management, Robert A. Gable

Information Superhighways Revisited: The Economics of Multimedia, Bruce Egan

Integrated Broadband Networks, Amit Bhargava

International Telecommunications Management, Bruce R. Elbert

International Telecommunication Standards Organizations, Andrew Macpherson

Internetworking LANs: Operation, Design, and Management, Robert Davidson and Nathan Muller

Introduction to Document Image Processing Techniques, Ronald G. Matteson

Introduction to Error-Correcting Codes, Michael Purser

An Introduction to GSM, Siegmund Redl, Matthias K. Weber, Malcom W. Oliphant

Introduction to Radio Propagation for Fixed and Mobile Communications, John Doble

Introduction to Satellite Communication, Bruce R. Elbert

Introduction to T1/T3 Networking, Regis J. (Bud) Bates

Introduction to Telecommunication Electronics, Second Edition, A. Michael Noll

Introduction to Telephones and Telephone Systems, Second Edition, A. Michael Noll

Introduction to X.400, Cemil Betanov

LAN, ATM, and LAN Emulation Technologies, Daniel Minoli and Anthony Alles

Land-Mobile Radio System Engineering, Garry C. Hess

LAN/WAN Optimization Techniques, Harrell Van Norman

LANs to WANs: Network Management in the 1990s, Nathan J. Muller and Robert P. Davidson

Minimum Risk Strategy for Acquiring Communications Equipment and Services,
Nathan J. Muller

Mobile Antenna Systems Handbook, Kyohei Fujimoto and J.R. James, editors

Mobile Communications in the U.S. and Europe: Regulation, Technology, and Markets, Michael Paetsch

Mobile Data Communications Systems, Peter Wong and David Britland

Mobile Information Systems, John Walker

Networking Strategies for Information Technology, Bruce Elbert

Packet Switching Evolution from Narrowband to Broadband ISDN, M. Smouts

Packet Video: Modeling and Signal Processing, Naohisa Ohta

Personal Communication Networks: Practical Implementation, Alan Hadden

Personal Communication Systems and Technologies, John Gardiner and Barry West, editors

Practical Computer Network Security, Mike Hendry

Principles of Secure Communication Systems, Second Edition, Don J. Torrieri

Principles of Signaling for Cell Relay and Frame Relay, Daniel Minoli and George Dobrowski

Principles of Signals and Systems: Deterministic Signals, B. Picinbono

Private Telecommunication Networks, Bruce Elbert

Radio-Relay Systems, Anton A. Huurdeman

RF and Microwave Circuit Design for Wireless Communications, Lawrence E. Larson

The Satellite Communication Applications Handbook, Bruce R. Elbert

Secure Data Networking, Michael Purser

Service Management in Computing and Telecommunications, Richard Hallows

Smart Cards, José Manuel Otón and José Luis Zoreda

Smart Highways, Smart Cars, Richard Whelan

Super-High-Definition Images: Beyond HDTV, Naohisa Ohta, Sadayasu Ono, and Tomonori Aoyama

Television Technology: Fundamentals and Future Prospects, A. Michael Noll

Telecommunications Technology Handbook, Daniel Minoli

Telecommuting, Osman Eldib and Daniel Minoli

Telemetry Systems Design, Frank Carden

Teletraffic Technologies in ATM Networks, Hiroshi Saito

Toll-Free Services: A Complete Guide to Design, Implementation, and Management, Robert A. Gable

Transmission Networking: SONET and the SDH, Mike Sexton and Andy Reid

Troposcatter Radio Links, G. Roda

Understanding Emerging Network Services, Pricing, and Regulation, Leo A. Wrobel and Eddie M. Pope

Understanding GPS: Principles and Applications, Elliot D. Kaplan, editor

Understanding Networking Technology: Concepts, Terms and Trends, Mark Norris

UNIX Internetworking, Second edition, Uday O. Pabrai

Videoconferencing and Videotelephony: Technology and Standards, Richard Schaphorst

Wireless Access and the Local Telephone Network, George Calhoun

Wireless Communications in Developing Countries: Cellular and Satellite Systems, Rachael E. Schwartz

Wireless Communications for Intelligent Transportation Systems, Scott D. Elliot and Daniel J. Dailey

Wireless Data Networking, Nathan J. Muller

Wireless LAN Systems, A. Santamaría and F. J. López-Hernández

Wireless: The Revolution in Personal Telecommunications, Ira Brodsky

Writing Disaster Recovery Plans for Telecommunications Networks and LANs, Leo A. Wrobel

X Window System User's Guide, Uday O. Pabrai

For further information on these and other Artech House titles, contact:

Artech House
685 Canton Street
Norwood, MA 02062
617-769-9750
Fax: 617-769-6334
Telex: 951-659
email: artech@artech-house.com

Artech House
Portland House, Stag Place
London SW1E 5XA England
+44 (0) 171-973-8077
Fax: +44 (0) 171-630-0166
Telex: 951-659
email: artech-uk@artech-house.com

WWW: http://www.artech-house.com